Voice & Data
Internetworking
Third Edition

ABOUT THE AUTHOR

Gil Held is an award-winning author and lecturer who specializes in the application of computer and communications technology. He is the author of more than 40 books on personal computers and data communications.

Voice & Data Internetworking
Third Edition

GIL **HELD**

Osborne/**McGraw-Hill**

New York Chicago San Francisco
Lisbon London Madrid Mexico City
Milan New Delhi San Juan
Seoul Singapore Sydney Toronto

Osborne/**McGraw-Hill**
2600 Tenth Street
Berkeley, California 94710
U.S.A.

To arrange bulk purchase discounts for sales promotions, premiums, or fund-raisers, please contact Osborne/**McGraw-Hill** at the above address. For information on translations or book distributors outside the U.S.A., please see the International Contact Information page immediately following the index of this book.

Voice & Data Internetworking, Third Edition

1234567890 1FGR 1FGR 01987654321

ISBN 0-07-213183-7

Publisher
  Brandon A. Nordin
Vice President & Associate Publisher
  Scott Rogers
Acquisitions Editor
  Steve Elliot
Project Editor
  Madhu Prasher
Acquisitions Coordinator
  Alex Corona
Technical Editor
  Ray Sarch

Copy Editor
  Claire Splan
Proofreader
  Anne Friedman
Indexer
  Valerie Robbins
Computer Designers
  Carie Abrew
  Dick Schwartz
Illustrators
  Michael Mueller
Series Design
  Peter F. Hancik

This book was composed with Corel VENTURA™ Publisher.

AT A GLANCE

CONTENTS

PREFACE

In the previous edition of this book I asked readers to imagine the cost of a long-distance or even an international call being billed at a penny a minute! At this price, even the popular "dime lady" who appeared in commercials for a leading long-distance communications carrier could not compete. Calls at a penny a minute or less with a minimum amount of organizational effort to achieve this is the driving force behind voice over data, an emerging technology that has the potential to revolutionize the manner in which both small and large corporations and government agencies construct and operate their communications networks.

Since the first edition of this book Voice over IP and Voice over Frame Relay technologies have considerably expanded. Today, many Web sites have "call buttons," which allow Web surfers to converse with sales personnel and customer representatives. The cost of long distance has significantly decreased, with conventional communications carriers now in competition with free and near free Internet calling.

The technology providing the capability to transport voice over data networks represents a natural evolution of the quest of network managers and LAN administrators to economize on the cost of organizational communications. Until recently, most organizations looked in the opposite direction, attempting to transmit data over existing voice transmission facilities. While data over voice will continue to play an important role in the creation of corporatewide integrated networks, voice over data represents a networking technology that can provide an outstanding return on your investment. In addition, unlike the use of ATM, which can require a complete change in an organization's infrastructure and severely tax existing operations during a changeover period, the switch to voice over data can be so mild that most persons are not aware of the additional capability until they've been informed of it.

Although the ability to transmit voice over networks originally designed to transport data offers significant economic advantages, those advantages are meaningless if the destination party cannot understand the spoken word. Unfortunately, many of the discount voice calling methods as well as the use of certain products under different operating environments will result in Mickey Mouse sounding like Minnie. Fortunately, there are many methods we can use to create a Voice over IP or Voice over Frame Relay solution for our organization that can provide the level of voice quality our customers rightly expect.

As a third edition, this book builds upon the prior edition to include experiences obtained by testing the technology. Because the ability to transmit voice over an IP or frame relay network is very time-dependent, numerous tips and techniques are described in this new edition that may enable you to make seemingly unworkable technology work. In teaching several seminars on the topic of this book in Europe, Israel, South America, and for Interop, I became noted for the expression "save a millisecond here and a millisecond there and the technology works!" Thus, throughout this new edition you will encounter "application notes" that provide tips and techniques whose implementation could save the milliseconds necessary to successfully implement the new technology.

The advantages of voice over data are considerable. However, as with any new technology, its implementation requires appropriate planning, which is best accomplished by understanding the technology, its strengths and limitations, the options to consider, and the methods by which the technology can be implemented—all topics covered in this book. So relax, grab a Coke or a cup of coffee, and pace yourself as we explore this relatively new technology, which, as you will learn, is based on a mixture of voice and data networking technologies.

As a professional author, I highly value reader feedback. Please feel free to contact me via my publisher, whose address is at the front of this book, or send e-mail to me at gil_held@yahoo.com. I welcome your comments and suggestions. For example, should a fourth edition contain greater detail in certain areas? Should I consider adding or perhaps removing some topics? Are there other issues you wish to bring to my attention?

Gilbert Held
Macon, Georgia

ACKNOWLEDGMENTS

Although it is a relatively easy process to select a book from a catalog or shelf in a store, the actual effort involved in its preparation and publication can be quite complex and can extend over a significant period of time. As a professional author, I learned long ago that the creation of a written work is a team effort requiring the cooperation and assistance of many persons. Thus, I would be remiss if I did not thank those who made this book possible.

First and foremost, I must thank my family for lost evenings and weekends while I worked on this new edition. It takes a considerable amount of patience and understanding to tolerate the efforts of a technical author who uses the fax machine at odd hours, fills bookshelves with reference material, and receives express packages early on Saturday mornings.

As an old-fashioned author who frequently travels to locations where even the best gadgets fail to provide the ability to recharge a notebook in hotels, I decided long ago that a pen and paper eliminates electrical-outlet incompatibilities. In addition, pens and paper are products that can be easily acquired worldwide and used on long international flights without fear of battery failure. Using pen and paper while flying through air turbulence results in some interesting examples of handwriting. Thus, once again I am grateful for the fine effort of Linda Hayes in converting my handwritten notes and drawings into a manuscript suitable for my publisher to work with.

The role of an acquisitions editor is most important in backing a writing project from its proposal and development on through the book production process. Thus, I would also like to again thank Steve Elliot, who became a father each time two of my books were published by McGraw-Hill. Now that a third edition is in print, Steve and his wife can work on keeping up with this book!

Last but not least, the cooperation and assistance of two equipment manufacturers deserves special mention. I sincerely appreciate the efforts of Mike Vizzi of ACT Networks and Eric B. Kirsten of Nuera Communications. Both gentlemen shared detailed information that enables me to discuss the operation of their firms' voice over frame relay products and to provide extensive data concerning the cost per minute of voice transport—an economic analysis that, for many readers, will fully justify implementing this relatively new technology.

CHAPTER 1

Introduction

During the late 1940s, an Englishman with a fondness for cigars quoted the classic and often repeated phrase, "Jaw, jaw, jaw is better than war, war, war." Approximately 50 years later, Winston Churchill, if he could view current events from heaven, would more than likely be amazed at the amount of jawing being performed.

Today, both the telephone and the personal computer are ubiquitous office and home products most of us use throughout the day, every day. The telephone is based on analog technology since it must support the analog waveforms generated by human speech. In comparison, the personal computer is based on digital technology since data is encoded and manipulated in terms of strings of binary 0s and 1s.

Until recently, the conventional wisdom associated with integrating voice and data transmission focused on two areas: integrating data transmission requirements into existing voice networks and constructing asynchronous transfer mode (ATM) networks that were designed to support the transfer of voice, data, video, and image information over a common network infrastructure. The integration of data transmission onto networks primarily constructed to transport voice calls between geographically separated organizational locations is anything but a recent phenomenon. The first generation of T1 multiplexers marketed during the 1980s provided this capability, and considerable improvement in voice digitization technology and the development of a new series of multiplexers enables corporate networks to use T1 and T3 transmission facilities and their fractional equivalents to carry router-to-router communications along with a significant amount of voice conversations plus faxes and videoconferencing. While data over voice will continue to be a viable mechanism for transporting information on a common network infrastructure, it is primarily applicable for organizations that have a significant requirement for voice communications between multiple locations.

Although ATM was proposed as a unifying technology developed to support voice, data, video, and imaging applications over a common network infrastructure, as with many technologies its hype is greater than its implementation. The scalability of ATM, which enables LAN-based data operating at 25 or 155 Mbps to be transported as a uniform flow of cells at a T1 operating rate of 1.544 Mbps to a communications carrier's central office and at optical carrier rates up to 10.0 Gbps (and perhaps higher by the time you read this book) between carrier offices, is an admirable concept; however, competitive LAN technology at a significantly lower cost resulted in many organizations postponing ATM to the desktop. Instead, ATM has gained a high degree of acceptance by communications carriers as a mechanism for an assortment of other transports, ranging in scope from traditional digitized voice carried on T1 lines to frame relay and TCP/IP data networks formed literally on top of ATM. Thus, network managers and LAN administrators looking for a practical, efficient, and low-cost mechanism to transport voice and data turned to a third option, which is the focus of this book: voice-over-data networks.

As we turn our attention to transporting voice over IP and voice over frame relay in this book we will note that in many situations we may be able to take advantage of the ATM infrastructure used by many Internet service providers (ISPs) and communication carriers. Because ATM can provide true Quality of Service (QoS) that can guarantee bandwidth and latency or delay, if we can map IP and frame relay to ATM we can take advantage of the capability of ATM. Thus, while voice-over-data networks provide an option to ATM they can also take advantage of an existing ATM infrastructure.

Concerning QoS, as we will note many times throughout this book, the ability to minimize the delay associated with the transport of packets through a network is the key to a successful implementation of a voice-over-data network. Because it is difficult to argue with success, we will describe and discuss application notes at relevant points throughout this book whose implementation can provide you with the edge concerning the successful implementation of a voice-over-data network.

OVERVIEW

Voice-over-data networks represent a technology developed to satisfy one of the most fundamental aspects of network management requirements: the necessity to integrate the transportation of voice and data in a cost-effective manner through the acquisition of equipment whose installation and operation limits potential disruption to ongoing organizational activities. That said, many readers may have the impression that the development of equipment to provide a voice-over-data network transmission capability represents a radical advance in technology. Although it is an advance in communications technology, most of the underlying technology has existed for several years. Basically, voice-over-data networks can be described as the application of voice digitization and compression schemes through a variety of hardware and software products to enable voice to be transported on networks originally developed to transport data. Two of these networks, which are the primary focus of equipment developers and whose use for voice over data is covered in detail in this book, are Internet Protocol (IP) and frame relay networks. As we will note later in this book, the basic concepts concerning the use of hardware and software to obtain a voice-over-data networking capability are applicable for both public and private networks. Thus, the information presented in this book will be applicable for voice over data occurring on the public Internet, private intranets, and public and private frame relay networks. Now that we have an overview of how voice over data fits into the methods of integrating voice and data networks and have had a very brief discussion of the underlying technology and its general benefits, let's broaden our horizon by turning our attention to the reasons that voice-over-data networking is one of the hottest technologies of the new millennium. Since this technology is similar to other communications technologies in that there are certain constraints and limitations associated with its use under different networking scenarios, we will also focus on potential networking problems that can adversely affect the transport of voice on data networks, and we'll discuss existing and evolving methods that may provide you with a mechanism to overcome many of these problems. To obtain an indication of where the industry is headed we will briefly review a chronology of events that occurred over the past few years. While the past may not be a totally valid indication of future events, we can obtain an appreciation for the general direction of the voice-over-data network field by examining some of these events. Last but not least, we will conclude this chapter with a preview of succeeding chapters. This will allow you to decide if you prefer to go directly to a chapter that has information specific to your needs or if, as a newcomer to the use of voice-over-data networks, you should proceed sequentially by chapter to build your overall knowledge.

RATIONALE

Today many organizations operate separate voice and data networks based on a traditional separation of the two technologies and an initial separation of equipment developed to support voice and data networking. Other organizations integrated all or a portion of their voice and data networks by adding equipment that enables data to be transported over circuits originally installed to support voice networking, resulting in a data-over-voice network infrastructure. When organizations failed to integrate their voice and data networks, two of the primary reasons were economics and technology. In some situations it made sense to maintain separate networks, as the cost associated with integrating the two might exceed the potential savings or require a period of time that would result in a relatively poor return on investment. As for technology, until recently, equipment was not available to effectively and efficiently transport voice over IP and frame relay networks, thus requiring consideration of other techniques, such as time division multiplexing, that would have a permanent effect on the ability to transport both voice and data regardless of the bandwidth requirements of each.

Classical Data-over-Voice Constraints

Figure 1-1 illustrates the classical method for integrating voice and data via time division multiplexing. In examining Figure 1-1, note that each multiplexer frame is configured to provide a static allocation of bandwidth between voice and data sources. For example, if the PBX was configured to provide 20 PCM-encoded voice conversations, each operating at 64 Kbps, then each frame would consist of 1.28 Mbps of bandwidth allocated to the PBX (64 Kbps × 20) and the remaining bandwidth of 256 Kbps to the data sources. Since the slots in the frame are fixed by time, if the data sources became inactive, the PBX could not take advantage of this fact to transmit additional PCM-digitized voice conversations. Similarly, if one or more of the 20 voice conversations supported by the multiplexer became inactive or terminated, the multiplexer could not allocate additional bandwidth to the data sources.

Another constraint associated with classical TDM operations is the requirement for organizations to construct their own networks via the installation of a series of point-to-point digital leased lines. Although this type of network is available for the exclusive use of the organization, that exclusivity is not without a price. In comparison, the cost associated with the use of packet networks that support the transmission requirements of a virtually unlimited number of different organizations can be considerably less expensive due to the sharing of the cost of the transmission facilities used to construct the packet network.

The reliability of a private leased-line network represents another constraint that must be considered. The most common way to enhance reliability is by installing additional circuits

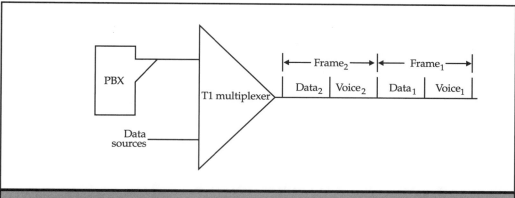

Figure 1-1. The classical method for integrating voice and data is via time division multiplexing, in which fixed segments of bandwidth are allocated to voice and data.

with diversity routing to ensure each circuit is routed through different central offices between source and destination, so the additional cost can be considerable—an organization would be charged for two circuits between each network location as well as a monthly fee for the diversity routing of the pair of circuits routed between locations.

Advantages of Voice-Over-Data Networking

The key advantages associated with the use of a packet network for the transmission of digitized voice can be considerable. Following are the advantages associated with transmitting voice over a packet-shared network:

- ▼ Bandwidth allocation efficiency
- ■ Ability to use modern voice-compression methods
- ■ Ability to take advantage of characteristics of a voice conversation
- ■ Sunk costs
- ■ Ability to use a single interface
- ■ Enhanced reliability of packet networks
- ▲ Economics associated with shared network use

Bandwidth Allocation

Both IP and frame relay represent packet-shared networks for which bandwidth is consumed only when transmission occurs. Thus, these networks remove the fixed bandwidth allocation associated with the first generation of T1 multiplexers. However, these networks were originally developed to transport data over relatively low-speed transmission facilities, such as 56- and 64-Kbps digital lines, with access to both of these networks only recently increased to T1 and for IP networks to a T3 operating rate, while a T3 connection rate to frame relay was standardized as this book was written.

Modern Voice-Compression Techniques

Since a PCM-digitized voice conversation requires a 64-Kbps operating rate, it would make no sense to attempt to transmit the digitized conversation over a 56- or 64-Kbps digital link to an IP or frame relay network, as doing so would preclude the ability to concurrently transmit data. Recognizing this problem, equipment developers incorporated a new generation of voice-compression algorithms into their products. Today, you can consider using some frame relay access devices (FRADs) that incorporate voice-digitization modules that lower the operating rate of a voice conversation to 4 or 8 Kbps. This means you can transmit between 8 and 16 voice conversations on one 64-Kbps frame relay connection or a mixture of voice and data over a single frame relay connection.

Human Speech Characteristics

Unless both parties to a conversation get quite animated and shout at one another, a human conversation is *half-duplex*. This means that as one party speaks, the other party operates in a civilized manner and listens. In addition, unlike the actor in a TV commercial for Federal Express who appeared to talk nonstop, we periodically pause as we talk.

When voice is digitized and transported via packet networks, small digitized samples of our conversation are transported via packets. Because the other party normally listens when we speak and we periodically pause for air, the packet network obtains the ability to use those gaps to transmit data, digitized voice, video, and other packets on its backbone network. In comparison, on the switched telephone network 64 Kbps of bandwidth is allocated through the telephone network for the duration of a conversation, regardless of the fact that voice is normally half-duplex and we periodically pause during a conversation. Thus, the characteristics of human speech are well suited for their transportation over a network.

Sunk Costs

The incorporation of voice compression and the development of techniques to enable compressed voice to flow effectively and efficiently through a packet network are two key mechanisms that resulted in the acceptance of the technology as a practical and economically viable mechanism for transporting voice over data networks. A similar incorporation of voice-digitization algorithms into IP equipment permits voice conversations to be transmitted using a very low amount of bandwidth on IP networks.

Since a major portion of the cost associated with the use of a frame relay network involves the monthly recurring fee for an access line and network port at the entrance to the frame relay network provider, any method that permits additional data sources to flow over a common connection becomes more cost-effective. Similarly, if your organization has a connection to the Internet or you operate an internal IP network referred to as an *intranet*, the ability to use existing transmission facilities represents transmission over sunk costs. That is, you are already paying for such equipment as router ports, channel service units, and transmission facilities. Hence, transporting digitized voice without incurring additional expenses other than for voice digitization can result in a very low cost per call minute.

The ability to transmit digitized voice over frame relay- and IP-based networks can provide several key advantages. Those advantages include the ability to transmit to and receive information from multiple locations via a single connection to a packet network, enhanced transmission reliability, and the economies of scale associated with the use of packet networks shared by many organizations.

Multiple Access via a Single Network Connection

A packet network permits the flow of multiple logical connections over a single physical circuit, thus allowing an organization to support transmission to multiple locations via a single network connection. Figure 1-2 illustrates this concept, showing an intermix of packets being transmitted to locations A and B from location C via the use of a packet network access device and a single network connection from that device to the packet network. In comparison, the use of a leased-line-based network would require an organization to install separate leased lines from location C to locations A and B. This in turn would require the networking device used with the leased-line network, such as a router or multiplexer, to have two ports, since one port would be required to support the connection to each leased line. This in turn would add to the cost of establishing a private leased-line-based network, since router and multiplexer ports can cost between $500 and $1000 or more per port. In addition, when your networking requirement expands such that you must consider the interconnection of a large number of locations, the cost of individual point-to-point networking connections and a large number of router or multiplexer ports per location can become prohibitive. In such situations, you would probably employ a different network structure—perhaps interconnecting several locations within distinct geographical areas to a single hub location and then interconnecting the hubs. Although this type of networking structure can be more economical than leased lines and routers or multiplexers, it is more appropriate for supporting a large voice-transmission internetworking requirement and a significant number of data sources than for supporting a moderate volume of voice over a data network. For the latter situation, the cost associated with using a frame relay or IP network for transmitting voice and data will usually prove to be more economical. This in turn opens up a vast number of networking possibilities, such as connecting national or international locations together via a packet network or even using a FRAD to support voice on a leased line between two locations that are required to transmit only a few data sessions of activity between locations, since the cost of adding voice support to some vendor products can be accomplished for a minimal one-time fee.

Enhanced Reliability

The backbone infrastructure of packet networks is commonly constructed based on a mesh topology. This topology, illustrated in Figure 1-3, commonly provides two or more routes or paths between network nodes. Depending on the number of network nodes between two network access points, the mesh structure can provide a large number of

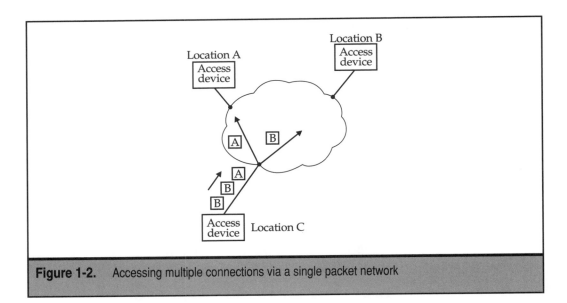

Figure 1-2.   Accessing multiple connections via a single packet network

alternative routes between network access points. For example, if the direct line connection between locations A and C should become inoperative, transmission between those two locations could continue on paths A-B-C or A-D-C. Thus the backbone mesh structure of packet networks provides a built-in alternative routing capability whose cost to support is shared by the large number of individuals and organizational users of the network. In comparison, it could be extremely expensive for a single organization to develop a mesh network infrastructure to improve network reliability.

Economics of Use

In our prior discussion of the advantages associated with the use of packet networks, we noted that they can provide access to multiple locations via a single network connection and that their mesh infrastructure to enhance reliability would be expensive to duplicate

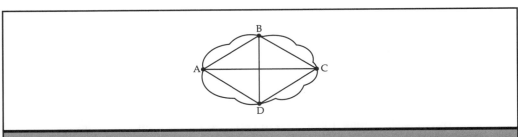

Figure 1-3.   Most packet networks use a mesh structure to connect access locations, providing an enhanced level of redundancy that improves network reliability.

for a private-line-based network. Both of these advantages provide an economic edge to the use of packet networks. However, a more substantial economic advantage can result from the cost structure associated with the use of different frame relay networks and the Internet. For example, the usage charge associated with some frame relay networks is based on a flat monthly fee determined by what is referred to as the *committed information rate* (CIR), which is the transmission rate the frame relay network provider guarantees will be serviced. The CIR can be less than or equal to the operating rate of the access line and is described in detail later in this book. This means that if you are not using the full CIR, you can transmit additional information in the form of conventional data or digitized voice without incurring any additional transmission charges other than a one-time expense associated with the acquisition of equipment required to transmit voice over a frame relay network. For the use of an IP network, most Internet service providers charge organizations a monthly fee based on the operating rate of the leased line connecting an organization's location. Thus, transmission of additional information in the form of conventional data or digitized voice over the Internet may not alter the organization's communications cost (other than the one-time expenditure for equipment to support the transmission of digitized voice). If the IP network represents an organization's internal private network, the addition of voice transmission over that network can also result in significant savings. This is because the use of modern voice-compression methods can enable voice packets to flow over the excess bandwidth available on many networks. Once again, this flow of digitized voice packets does not alter the monthly cost of the organization's transmission facilities and is normally obtained as a result of the one-time expenditure for equipment to digitize and packetize voice so it can flow over an IP network. Now that we have an appreciation for some of the advantages associated with the transmission of voice over data networks, let's turn our attention to some of the networking problems that may occur. By understanding some of the potential pitfalls of transporting voice over data networks, you can make more intelligent decisions about implementing this networking technique based on your organization's current and evolving communications requirements, the existing network infrastructure used by your organization, and the potential use of different types of packet networks.

POTENTIAL PROBLEMS TO CONSIDER

The transmission of voice over data networks includes a degree of risk associated with the use of the technology. Potential problems that network managers and LAN administrators should consider can be classified into five general areas: reliability, predictability, security, standards, and the cost of the switched telephone network.

Reliability

For the purposes of this book, we will define *reliability* as the ability of a packet to reach its destination. The transmission of voice over packet networks originally developed to transmit data entails a degree of risk concerning the reliable delivery of voice-encoded packets. Some packet networks include a discard mechanism by which packets can be figuratively routed to the "great bit bucket in the sky" when the level of utilization of the network reaches a predefined level. While performance is not significantly altered by data packets transporting e-mail messages, carrying interactive query-response data, or retransmitting files of a dropped packet (or sequence of packets), the same is *not* true when digitized voice is being transported. Speech has to be reconstructed so it sounds natural, and this requires an absence of extended delays during which dropped packets are retransmitted. Thus, the ability to effectively transport voice over data networks requires a mechanism to ensure the reliable delivery of packets. As we describe and discuss the operation of certain types of equipment later in this book, we will note that vendors have increased the potential for reliable delivery of voice-encoded packets primarily by limiting the length of such packets, lessening the probability that they will be dropped by a packet network. Because many router queuing algorithms allow priority to be given to packets based upon packet length, this technique can work reasonably well. However, as more and more short-length packets begin to flow over a network the potential for reliable delivery based upon packet length will lessen. Thus, a technique that may work reasonably well when your organization is among the first to use a new technology may result in unexpected problems as other users migrate to the technology.

Another associated reliability problem with respect to transmitting voice over a data network is the manner by which a receiver performs when a voice-carrying packet is dropped. Most equipment simply does nothing, resulting in a period of silence which, if packet dropping is only occurring rarely, is more than likely imperceptible to the human ear. However, if several packets are transmitting digitized voice flow to the great bit bucket in the sky, the result is an audible gap of silence that becomes noticeable to the human ear. To overcome this problem, some equipment vendors generate a bit of noise, which is not as noticeable as a gap of silence.

Predictability

Although many persons might consider predictability to be similar to reliability, they actually represent two separate problems associated with the transmission of voice over data networks. *Reliability* refers to the ability of packets containing digitized voice to reach their intended destination without being dropped by the network. *Predictability* refers to the delivery of those packets without an excessive amount of delay that would result in the reconstruction of the transported conversation sounding awkward. Depending on the type of voice being transported, both reliability and predictability may or may not be an issue. To understand this, let's examine the two primary categories of voice your organization may wish to transport over a data network.

Effect on Voice Applications

There are two general categories of voice applications you can consider for transporting voice over a data network: real-time and non-real-time. A real-time voice application, such as a telephone call, requires both reliability and predictability. In comparison, the attachment of a voice message to an e-mail or the transmission of prerecorded voicemail from one location to another requires neither reliability nor predictability.

Concerning the attachment of a prerecorded voice message, reliability with respect to minimizing the dropping of packets transporting digitized voice is not required, since the eventual retransmission of dropped packets ensures the conversation arrives at its destination, where it will be listened to after the pieces of the voice message arrive and are stored as an entity. Similarly, predictability is no longer an issue, as packets arriving with random delays are not listened to in real time. Instead, the packets are assembled as an entity prior to being listened to, which removes gaps associated with their transmission over a packet network.

Techniques to Improve Predictability

There are two basic techniques that can be used to increase the predictability of packets arriving at their intended destination. The first technique involves providing a QoS function to reserve network resources for the transmission of the sequence of voice packets with a guaranteed minimum delay. Although QoS is built into ATM, it has only recently been developed as a mechanism for incorporation indirectly into IP networks via the Resource ReSerVation Protocol (RSVP). Unfortunately, RSVP requires the use of RSVP-compatible devices throughout the route of packets flowing through an IP network to obtain the ability to reserve network resources from source to destination. On a public IP network like the Internet, it may be several years before a significant amount of vendor equipment is upgraded to RSVP to enable a small fraction of RSVP requests to be fully honored. Of course, if your organization has an internal IP network, you have a far greater ability to control the upgrading of equipment to support RSVP.

In addition to RSVP there are other components of what this author likes to refer to as the QoS puzzle. Those components range in scope from the Institute of Electrical and Electronic Engineering (IEEE) 802.1p standard for prioritizing traffic in a switched LAN environment to the Internet Engineering Task Force (IETF) Request for Comments (RFCs) covering Differentiated Service (DiffServ) that expedites traffic into a network, and Multi-Protocol Label Switching (MPLS) that expedites traffic through a network.

Recognizing that RSVP may be several years away from being useful, that the interoperability of 802.1p, DiffServ, and MPLS remains to be seen, and that an equivalent method for frame relay is lacking, equipment developers have attacked the need for predictability via the control of voice packets. In doing so, equipment vendors have introduced techniques that fragment voice packets into smaller-size packets to enhance their ability to be transported over a network, developed schemes for the prioritization of the transmission of voice packets over data so voice packets reach the network first, and initiated other techniques that are described in detail later in this book. Although such techniques do not actually guarantee that voice packets will arrive at their destinations in a timely and predictable manner, they greatly enhance the probability that they will do so.

Security

Unless you are working for the CIA, DIA, or another spook agency, chances are high that you will never have a second thought concerning the possibility of your conversation being overheard when you use the Public Switched Telephone Network (PSTN) or an internal corporate voice network. Although the probability remains slim that a voice conversation will be overheard when you use a voice-over-data network transmission method in which voice packets or frames flow over a public packet network, the use of a public packet network that is connected to an internal corporate network results in a new security risk—a risk with far greater potential damage to your organization than overheard conversations about a planned trip, travel arrangements, or even the potential bid on a construction project. That potential damage results from the connection of an organization's internal private network to a public packet network, which opens the internal network to access from anyone who can connect to the public network.

Although an organization will always have a degree of exposure to the contents of digitized voice packets being inadvertently or intentionally read as they are routed through a public packet network, a perhaps more significant problem is the fact that every hacker and cracker can attempt to access your organization's network once that network is connected to a public packet network. What took burglars and swindlers hours or days to accomplish, a hacker can now do in minutes using a personal computer. From a security perspective, there are three issues to be considered: barring uninvited persons from accessing your network, verifying the identity of those who access your network, and encoding the contents of packets so they are not readable. These issues involve access control, authentication, and encryption with respect to transmission over a public packet network, as illustrated in Figure 1-4.

Access Control

Access control represents a mechanism to enable or disable transmission between a public and private network, between two public networks, or between two private networks based on some predefined metric, such as the source address of a packet. One of the most common methods used to implement access control is through the construction of an access list programmed into a router, assuming the router incorporates this feature.

One of the least understood problems associated with access control lists is the trade-off between security in a voice-over-IP environment and delay in the processing of the extended list that can adversely affect the intelligibility of the reconstructed piece of voice at its destination. Because the use of TCP and UDP ports for voice are not standardized, this means that if your organization supports several vendor products and uses a router access control list for security, your list more than likely includes several voice-over-IP-related statements. Because access lists are examined from the top down, sequentially, placing your voice-related statements at the bottom of the list can result in a few extra milliseconds of delay.

APPLICATION NOTE: If using access control lists, move your voice-related statements toward the top of the list.

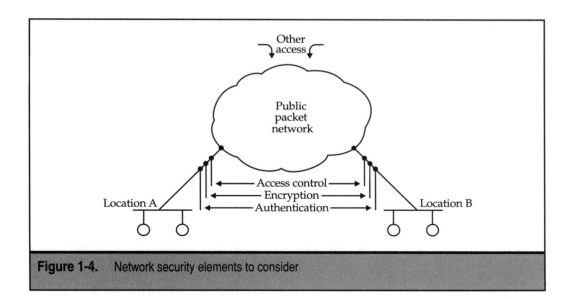

Figure 1-4.   Network security elements to consider

To gain a few milliseconds, you should consider placing your voice-related statements near the top of your access control list. Since most lists begin with statements that disallow inbound packets containing source addresses from your network and RFC 1918 addresses to prevent address spoofing, a good technique is to place your voice-related statements right after your antispoofing statements in your access control list.

While the relation of voice-related statements to the top of a router's access control list may appear to open up a vulnerable area in your network, let's consider the worst that can happen. Suppose at the top of the list you include two statements, one allowing TCP on the port used by your organization's voice gateway for call control and a second statement that permits voice packets transported by UDP to the voice gateway. If your organization's voice gateway provides connectivity to a PBX for local calls, the best a hacker might accomplish, assuming he figures out the type of voice gateway used by your organization, would be to route calls to your employees. Distracting, yes, and you would then readjust the order of your access control list statements.

In examining the three elements of network security illustrated in Figure 1-4, note that access control is shown as the innermost layer with respect to the relationship between two organizational networks (A and B) communicating via the use of a public packet network. Access control is placed as the innermost layer because the most popular access control device, a router, can be considered to represent the demarcation between the public and private network. A second access control device that has gained widespread popularity is the firewall. In addition to providing an access control capability through the creation of access lists, a firewall can add a significant number of additional security features, including authentication, encryption, message alert generation, and proxy services. Both the use of router access lists and firewalls are discussed later in this book.

Encryption

Since encryption usually occurs based on the operation of a device behind a router, it is shown as the middle layer of security elements depicted in Figure 1-4. However, readers should note that encryption can be one of the optional features provided by many firewall vendors along with authentication. Thus it is entirely possible for encryption and authentication functions to be colocated on a common device. Similarly, a firewall could be used to provide authentication while a private network device, such as a program operating on a computer located on the private network, performs encryption. In this situation, authentication would be performed prior to encryption, which from a practical standpoint makes more sense when receiving packets, because the productivity of equipment would be adversely affected if, after decrypting a packet, its originator could not be authenticated and the packet was subsequently rejected.

Today you have a range of products that can be considered for use for encryption of packets. However, unlike point-to-point circuit utilization where an entire packet can be encrypted and decrypted without affecting its ability to be routed, the application of encryption for data flowing over a public packet network requires the use of an intelligent encryption device. That device must recognize the separation of the information field from the header and trailers of a packet, operating only on the information field. Otherwise the packet would not be routable through the network.

The use of encryption to hide the meaning of packets results in another delay, which by itself may not be meaningful, but in conjunction with other factors when moving voice over a data network can adversely affect the intelligibility of reconstructed voice. Because a hacker must not only intercept your voice packets but in addition must know the voice-compression algorithm, it is not a simple process to hear an unauthorized conversation. Because of this, in many situations you may wish to carefully consider the benefit of encryption prior to its use.

APPLICATION NOTE: Encryption adds to the delay of packets transporting digitized voice and should be employed only when absolutely necessary.

Authentication

The third major element of network security is verifying that the originator of a packet is the person he or she claims to be, a process more formally referred to as *authentication*. Although the sound of voice in a personal conversation may suffice to verify the originator of a call, when you connect an internal corporate network to a public packet network, there may be some data access requirements that will require authentication. If so, you will then have to consider the acquisition of equipment that can distinguish between the information transported in different types of packets and require authentication only for packets attempting to access predefined services on the private network. Now that you understand the rationale for transmitting voice over data networks and some of the potential problems, let's turn our attention to some of the key reasons you would consider adding a voice-over-data networking capability to your organization's network infrastructure (in addition to the economics associated with the technology). We'll look at a few of the applications that may by themselves justify adding this capability to your network infrastructure. However, prior to doing so, a brief discussion of regulatory considerations and standards is in order.

Regulatory Considerations

While the transmission of voice over data networks provides a variety of productivity and economic advantages that make it a very desirable technology, readers should be aware of certain regulatory issues that can affect its implementation. These include national and international efforts aimed at precluding individual organizations from transmitting voice over their internal networks and preventing voice resellers that provide a telephone service via the use of gateways from competing with communications carriers.

Rationale

Economics is the primary reason for national and international regulations attempting to prevent or limit the ability of organizations to transmit voice over data networks. Over the past century, communications carriers invested hundreds of billions of dollars globally in technology based on the use of 64-Kbps time slots to transport digitized voice. The ability to transport voice at data rates as low as 2.4 Kbps is more than competitive; it provides the eventual capability of relatively newly formed alternative communications carriers to put long-established organizations out of business. It is sort of like one widget manufacturer being able to undercut the cost of another by a factor of 8 or 16 or more!

National Events

At the national level, the Association of Telecommunications Carriers, an organization representing thousands of small telephone companies, petitioned the Federal Communications Commission (FCC) to prevent the establishment of voice calling over data networks, as their ability to earn long-distance revenue is curtailed while gateways tie up their local exchanges.

In addition, another interesting aspect of regulation involves the desire of local telephone companies to charge for the delivery of long-distance calls that are delivered to a subscriber of an Internet connection. If a friend or business associate uses Worldcom, AT&T, Sprint, or another long-distance communications carrier to dial a telephone number where local access is provided by another carrier, the carrier that provides local access is rightly compensated for completing the call. Currently, in the wonderful world of Internet telephony, there is no mechanism for local access providers to be compensated, which rightly makes them angry. At the time this new edition was prepared, the FCC ruled that Internet calls represent long-distance calls; however, it remains to be seen what effect this will have on the ability of local access providers to be compensated for delivering such calls.

To understand the potential effect of charges for delivery, assume a local access provider bills calls at a nickel per minute. Vendors currently charging 4.9 cents per minute for Internet calls would go out of business because they could never earn a profit. However, for private networks based on the use of leased lines, the preceding local access delivery charge would not be applicable. Thus, a corporation that uses leased lines to form a private network or gain access to a public packet network would not be subject to local delivery charges.

The amount of money provided by long-distance carriers to local exchange carriers for "last mile" delivery of telephone calls is anything but nominal on a cumulative basis. According to trade reports the big three inter-exchange carriers spent tens of billions of dollars paying for local exchange carrier call delivery. This explains why AT&T spent over $100 billion purchasing cable TV systems to gain access into homes, which allows phone calls over cable to bypass the local carrier. Because numerous lobbyists are involved on each side of phone delivery, it is difficult to predict what decision, if any, will occur concerning the bypass of local exchange carriers by voice-over-IP and voice-over-frame relay transmission systems. However, it is worth noting that a change could render your economic model obsolete.

International Events

At the international level, several countries have blocked access to Web sites that offer software for making international calls over the Internet. The latter situation occurred during June 1997 when the Czech Republic, Hungary, Iceland, Portugal, and Palau (an island in the South Pacific) made it unlawful for customers to access a Web site that provides software enabling voice calls to be placed over the Internet. In March 1999, Lebanon added a new wrinkle to the prevention of Internet telephony when it officially banned Internet service providers from offering overseas Internet calls to their subscribers. Since the software enables calls to be made at a cost of between 10 and 30 cents per minute, while long-distance companies in those countries charge between $2 and $4 per minute, the preceding actions could be considered one of self-preservation for the international long-distance service offered by certain national communications carriers. One vendor affected by this action asked the FCC and White House to intervene on the issue. While the potential effect of national and international actions against voice-over-data network applications is essentially anyone's guess, it is this author's opinion that restrictive regulations will eventually be self-defeating. Although voice-over-data networks are in their infancy, they represent a relatively easy-to-apply technology that is difficult to bar by the regulatory process. Instead of attempting to fight this technology, national and international communications carriers should be looking at methods to integrate the technology into their communications infrastructure. They might then be able to develop schemes to price the transmission of low-bit-rate digitized voice at a substantial discount from their regular tariff and thus encourage persons to use their facilities.

Standards

Standards can be considered the glue that enables different vendor products to interoperate. In the area of standards, voice over IP is still in its infancy, with many standards clearly lacking, thus forcing vendors to use other standards that may not be the most suitable for operation. Although voice over frame relay is presently more standardized due to the efforts of the Frame Relay Forum in developing Implementation Agreement (IA) (covering numerous aspects of transporting voice over frame relay), there are still other areas that lack standardization. For example, frame relay providers have not standardized how their prioritization occurs or queuing methods used to support prioritization.

Currently, popular IP standards include the Real-Time Protocol (RTP), which involves the time-stamping and sequencing of packets; RSVP, which involves the reservation of bandwidth; and H.323, which represents an umbrella standard originally developed for conferencing but now embraced by many vendors because of control bandwidth utilization. Other important standards include the Session Initiation Protocol (SIP) and the Media Gateway Control Protocol (MGCP) as well as such quality of service standards as DiffServ and MPLS. In this book we will focus our attention upon each of those standards.

In addition to the previously mentioned standards, there are two evolving standards that, while extremely important for the success of Internet telephony, may not be necessary for an organization implementing a voice-over-IP capability. These two standards are the Open Settlement Protocol (OSP), which is designed to handle authorization, call routing, and call detail billing between ISPs; and iNOW!, promoted by several vendors to obtain interoperability among IP telephony platforms produced by different vendors.

Switched Network Cost

When the first edition of this book was written, the Sprint "Dime Lady" was a popular fixture on television commercials. During those commercials television viewers were provided the opportunity to change their long-distance provider and have their long-distance calls made from 7:00 P.M. through 7:00 A.M. weekdays and all day during the weekend billed at a dime per minute. When this new edition was being written a few years later, Sprint was advertising a calling plan that gave subscribers 1,000 minutes of long distance per month at a cost of $20, or two cents per minute.

The reduction in the cost of toll-quality long distance represents an important criterion if you are considering implementing a voice-over-data facility for your organization. Because the cost of long distance is dropping faster than a rock in a lake, it is important not only to compare the cost of a voice-over-data network solution against the current cost of long distance, but, in addition, against the potential cost of long distance over the next few years.

APPLICATIONS

When discussing potential and existing voice-over-data network applications, we are literally moving into a new dimension—akin to the way in which the advent of three-dimensional graphics revolutionized that industry. Following is a list of seven examples of voice-over-data network applications:

▼ Remote telecommuter support

■ Document conferencing

■ Help desk access

- Integrated call management
- Order placement
- Web-enabled call center
- ▲ Unified messaging

We'll discuss them briefly to illustrate the potential of this relatively new technology.

Remote Telecommuter Support

If the 1980s can be characterized as the decade of the PC and the 1990s as the decade of the LAN, the first decade of the new millennium will probably be characterized as the decade of the telecommuter. The ability to work at home is beginning to be recognized as a valuable mechanism to hold onto workers in a tight labor market. Unfortunately, many workers have only a single telephone line in their home. This means they must disconnect any remote access operation they are performing in order to call the office or another location if they have a question. This obviously decreases the productivity of the telecommuter.

Now imagine a person working at home and accessing the corporate database when a question arises concerning the information being viewed. Suppose the remote telecommuters now have the ability to click on an icon located on their screens and dial a telephone number. While viewing the database they obtain the ability to converse with a coworker who answers their question. As you might imagine, the productivity of the remote telecommuter has increased.

If the preceding sounds far-fetched, it isn't. As we will shortly note in this section, several Web sites now provide a no-cost or low-cost local and long-distance voice-over-IP capability that enables subscribers to use their PC to make calls.

Document Conferencing

Document conferencing enables multiple parties to verbally communicate while viewing the same document on their computer screens. Until recently document conferencing was based on the use of the public switched telephone network for voice communications, while software on multiple computers used the corporate network or the Internet to provide real-time document-editing capability. With the ability to transmit voice over data, it becomes possible to integrate voice and data communications requirements onto a common network infrastructure. This not only makes it simpler to coordinate software, but also eliminates the use of the PSTN, which could result in considerable savings, especially if document-conferencing audio and data are transported between international locations.

Help Desk Access

Imagine that a customer just bought your firm's latest software or hardware product. Suppose that during its installation the customer encounters a problem or has a question. Now further suppose that the installed hardware or software product generates an interface to the customer's favorite browser and, upon recognition that its PC is multimedia-ready, generates a screen displaying a message such as "click here to talk to our customer service representative." By clicking on an icon, the user's browser establishes a connection to the product help desk operated by your organization, which supports a digitized telephone application that enables the customer to verbally define the problem and seek assistance.

The preceding help desk operation represents a voice-over-data network application that makes good sense from both a customer satisfaction and an economic perspective. Such applications could forestall toll-free calls to your organization's help desk facility. In addition, since many organizations do not provide international toll-free access to their customer assistance or help desk, providing voice support via the Internet could be a marketing ploy that makes one product more valuable than an equivalent product whose manufacturer does not offer this feature. Thus, marketing and economic advantages accrue by providing voice access via a data network to an organizational help desk.

Integrated Call Management

Call management represents the routing of inbound and outbound calls using the most appropriate technique for the call to reach its intended destination. When you add the ability to transmit voice over data networks, you can integrate this capability with existing voice applications and transmission methods to obtain an integrated call management capability. For example, a PBX could be programmed to recognize the dial prefix 6 as the code to route calls from the local PBX to another location via an internal private IP network, a frame relay network, or the Internet.

Another call management technique that provides an interesting discussion topic is international callback. If you travel abroad and read the *International Herald Tribune* or scan some foreign newspapers you will more than likely note advertisements proclaiming "Cheap Callback," "Save 50% or more on international calls," and similar headings designed to get your attention. Although you might expect some sort of illicit scheme is in progress, vendors offering rates 50 percent or more below international calling rates are legitimate. They use a callback feature that lets subscribers dial their facility in the United States, which in many cases does not answer their call but reads their caller ID telephone number and dials back using much lower rates. On receiving the callback, the subscriber enters an access code and the desired telephone number, and the equipment in the United States generates a second telephone call to the destination. Since the cost of *both* U.S.-originated calls is significantly less than the cost of *one* international call originated in some overseas locations, a portion of the savings is passed along to the subscriber to encourage the use of callback.

To economize even more on the cost of international calls, many callback operators recently turned to voice-capable FRADs, using them on international circuits to transport up to 16 simultaneous calls over one 64-Kbps frame relay access line or on a leased 64-Kbps circuit. In fact, ACT Networks reported high sales of their voice-capable FRADs to callback operators due to the ability of their equipment to significantly reduce the cost of long-distance international calling.

Order Placement

Although in its infancy, the use of the Internet for ordering products ranging from books to bagels is literally exploding. The ability to communicate verbally with an order desk adds a new dimension to selling products over the Internet. Let's consider the manner in which products are currently sold and compare it to how they *could* be sold, using as an example the sale of coffee makers. Say you locate a World Wide Web page that displays a picture and description of a 12-cup coffee maker. Let's assume you have a question concerning its filters. Normally, you might click on the e-mail icon that would generate a form with the company's e-mail address included in the "to" line. After filling out a message and clicking on the Send button, you would probably go on to check other sites, since you know the response to your query will be far from instantaneous. Now suppose that instead of clicking on an e-mail icon, you could click on a voice icon and directly communicate with a customer service representative who immediately answers your query and concludes with, "Can I process your order now?" This is obviously a more effective method for obtaining a sale.

Another possible variation of the use of voice for Internet order placement for potential customers who do not have an appropriate voice capability involves integrating an organization's call management capability with Web browsers' activity. For example, instead of clicking on a voice icon, the potential customer might also be offered the option to click on a Help icon that enables the user to enter his or her name and telephone number on the WWW page. This information is passed to the vendor, which results in the call management system dialing the number and connecting a customer representative to the dialed connection. For both methods discussed in this section, the end result is the use of voice in conjunction with data to assist potential customers and to hopefully improve the probability of closing a sale.

In the event the preceding sounds like fantasy, your attention is directed to the Surf&Call Web site whose home page is shown in Figure 1-5. Surf&Call is a relatively new business venture from VocalTec, a pioneer in the development of voice over IP.

Surf&Call is a Web plug-in that appears as a button on a Web page. Web surfers can download a client program from the Surf&Call Web site **www.surfandcall.com** that provides them with the ability to communicate with employees at a Surf&Call-enabled site. If the client PC has a microphone, sound card, and speakers, the Web surfer obtains the ability to have real-time voice connectivity with any Surf&Call-enabled Web site.

The initial availability of Surf&Call dates to 1997. However, it wasn't until a few years later that the program became more practical for use as Internet service providers upgraded their backbone infrastructure that reduced the latency or delay of packets flowing through the Internet.

Recognizing that many but not all PCs have microphones, sound cards, and speakers, Surf&Call provides other options that indicate why businesses must carefully consider the universe of potential customers and their hardware and software. For example, Surf&Call also supports regular telephone calls to Web site operators and synchronizes

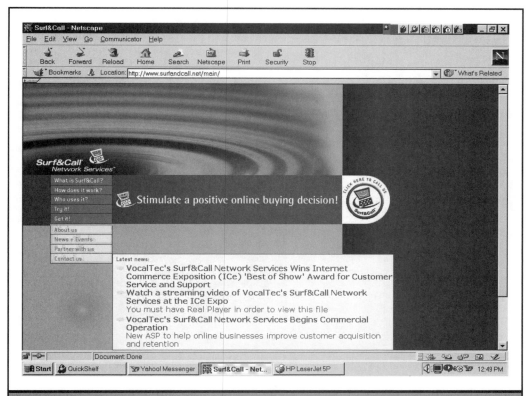

Figure 1-5.   The Web page of Surf&Call, a company that provides a Web plug-in that appears as a button on a Web page and supports real-time voice connectivity between Web surfers and Web server operators

client operations with a call center agent. The software also supports real-time messages, which permits three methods of conversation between customers and Web site operators in addition to e-mail and form submittal that are typically anything but real-time.

When this book was prepared the folks at Surf & Call were beginning to promote their product using full-page advertisements in different trade magazines. In addition to Deutsche Telekom A.G. of Bonn, Germany, which is the world's third largest communications company, Surf & Call has clients ranging from a computer rental firm to a vacation firm. Figure 1-6 illustrates the Web home page of Readyserve, a computer rental firm. In examining that home page note the Surf & Call button in the lower-left portion of the screen as well as a second button protruding into view in the lower-center of the screen. By clicking on either button the Web surfer can initiate a real-time conversation with an employee of Readyserve.

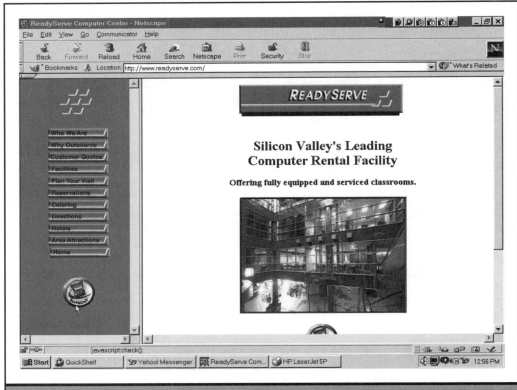

Figure 1-6. The Readyserve Web home page contains two Surf & Call buttons which enable a real-time conversation to be initiated with an employee of the Web site.

Web-Enabled Call Center

When you make an airline reservation, request a brochure, or place an order from a catalog by phone you are more than likely having your telephone call query serviced by a call center. Call centers date to the 1970s, although multivendor call centers are a more recent version that provides significant economies of scale to conventional call center operations.

A multivendor call center services calls for different vendors. Calls dialed using different 800, 877, and 878 toll-free numbers are routed to the call center, and sent to the first available operator and integrated with a computer system that displays the name of the organization being dialed. This allows the operator to greet the caller appropriately when the call is answered and makes the caller believe they are conversing with an employee of the organization on the advertisement that listed a toll-free number for obtaining further information.

In an Internet environment the economies of scale associated with conventional call centers are applicable. Thus, it is both possible and probable to expect call centers to be established that will contain operators that will provide real-time conversational support for different Web sites. However, instead of a computer at the call center displaying a vendor name based upon the dialed telephone number the call center computer will use the destination IP address as the vendor identifier.

Unified Messaging

Unified messaging refers to the use of a system to manage all methods used to convey information, such as voicemail, e-mail, fax, and real-time voice. The addition of voice-over-data networks into a unified messaging system enables employees to make more effective use of all methods used to convey information. This is because it enables a common graphical user interface (GUI) to be used and learned, and eliminates the need for users to spend additional time locating and retrieving messages from nonunified separate systems. In addition, by integrating voice-over-data network applications into a unified messaging system, the system can facilitate the use of voice application, such as real-time voice or voicemail that employees may use instead of dialing over the PSTN, resulting in both enhanced productivity and the cost avoidance associated with not having to leave real-time voicemail messages when the party called using the PSTN is not available.

One example of the potential of unified messaging is the evolution of various Internet messenger products, such as Yahoo! Messenger or AOL's ICQ. Because this author is a frequent user of Yahoo! Messenger, let's look at an example of unified messaging using a beta version of the Yahoo! product that was available during late 2000.

The Yahoo! Messenger for Windows-Beta version used by this author added the ability to make free PC-to-phone calls anywhere in the United States to Yahoo! Messenger as well as providing other improvements, such as a new chat client and a toolbar that facilitates the sending of emoticons, changing text colors, and modifying other visual effects.

The left portion of Figure 1-7 illustrates the new beta version Yahoo! Messenger window. Note the call button located just under the Friends menu option. Clicking on the call button results in the display of the telephone dialing window referred to by Yahoo! as the

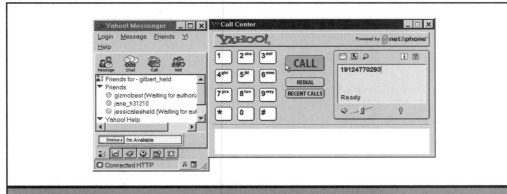

Figure 1-7. A new feature added to Yahoo! Messenger and similar programs is the ability to make free long distance and local calls anywhere within the United States.

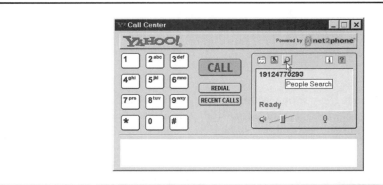

Figure 1-8. The Yahoo! Call Center provides the ability to retrieve information from a user's address book or from Yahoo! Yellow Pages and Yahoo! People Search by clicking on an appropriate button.

"Call Center." The right portion of Figure 1-7 illustrates the Yahoo! Call Center display. As you click on each telephone button the corresponding number is displayed. Once a number is entered, clicking on the Call button initiates a call via the Internet to a voice-over-IP gateway connected to the PSTN at a location near the dialed destination telephone number. The gateway converts the call's voice digitization scheme to PCM, delivering the call to its destination.

If we focus our attention upon the three icons above the number entered to be called in Figure 1-7, we can obtain an appreciation for the manner by which Yahoo! is integrating telephone calling within its Messenger product. The three icons provide access to a subscriber's Address Book, Yahoo!'s Yellow Pages and Yahoo!'s People Search, respectively. An example of the use of these icons is illustrated in Figures 1-8 and 1-9. In Figure 1-8 the cursor is located over the People Search button. Once you click on that button your browser is automatically directed to the Yahoo! People Search facility. This is illustrated in Figure 1-9 in which this author entered his name for a telephone search.

Rather than leave readers in suspense concerning this author's telephone number, let's proceed to locate it. As a result of the limited search criteria consisting of a name and state, this author was able to locate his telephone number. The result of the search initiated in Figure 1-9 is illustrated in Figure 1-10.

In examining Figure 1-10, note that above the author's retrieved phone number is the legend "Click to call." While Yahoo! supports a second method for obtaining free calling within the United States, that method is not integrated into its Messenger program and requires a separate program which, interestingly enough, is provided by Net2Phone, the company that added the calling feature to Yahoo! Messenger.

Figure 1-11 illustrates the screen display resulting from this author attempting to make a call to his telephone from the Yahoo! People Search facility. Instead of recognizing that Yahoo! Messenger is active, the browser page code took this author to another page

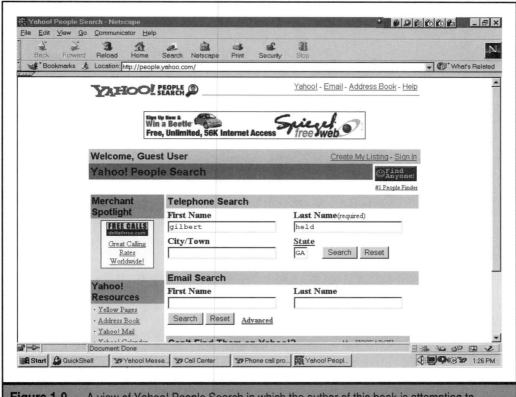

Figure 1-9.   A view of Yahoo! People Search in which the author of this book is attempting to determine if his telephone number can be reached.

that would require a different program to be downloaded. While Yahoo! should be commended for its effort toward providing a unified messaging capability, there is obviously additional effort required to be placed on a "to do" list.

Earlier in this chapter we noted that one of the problems associated with an organization implementing a voice-over-data network solution by acquiring hardware is the rapid decline in the cost associated with using the PSTN. In addition to conventional telephone operators significantly reducing the cost of long-distance calls you must also consider vendors that are providing telephony services over public and private IP networks. For example, Net2Phone was providing 200 minutes of free PC phone calls to persons who registered with them in late 2000. In addition, that company was billing calls within the U.S. at 1 cent per minute and offered international rates as low as 3.9 cents per minute. While these rates make it difficult to justify an internal organizational voice-over-data networking solution, it is also important to note that you may have better success with an internal network solution with respect to controlling voice quality. This is especially true

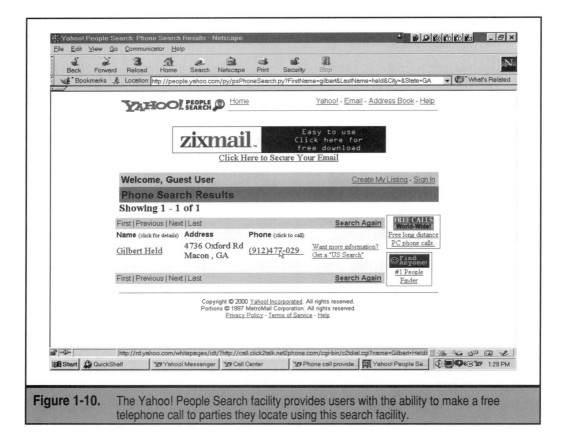

Figure 1-10.   The Yahoo! People Search facility provides users with the ability to make a free telephone call to parties they locate using this search facility.

in a business environment where many organizations rightfully wish to prioritize voice quality over cost. As we probe deeper into the characteristics of voice over data networks we will investigate many techniques we can employ to control the fidelity of reconstructed voice.

CHRONOLOGY OF EVENTS

To obtain a reasonable understanding of where voice over IP and voice over frame relay is headed, we can note some of the events that occurred over the past few years. In doing so it is important to remember that during 1996 most of us were paying a quarter or more per minute for a long distance call. Thus, at the very least, competition from voice-over-data network operators to include Internet Telephony Service Providers (ITSP) has provided both actual and potential competition that lowered the cost associated with the use of the public switched telephone network for long-distance calls.

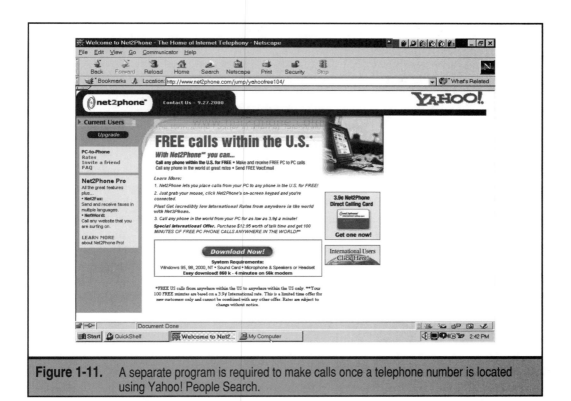

Figure 1-11.   A separate program is required to make calls once a telephone number is located using Yahoo! People Search.

During 1997 there were two significant events that marked the beginning of the use of the Internet as a voice transport mechanism. In August 1997 VocalTec announced its NextGen telephony program and a joint marketing program with several ITSPs. This program allowed PC users running the vendor's Internet Phone Service to dial up an Internet connection and access an ITSP gateway to route a long-distance call over the Internet onto the PSTN for local delivery. A second milestone during 1997 occurred in October of that year when VocalTec introduced its Surf&Call Web-to-telephone browser plug-in. In a few years' time when we look back at 1997 we will more than likely recognize these two events as pioneering efforts that gave birth to an explosive new industry.

During 1998 several large communications vendors announced both products and marketing agreements. In March of 1998, 3Com and Siemans announced a joint venture to develop an IP voice gateway to connect an Ethernet switch and PBX, enabling corporate users to run voice traffic over their IP networks. The gateway was designed to convert a call into packets transported via IP, which are passed to a 3Com switch. The switch would direct the packets to a router port. Also during 1998, ICG Communications announced the deployment of a nationwide Internet telephone service that was scheduled for 166 cities in the United States. Calls originating and terminating on ICG's network would be billed at 5.9 cents per minute, while calls terminating on the PSTN would be billed at 7.2 cents per minute.

During 1999 Netscape, now a subsidiary of AOL, announced that future versions of its Communicator and Netcenter products would feature an icon for IDT Corporation's Net2Phone service. Also during 1999, AT&T added VocalTec equipment for a second gateway for IP telephony via its Global Clearinghouse.

Another notable event that occurred during 1999 was Microsoft's announcement that it was incorporating emerging QoS standards into Windows 2000. In Windows 2000 Microsoft included an extension to the IETF Resource Reservation Protocol (RSVP), which combines RSVP with DiffServ. The extension, referred to as D Class, is also backed by Cisco Systems and other vendors and enables an application to reserve bandwidth, which, when granted, results in host systems marketing packets with applicable bits in the DiffServ byte.

As we entered the new millennium the number of minutes of voice traffic being transported over the Internet and private IP networks was growing near exponentially. For example, ITXC Corporation announced at the end of September 2000 that that month was its first month with over 100 million minutes of Internet telephony traffic. Progress in voice digitization techniques coupled with a substantial increase in Internet backbone bandwidth permitted voice-over-IP quality of reproduced voice to substantially increase. In fact, in September 2000 VocalTec's Surf&Call Network Services, which was introduced in 1997, began commercial operation.

While significant advances in voice over IP and voice over frame relay have occurred in just four years, we are not safely out of the woods. Many times it is difficult, if not impossible, to achieve the quality of reconstructed voice we take for granted when making a call over the PSTN. Many application notes throughout this book detail techniques that, when implemented, can result in a viable and practical implementation of available technology.

TECHNOLOGICAL SUCCESS

Voice-over-data networking is similar to any new technology in that to be successful it must provide a certain level of customer satisfaction. We can gain an appreciation of that level of customer satisfaction by comparing telephone company circuit switching to packet switching and then turning our attention to what I call the *Rosetta stone for success*—the maximum delay you should strive to undercut prior to constructing a voice-over-data network transmission facility.

Circuit Switching vs. Packet Switching

Table 1-1 provides a comparison of circuit switching performed by the traditional telephone companies and packet switching. In examining the entries in Table 1-1, note that the pulse code modulation (PCM) used by telephone companies represents a "toll-quality" voice-digitization method that results in high-quality reconstructed

Parameter	Circuit Switching	Packet Switching
Dedicated bandwidth	Yes	No
Quality of service	Yes	No
Voice quality	Toll-quality	Non-toll-quality
Delay latency	Minimal	Variable
Utilization level	Poor	High
Economics of utilization	Low	High
Call management features	Numerous	Few

Table 1-1.   Comparing Circuit Switching and Packet Switching

voice. In comparison, most voice-over-data network equipment involves the use of low-bit-rate coders that may provide near-toll-quality reconstructed voice. However, because delays occur randomly on a packet network, the reconstruction of voice is subject to variable delays.

Although not indicated in Table 1-1, it is important to note that once you make a connection over the switched telephone network there is no nonplanned disconnection. In comparison, under periods of congestion, both routers and frame relay switches drop packets. Concerning the last entry in Table 1-1, call management features can include call waiting, call forwarding, caller ID, and unified billing. While most telephone companies market a full range of call management features, most voice-over-data network solutions presently provide few such features. In my opinion, successful packetized voice requires characteristics and features similar to those offered by the legacy telephone company.

The Rosetta Stone of Delay

If we follow standards to the letter we would note that the International Telecommunications Union (ITU) Recommendation G.114 places a limit on telephone call round trip delay. That limit is 300 milliseconds (ms), or assuming each direction provides the same delay, 150 ms from one location to another. While the G.114 Recommendation should not be dismissed, from a practical standpoint it is well known that most humans can endure approximately 250 ms of delay before a voice conversation becomes awkward and perhaps intolerable. When voice is transmitted over a packet network, there are both fixed and variable delays that have to be considered. By understanding where these delays occur, you can consider different actions to ensure that latency on an end-to-end basis is below 250 ms; if you are not able to provide this capability, you would then want to consider temporarily abandoning your voice project until such time as you can ensure a reasonable degree of latency. Thus, understanding congestion and delay results in the Rosetta stone of implementing a voice-over-data network project.

Table 1-2 indicates the general range of fixed and variable packet network delays. In examining the entries in Table 1-2, note that the interprocess delays represent handoffs at routers and do not consider extra delays resulting from access list processing or encryption. Also note that network access and egress delays are based on the line operating rate. For example, a relatively small voice packet might require 2.5 ms when access and egress occurs on a T1 line operating at 1.544 Mbps, while the delay could expand to 7 ms when access and egress occurs on a 56-Kbps line.

APPLICATION NOTE: Understanding the delays associated with access and egress lines, different voice-compression methods, and the network provides the ability to consider different techniques to reduce overall latency.

Now that we have an appreciation for the constraints that govern our ability to successfully implement voice-over-data network technology, let's turn our attention to the actual methods used by the technology. Since these methods are the focus of the remainder of this book, we will conclude this chapter with a preview of the topics covered in succeeding chapters.

Cause	Delay in ms
Fixed Delays	
Compression (voice coding)	20–45
Interprocess at origin	10
Network access at origin	0.25–7
Network delay	20–100
Network egress at destination	0.25–7
Interprocess at destination	10
Jitter buffer (configurable)	10
Decompression (voice)	10
Fixed delay	90–199

For worst case a call can tolerate 250 - 80 to 250 - 199 or between 170 and 51 ms of variable delay.

Table 1-2.   Fixed and Variable Packet Network Delays

TOPIC PREVIEW

You can use the information presented in this section by itself or in conjunction with the index in this book to focus your attention on a particular topic of interest. You can read topics in the order presented, or you can tailor your reading by going directly to the material that corresponds to your specific requirements.

The Internet Protocol

In Chapters 2 through 4 we will turn our attention to obtaining a basic understanding of IP and frame relay networks, POS, and IP telephony-related protocols. Some readers may find they are already familiar with the information in these chapters. If this applies to you, you may wish to skip or simply skim their contents. For other readers, the information presented in these chapters will provide a basic foundation concerning the operation of the two key types of packet networks for which the transmission of voice is both economical and practical based on the recent development of a variety of hardware and software products covered in this book.

In Chapter 2, we will discuss TCP/IP, including how the IP stack is constructed, applications that reside on the stack, and why basic IP operations lack predictability. In doing so we will examine some of the major protocols in the TCP/IP protocol suite, to include ICMP, TCP, and UDP. We will also examine the dataflow in a TCP/IP network, why bottlenecks occur, and the need for a QoS capability.

QoS and IP Telephony-Related Protocols

In Chapter 3 we will turn our attention to two important topics that provide the building blocks for a successful voice over IP implementation—QoS and IP telephony-related protocols.

Concerning QoS, we will first examine the IEEE 802.3p standard at layer 2 of the protocol stack. Moving up the protocol stack we will turn our attention to the Type of Service (ToS) byte in the IP header and its revision into the DiffServ byte. We will also discuss how the dataflow into a network can be controlled by router queues and the expedited flow of traffic based upon the use of Multi-Protocol Label Switching (MPLS). In concluding this chapter we will examine such "modern" protocols that provide a mechanism for implementing voice over IP as the Real-Time Protocol (RTP), the ReSerVation Protocol (RSVP), the H.323 standard, and the Session Initiation Protocol (SIP).

Frame Relay

Chapter 4 will examine the rationale for frame relay with an overview of basic X.25 packet switching, noting strengths and weaknesses that have resulted from the rapidly growing market for frame relay products and services. Following the overview, we will focus our attention directly on frame relay. We'll learn how private networks are connected to this public network, define various frame relay parameters, and discuss the conventional flow of information through this type of network, including how frames can be lost and how higher layers in the protocol stack operate to ensure that frame loss does not translate into the actual loss of data.

Voice Basics

No book on voice-over-data networks would be complete without a detailed discussion of the characteristics, digitization, and encoding of voice, which is the focus of Chapter 5. In this chapter, we will examine the characteristics of human speech, which will serve as a foundation for discussing how voice can be analyzed and synthesized. As we discuss different voice-digitization methods, we will also cover different international standards, laying the foundation for information presented in the next three chapters.

Telephone Operations

The transmission of voice over a data network requires the selection of equipment that supports the signaling method used by organizational PBXs. Thus, it is important to obtain an appreciation of how a telephone call is routed between PBXs or over the switched telephone network and the different types of signals required to establish a call. This information is presented in Chapter 6.

Voice over IP

As previously mentioned, there are two types of packet networks applicable for transporting voice. One type of packet network is based on the IP protocol and can range in scope from private IP networks to the mother of all public IP networks—the Internet. In Chapter 7, we will turn our attention to the transmission of voice-over-IP-based networks, examining different techniques used to accomplish this function.

In doing so we will examine how popular vendor equipment can be configured to expedite traffic through an IP network. Because Cisco Systems has a significant share of the market for routers, many of our examples will be oriented toward Cisco's Internetwork Operating System (IOS). We will conclude this chapter with a discussion of several traffic control methods that can facilitate your voice-over-IP application becoming a reality.

Working with Voice-over-IP Gateways

The key to the successful operation of many voice-over-IP applications is obtained through the configuration of a voice-over-IP gateway. In Chapter 8 we will turn our attention to the configuration, operation, and utilization of several gateway products.

Voice over Frame Relay

Continuing our practical examination concerning the transmission of voice over data networks, Chapter 9 has a discussion of the transmission of voice over frame relay networks. After reviewing different techniques developed to perform this operation with some degree of network predictability, we will turn our attention to the operation of appropriate vendor equipment, the economics associated with this technology, and the various network integration issues you should consider. This chapter will give you an understanding of the practical issues associated with moving voice over frame relay networks, including the use of currently available equipment and its integration into existing and planned private networks.

Voice over ATM

As previously mentioned in this book, ATM provides a transport mechanism primarily used by communications carriers. In addition, ATM provides several classes of service to include a constant bit rate (CBR) capability that provides a true QoS capability. Due to this, ATM can be used as a backbone to interconnect geographically separated IP networks, permitting voice over IP to be transmitted via ATM on an assured level of service. In Chapter 10 we will first obtain an overview of ATM and then focus on its use as a mechanism to support both IP and frame relay traffic to include voice over each of those transport facilities.

Management

No book covering the transmission of voice over IP and frame relay would be complete without considering management tools and techniques. At the conclusion of this book we will examine several management issues and evaluate the use of different tools to help you decide prior to investing in equipment and/or line facilities whether the technology will work at an acceptable level.

CHAPTER 2

IP and Related Protocols

Any discussion concerning the transmission of voice over IP networks requires a degree of knowledge concerning the Internet Protocol. Since IP represents one component of the TCP/IP family, it is difficult to discuss IP as a separate entity unto itself. Recognizing this fact and the fact that the actual implementation of voice over IP requires knowledge about setting IP addresses and subnet masks and pointing your computer to the correct gateway and domain name server, this chapter is designed to provide readers with practical information necessary to configure and operate any IP device to include voice-over-IP products. In addition, because knowledge of the operation of the TCP/IP protocol stack provides us with the ability to understand how certain activities adversely affect the flow of data through an IP network, we will be able to note the effect of certain network-related operations upon latency. This in turn will provide us with the ability to consider certain network adjustments that may make an otherwise unworkable voice-over-IP transmission method workable.

In this chapter, we will focus our attention on IP; however, since it represents a component of the TCP/IP family, we will first review the layered structure of that family with respect to the International Standards Organization (ISO) Open System Interconnection (OSI) Reference Model. Once this is accomplished, we will turn our attention to network and data link layer addressing. In doing so, we will also examine the method by which Ethernet and Token Ring frames are transmitted on local area networks and the manner in which TCP/IP uses 32-bit IP addresses. This information will then be used to discuss the need for a mapping or translating mechanism between data link and network addresses, which is accomplished through the Address Resolution Protocol (ARP). Because it is important to obtain an understanding of the manner by which a TCP/IP network is operating, the Internet Control Message Protocol (ICMP) provides data concerning network errors, network congestion, and timeouts. In addition, ICMP supports an Echo function that can be used to determine if a distant host is operational as well as a Ping utility that uses ICMP Echo and Echo Reply messages to determine round-trip delay. Due to the usefulness of ICMP we will also examine this protocol in this chapter. We will discuss the role of the domain name server and its use, as well as the structure of the Internet Protocol. In addition, since IP does not operate by itself, we will move up the protocol stack, turning our attention to the transport layer and the role of the Transmission Control Protocol (TCP) and the User Datagram Protocol (UDP).

THE TCP/IP FAMILY

The Transmission Control Protocol/Internet Protocol (TCP/IP) represents a family of protocols that evolved over a period of time to perform predefined tasks. The roots of TCP/IP can be traced to the U.S. Defense Department's Advanced Research Projects Agency (DARPA), which developed a series of communications protocols for transporting data between geographically separated networks via a common network infrastructure known as the Advanced Research Projects Agency Network (ARPANET). ARPANET research formed the basis for the development of the *Internet*, which, when used with a capital "I," refers to a collection of interconnected networks.

Although ARPANET and the evolution of the Internet predate the development of the ISO's OSI Reference Model, the TCP/IP protocol family is a layered network architecture that is very similar in structure to the OSI Reference Model. Thus, prior to discussing the TCP/IP protocol family, let's quickly review the OSI Reference Model.

The OSI Reference Model

The OSI Reference Model was developed as a mechanism to subdivide networking functions into logical groups of related activities referred to as *layers*. Each network layer was designed to represent a collection of independent tasks that would be self-contained from a programming point of view but that could easily interact with another layer developed by the same programmers or even persons from a different company who followed the rules associated with creating a layer's set of tasks. Thus, another goal of the OSI Reference Model was to create an open system architecture that would facilitate interoperability between different vendor products.

Figure 2-1 illustrates the general structure of the OSI Reference Model and the relationship of each layer in the model to the other layers in the model. The lower four layers in the model focus on tasks required for the transmission of data. Those tasks include the creation of fields to transport data so each entity can be correctly routed to its destination via the use of addressing information, as well as the use of cyclic redundancy checking to enable the contents of each entity to be verified. In comparison, the upper layers are concerned with the manner in which the application interface is presented to the user and are not concerned with the manner by which data gets to the application. Let's briefly examine the tasks performed at each layer.

Layer number	Layer
7	Application
6	Presentation
5	Session
4	Transport
3	Network
2	Data link
1	Physical

Figure 2-1.   The seven-layer OSI Reference Model

The Physical Layer

The lowest layer of the OSI Reference Model is the physical layer. This layer is responsible for the mechanical, electrical, functional, and procedural mechanisms required for the transmission of data. It can be considered to represent the physical connection or cabling of a device to a transmission medium. When we examine the TCP/IP protocol suite we will note that it does not define a physical layer. Because the TCP/IP protocol suite actually commences at the network layer, it depends on the data link layer associated with other protocols, such as Ethernet and Token Ring, for data delivery. Because such data link protocols include a physical layer, TCP/IP in effect uses the physical layer of the data link protocol for data delivery.

The Data Link Layer

The data link layer is responsible for the manner in which a device gains access to the medium specified in the physical layer. In addition, the data link layer is also responsible for the manner in which data is formatted into defined fields and the correction of any errors occurring during a transmission session. Common examples of data link protocols include the various "flavors" of Ethernet, including Fast Ethernet and Gigabit Ethernet as well as Token Ring.

The Network Layer

The network layer is responsible for the physical routing of data. To accomplish this task, the network layer performs addressing, routing, switching, sequencing of data packets, and flow control, with the latter used as a mechanism to prevent data from overflowing buffers and becoming lost during periods of network congestion. In the TCP/IP protocol suite, the IP represents a network layer protocol. IP version 4 (IPv4) and the emerging IP version 6 (IPv6) use 32- and 128-bit addresses to identify source and destination interfaces. In addition, all routing on an IP network occurs by routers examining the destination IP address, checking its table entries, and making a routing decision concerning which port to output a packet to with a particular destination IP address.

Although ICMP is normally considered a separate protocol, ICMP messages are delivered by the use of an IP header affixed to the ICMP message. Thus, ICMP is normally considered to represent a network layer protocol.

The Transport Layer

The transport layer is responsible for ensuring that the transfer of information occurs correctly once a route is established through a network. This means the transport layer is responsible for controlling the communications session between network nodes once a path is established by the network control layer. This control includes verifying that transmitted data matches the data received.

The TCP/IP protocol suite includes two transport layer protocols: TCP and UDP. As we will note later in this chapter, TCP is a connection-oriented protocol that requires handshaking prior to the transmission of data, while UDP is a connectionless protocol that operates on a best-effort delivery basis.

The Session Layer

The session layer represents the first of three upper layers of the OSI Reference Model. This layer is responsible for establishing and terminating data streams between network nodes. Since each data stream can represent an independent application, the session layer is also responsible for coordinating communications between different applications that require communications.

The Presentation Layer

The presentation layer is responsible for isolating the application layer's data format from the lower layers in the OSI Reference Model. To accomplish this task, the presentation layer provides data transformation, formatting, and syntax conversion, converting application data into a common format for transmission and reversing the process for inbound or received data.

The Application Layer

At the top of the OSI Reference Model, the application layer functions as a window through which the application gains access to all of the services of the model. In the reverse direction, the application layer displays received information in an appropriate format. In the TCP/IP protocol suite, applications can be considered to represent a combination of layers 5 through 7 of the OSI Reference Model. Prior to turning our attention to the TCP/IP family and its relationship to the OSI Reference Model, let's briefly review the movement of data within the protocol stack and how the attachment and removal of protocol headers facilitate the layering process.

Data Flow

Within an ISO network, each layer from top to bottom appends appropriate heading information to packets of information flowing within the network while removing the heading information added by a lower layer when the data flow is in the reverse direction. In this manner layer n interacts with layer $n - 1$ as data flows through the network. Figure 2-2 illustrates the data flow within an OSI Reference Model network for an outgoing frame. For a received frame, the process would be reversed, with headers stripped or removed as the packet flows up the reference model.

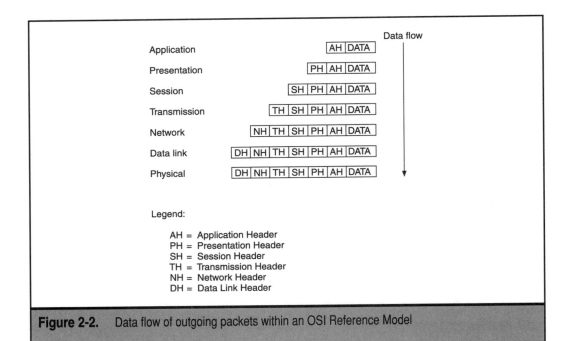

Figure 2-2.   Data flow of outgoing packets within an OSI Reference Model

It should be noted that when referring to the data link layer, the unit of transportation is referred to as a *frame*. At higher layers, the information field contained in several layer 2 frames may be combined into one larger packet. When transmission flows onto a local area network the contents of a packet whose information field exceeds the length of the information field of a LAN frame will be subdivided into segments, enabling multiple frames to transport the contents of a packet. In this book, we will refer to *frames* when a unit of transmission flows at the data link layer, while that same unit of information or grouping of units of information into a single entity flowing at and above the network layer will be referred to as a *packet*.

The TCP/IP Protocol Family

The TCP/IP protocol family was developed as a layered network architecture, even though its development predated the OSI Reference Model. In fact, the TCP/IP protocol suite represents the first major effort to develop a layered network architecture. As you might surmise, its development effort achieved a considerable degree of recognition, as it represents the only protocol suite used for transmission on the Internet.

Although TCP/IP predates the OSI Reference Model, at the lower layers there is a general one-to-one correspondence between the TCP/IP protocol suite and the OSI Reference Model. At the upper layers there is considerable divergence between the two, as most TCP/IP applications roughly correspond to the upper set of OSI Reference Model layers from layer 5, the session layer, through layer 7, the application layer.

Figure 2-3 provides a general indication of the correspondence between the TCP/IP family of protocols and the OSI Reference Model. At the data link layer, the referenced figure shows how TCP/IP resides on top of various data link layers that are not in actuality part of the TCP/IP family of protocols, enabling TCP/IP to transport data between local area networks that are located in close proximity to one another or geographically separated from one another by a few miles or by thousands of miles.

The protocols that reside above the transport layer and roughly correspond to the upper three layers of the OSI Reference Model represent some well-known applications as well as a special transport protocol mechanism developed to deliver audio and video packets. Thus, prior to focusing our attention on the lower layers of the TCP/IP protocol stack, let's briefly examine some applications and audio and video transport protocols that correspond to the upper layers of the OSI Reference Model.

FTP

The File Transfer Protocol (FTP) is a mechanism for moving data files between hosts via a TCP/IP network. FTP operates as a client-server process, with the client issuing predefined commands to the server to navigate its directory structure and to upload and

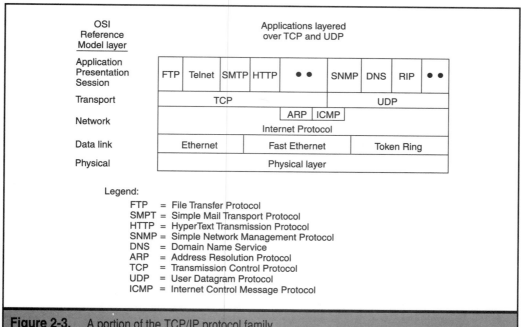

Figure 2-3.   A portion of the TCP/IP protocol family

download files to and from the server. Examples of client commands include GET to retrieve a file and MGET to retrieve a series of files that could be specified using a wildcard such as PAY.*, where the asterisk is used to specify "any extension."

Two types of FTP access are supported by servers: anonymous and via a previously established account. Anonymous access allows any person accessing the server to enter the USER ID "anonymous" to access the server or a predefined directory on the server. When "anonymous" is used as the USER ID, no password checking is employed, although many FTP servers request the user to enter his or her e-mail address in the password field. If access is via a previously established account, password checking is employed.

Telnet

Telnet represents another TCP/IP client-server application. This application is designed to enable a client to access a remote computer as though the client were a terminal directly connected to the remote computer. There are several versions of Telnet, with TN3270 providing the client with the ability to emulate different types of IBM 3270-type terminals to obtain access to IBM mainframes operating TCP/IP.

SMTP

The Simple Mail Transport Protocol (SMTP) provides the data transportation mechanism for electronic messages to be routed over a TCP/IP network. This protocol is completely transparent to the user since there are no user commands that govern the transfer of electronic mail via SMTP.

HTTP

The HyperText Transmission Protocol (HTTP) represents a relatively recent addition to the TCP/IP family in comparison to the previously mentioned protocols. HTTP is the protocol used by Web browsers to communicate with Web servers and vice versa.

SNMP

The Simple Network Management Protocol (SNMP) provides the mechanism to transport status messages and statistical information about the operation and utilization of TCP/IP devices. In addition, under SNMP, devices can generate alarms when certain predefined thresholds are reached. Under SNMP, the client-server processes are altered, with a server becoming a network manager that controls clients, referred to as *agents.* Although the previously described members of the TCP/IP family use TCP as a transport mechanism, SNMP uses the User Datagram Protocol.

DNS

The Domain Name Service (DNS) provides a very important service by enabling "near-English" host computer names (such as ftp.xyz.com to indicate an FTP server operated by the XYZ commercial firm) to be translated into a unique IP address that represents the physical address of the interface of the FTP server on a network (in this case, the

XYZ Corporation). Similar to SNMP, DNS uses the User Data Protocol as a transport mechanism.

RTP

The Real-Time Transport Protocol (RTP) represents a special type of protocol developed to support applications requiring the real-time delivery of data such as audio and video. RTP has evolved from the Visual Audio Tool (VAT) protocol, which was used to support the first working voice-conferencing program carried over the Internet. Although many Internet telephony products use proprietary audio coding techniques and protocols, a number of products use RTP as a "transport" protocol on top of UDP. In addition, other vendors, including Microsoft and Netscape, have committed to using RTP. Because of its evolving role in the transportation of audio and video information on TCP/IP networks, we will examine this protocol in some detail in Chapter 3 when we discuss POS and IP telephony-related protocols.

Transport Protocol Overview

Returning to Figure 2-3, you will note that the familiar TCP/IP applications correspond roughly to the upper three layers of the OSI Reference Model. You will also note that there are two transport protocols in the TCP/IP protocol suite: the Transmission Control Protocol (TCP) and the User Datagram Protocol (UDP). Some applications, such as FTP, Telnet, SMTP, and HTTP, were developed to use TCP, while other applications, such as DNS and RTP, were developed to use UDP. Thus, it is important to note the differences between each transport protocol.

TCP

TCP was developed to provide a reliable, connection-oriented service that supports end-to-end transmission reliability. To accomplish this task, TCP supports error detection and correction as well as flow control to regulate the flow of packets through a network. Error checking requires the computation of a cyclic redundancy check (CRC) algorithm at a network node based on the contents of a packet and a comparison of the computed CRC against the CRC carried in the packet. The CRC carried in a packet is created by the same algorithm applied to the contents of the packet by the originating node. If the two CRCs do not match, an error is presumed to have occurred and the packet is corrected by retransmission. The request to retransmit and the retransmission process delay the flow of the packet through a TCP/IP network. Thus, although TCP is used by FTP, Telnet, SMTP, and other applications where the integrity of data is a primary concern, it can result in unacceptable delays when transporting digitized voice and is rarely, if ever, used for voice transmission. In fact, some developers of voice-over-IP products prefer to drop a delayed packet containing digitized voice information and either use a period of silence or attempt to predict the contents of the packet based on a previously received packet rather than accept a delayed packet that results in a distortion to a portion of reconstructed speech.

Another characteristic of TCP that deserves mention is the fact that it is a connection-oriented protocol. This means a session between originator and receiver has to be established prior to data transfer being permitted. Although TCP is rarely, if ever, used to transport digitized voice, its connection-oriented capability results in its use for call control operations. This explains why just about all voice-over-IP products use TCP for call control, including the setup of a call, while UDP is used to transfer the digitized conversation. Now that we have a general appreciation for the operational characteristics of TCP, let's turn our attention to UDP.

UDP

UDP was developed to provide an unreliable, connectionless transport service. Before we are tempted to make a nasty comment concerning its unreliability, we should note that this is not necessarily bad and adds a degree of flexibility to the protocol family. That is, if reliability is required, a higher layer, such as the application layer, can be used to ensure that messages are properly delivered.

A second property of UDP that warrants a discussion is the fact that it is a connectionless protocol. This means that instead of requiring a session to be established between two devices, transmission occurs on a best-effort basis. That is, functions associated with connection setup and the exchange of status information as well as flow control procedures are avoided. While this removes a considerable amount of overhead, there is a price paid for obtaining this capability. To understand this price, we must digress a bit and look at the two methods in which the TCP/IP protocol suite routes data between network nodes: by the establishment of virtual circuits and the use of datagrams.

Virtual Circuit Transmission

When transmission occurs via a virtual circuit, a temporary path is established between source and destination locations. The establishment of a virtual circuit requires each network node to maintain a table of addresses and destination routes to enable a path to be established for the duration of the communications session. This fixed path, which is established for the duration of the transmission session, can be considered as a logical linking of nodes on a temporary basis. Once the communications session is completed, the previously established path is relinquished.

The key advantage associated with the transmission of data via a virtual circuit results from the use of the same path for all transmission. Although this precludes the use of an alternate route if a circuit outage occurs on the virtual path, it also precludes the need for data sequencing. This means that the possibility of duplicate data packets occurring is eliminated, resulting in an easier mechanism for the management of data flow between source and destination.

APPLICATION NOTE: Within an IP network, routers periodically transfer the contents of their routing tables, during which time the transmission of data, including voice-digitized packets, is suspended. Because edge routers commonly are connected only to an ISP router, you should consider configuring the edge router for static routing. This will preclude the transfer of router table entries and the resulting delays to traffic such table transfers cause.

Datagram Transmission

Datagram transmission results in the ability of transmission at the network layer to avoid the need to establish a fixed path between source and destination. Instead, packets are subdivided into units of data referred to as *datagrams.* Datagrams are transmitted via a broadcasting technique in which they are forwarded onto every port other than the port on which they were received. While this transmission technique results in duplicate traffic occurring on some network paths, it can considerably simplify network routing. This simplification results from the fact that since there are no fixed paths between nodes, there is no requirement to support a recovery method to reestablish a path if a circuit or intermediate node should fail.

Datagram transmission represents a connectionless or best-effort type of transmission. Transmitting datagrams onto all ports other than the port data is received on is a relatively easy process to implement. Figure 2-4 illustrates the transmission of a datagram on a three-node network from LAN A to LAN B. Note that Router A forwards datagrams to Routers B and C. Router C in turn transmits the datagram to LAN C as well as to Router B. Thus, two datagrams arrive at Router B. This means that duplicate datagrams will appear on LAN B and Router C, and LAN C will receive datagrams that are not destined for those locations, thus requiring a higher level of network utilization than if a virtual circuit method of transmission were used.

As previously discussed, most implementations of voice-over-IP networks use TCP to establish communications between network devices, while UDP is used for the flow of digitized data. Although UDP is used to transport datagrams, the actual transfer occurs via the use of virtual circuits as an IP header added to the UDP header for routing purposes. Although datagram transmission was popular during the 1960s, its generation of duplicate datagrams more than negates its ease of implementation, and all known network operations now avoid this transmission method.

THE INTERNET PROTOCOL

Now that we have an appreciation of TCP and UDP, let's turn to the network layer and examine the Internet Protocol. To understand the relationship of TCP and UDP to IP, let's examine Figure 2-5, which illustrates the formation of a series of headers as application data is transported via a TCP/IP network onto a local area network, with the latter result-

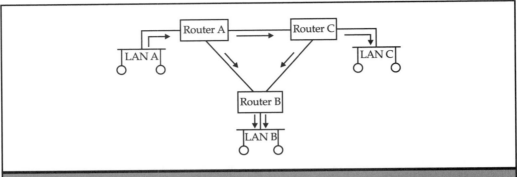

Figure 2-4. Datagram transmission can result in duplicate packets arriving at certain network locations.

ing in the use of a LAN frame as the transport mechanism. Each of the headers is employed to facilitate the transfer of information at the indicated network layer. For example, the TCP header will contain information that allows this layer 4 protocol to track the sequence of the delivery of datagrams so they can be placed in their correct order if they arrive out of sequence.

In examining Figure 2-5, note that the layer 4 protocol results in either a TCP or UDP header appended to application data. When passed to the IP layer, an IP header is added that includes the use of source and destination fields in the form of IP addresses, enabling data to flow from source to destination. At layer 2, which commonly represents a local area network transmission facility, a LAN header such as those formed by Ethernet or Token Ring will prefix the IP header, and a LAN trailer will be added as a suffix. The trailer typically consists of cyclic redundancy check characters that provide a mechanism for the LAN to determine if a frame of data is received without errors.

As we will note later in this section, LANs use 6-byte addresses assigned by the IEEE, which results in each station on a node having a unique 48-bit address. In comparison, IP uses a 32-bit address assigned by the Internet Network Information Center (InterNIC), whose composition, as we will shortly note, identifies a network and a station or an interface node on a network. Thus, there is no correlation between a data link address and an IP address, which means that when IP enables data to be correctly delivered to a LAN, another mechanism is required to enable data to reach a destination that uses a different addressing scheme. That mechanism is the Address Resolution Protocol, which we will discuss after we cover the composition of the IP header and examine IP addressing in detail.

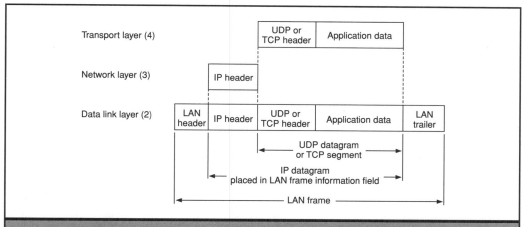

Figure 2-5.   The relationship of headers at the transport, network, and data link layers

Bytes vs. Octets

Prior to examining the IP header, a few words are in order concerning the use of the terms "bytes" and "octets." During the 1960s when computers were in a state of evolution the term "byte" was used to represent the number of bits a computer operated upon as an entity. Computers that had 5 through 9 bits per byte were common and the meaning of the term "byte" was vague.

Recognizing the fact that the term "byte" was nebulous, standards-making bodies turned to the term *octet* to reference eight bits operated upon as an entity. While the term "octet" is still used in standard documents, most if not all computers now use 8-bit bytes. Thus, in this book this author will use the terms "byte" and "octet" interchangeably, but will primarily use *byte* to reference a collection of eight bits operated on an entity.

The IP Header

Figure 2-6 illustrates the fields contained in the IP header. Note that the header contains a minimum of 20 bytes of data, and the width of each field is shown with respect to a 32-bit word. To obtain an appreciation for the functions performed by the IP header, let's examine the functions of the fields in the header.

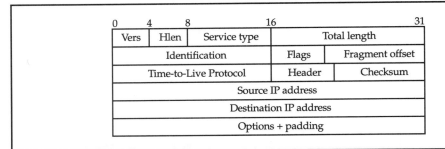

Figure 2-6. The IP header

Vers Field

The Vers field consists of 4 bits that identify the version of the IP protocol used to create the datagram. The current version of the IP protocol is 4 and the next generation IP protocol is assigned version number 6.

Hlen and Total Length Fields

The Hlen field is 4 bits in length. This field, which follows the Vers field, indicates the length of the header in 32-bit words. Care should be taken when interpreting the value of the Hlen field when using a protocol decoder. This is because this field has its value in terms of 32-bit words. For example, the shortest IP header is 20 bytes, which is 160 bits. When you divide 160 bits by 32 (160/32) a value of binary five (0101) will appear in the Hlen field. In comparison, the total length field indicates the total length of the datagram, including its header and higher-layer information. Since 16 bits are used for this field, an IP datagram can be up to 2^{16}, or 65,536, octets in length.

Service Type Field

The purpose of the Service Type field is to indicate how the datagram is processed. This field is also referred to as the Type of Service (TOS) byte and its composition is illustrated as follows:

7	6	5	4	3	2	1	0
R	TOS				Precedence		

where

R represents reserved.

Precedence provides eight levels, 0 to 7, with 0 normal and 7 the highest priority.

Type of Service (TOS) indicates how the datagram is handled:

0000	Default
0001	Minimize monetary cost
0010	Maximize reliability
0100	Maximize throughput
1000	Minimize delay
1111	Maximize security

As noted in the preceding illustration the Service Type field consists of two sub-fields: Type of Service (ToS) and Precedence. The ToS sub-field indicates how a datagram should be handled while the Precedence sub-field permits the transmitting station to indicate to the IP layer the priority for transmitting a datagram. Concerning the latter, a value of 000 indicates a normal precedence, while a value of 111 indicates the highest level of precedence and is normally used for network control.

The settings in both sub-fields are used to indicate how a datagram should be processed. To obtain an appreciation for the use of these sub-fields, let's assume an application is transmitting digitized voice that requires minimal routing delays due to the effect of latency on the reconstruction of digitized voice. Setting the Type of Service sub-field to a value of 1000 indicates to each router in the path between the source and the destination network that the datagram is delay-sensitive and its processing by the router should minimize delay. By providing a high precedence sub-field value, the computer forming the datagram informs the IP layer to prioritize its flow.

Although the use of the Service Type field provides a priority mechanism for the routing of IP datagrams, it is important to note that the IP standard does not mandate the specific actions that are caused by the values of the precedence bits. However, the Service Type field provides a mechanism for mapping LAN priority settings into an IP network, enabling end-to-end precedence to be obtained if routers along the path support the use of the field settings in a common manner.

When a router runs short of memory it will discard some datagrams. If an application sets the ToS sub-field to a value of 0010, the packets transporting the application will be less eligible for discard than other packets.

APPLICATION NOTE: For transporting voice over IP, you will normally be more concerned with the effect of delay than with that of periodic packet dropping, because real-time voice cannot be retransmitted. For this reason, you should configure your application to set the ToS sub-field in the Service Type field to a value of 1000, which minimizes delay.

Although a TCP/IP-based network is difficult to configure so that its reserved bandwidth is reliable and predictable enough to transport digitized voice, the Service Type field provides a mechanism to overcome certain shortcomings. For example, by mapping ToS sub-field values into ATM classes of service, it becomes possible to use IP to transport digitized voice over an ATM backbone with a quality of service that provides both reliability and predictability. When we cover voice-over-ATM networking in Chapter 10, we will examine the previously mentioned method of transporting digitized voice contained in IP datagrams over an ATM backbone.

Another item concerning the Service Type byte that warrants attention is the fact that the IETF was in the process of reusing the byte, naming it the Differentiated Service (DiffServ) byte. DiffServ is a traffic-expediting mechanism that was in the process of being standardized when this book revision occurred, and is described in Chapter 3.

Identification and Fragment Offset Fields

The Identification field enables each datagram or fragmented datagram to be identified. If a datagram is fragmented into two or more pieces, the Fragment Offset field specifies the offset in the original datagram of the data being transported. Thus, this field indicates where a fragment belongs in the complete message. The actual value in this field is an integer that corresponds to a unit of 8 octets, providing an offset in 64-bit units.

Time-to-Live Field

The Time-to-Live (TTL) field specifies the maximum time that a datagram can exist. This field is used to prevent a misaddressed datagram from endlessly wandering the Internet or a private IP network. Since an exact time is difficult to measure, it is commonly used as a hop count field, that is, routers decrement the value of this field by 1 as a datagram flows between networks. If the value of the field reaches 0, the datagram is discarded.

Flags Field

The Flags field contains two bits that are used to denote how fragmentation occurs, with a third bit in the field presently unassigned. The setting of one of the two fragmentation bits can be used as a direct fragment control mechanism, since a value of 0 indicates the datagram can be fragmented, while a value of 1 indicates it cannot be fragmented. The second bit is set to 0 to indicate that a fragment in a datagram is the last fragment, while a value of 1 indicates that more fragments follow.

Protocol Field

The purpose of the Protocol field is to identify the higher-level protocol used to create the message carried in the datagram. For example, a value of decimal 6 would indicate TCP, while a value of decimal 17 would indicate UDP.

Source and Destination Address Fields

The Source and Destination Address fields are both 32 bits in length. Each address represents both a network and a host computer on the network. Since it is extremely important to understand the composition and formation of IP addresses in order to correctly configure devices connected to an IP network, we will turn our attention to this topic. Once we understand IP addressing, we will then examine the address resolution process required to enable layer 3 packets that use IP addresses to be correctly delivered via layer 2 addressing.

IP Addressing

In this section, we turn our attention to the mechanism that enables TCP and UDP packets to be transmitted to unique or predefined groups of hosts. That mechanism is the addressing method used by the Internet Protocol, commonly referred to as IP addressing. The current version of the Internet Protocol is version 4. The next-generation Internet Protocol, which is currently being operated on an experimental portion of the Internet, is referred to as version 6 and noted by the mnemonic IPv6. Since there are significant differences in the method of addressing used by each version of the Internet Protocol, we will cover both versions in this section. First, we will focus our attention on the addressing used by IPv4. Once we have an appreciation for how IPv4 addresses are formed and used, we will turn our attention to IPv6. By first covering the addressing used by IPv4, we will have the ability to discuss address-compatibility methods that will allow IPv6 addresses to be used to access devices configured to respond to IPv4 addresses.

Overview

IP addresses are used by the Internet Protocol to identify distinct device interfaces such as interfaces that connect hosts, routers, and gateways to networks as well as to route data to those devices. Each device interface in an IP network must be assigned to a unique IP address so that it can receive communications addressed to it. This means that a multiport router will have one IP address for each of its network connections.

IPv4 uses 32-bit binary numbers to identify the source and destination addresses in each packet. This address space provides 2,294,967,296 distinct addressable devices—a number that exceeded the world's population when the Internet was initially developed. However, the proliferation of personal computers, the projected growth in the use of cable modems that require individual IP addresses, and the fact that every interface on a gateway or router must have a distinct IP address have contributed to a rapid depletion of available IP addresses. Recognizing that hundreds of millions of Chinese and Indians may eventually be connected to the Internet and also recognizing the potential for cell phones and even pacemakers to communicate via the Internet, the Internet Activities Board (IAB) in 1992 commenced work on a replacement for the current version of IP. Although the addressing limitations of IPv4 were of primary concern, the efforts of the IAB resulted in a new protocol with a number of significant improvements over IPv4, including the use of 128-bit addresses for source and destination devices. This new version of IP, which is referred to as IPv6, was finalized in 1995 and is currently being evaluated on an experimental portion of the Internet. Since this section is concerned with IP addressing, we will cover the addressing schemes, address notation, host address restrictions, and special addresses associated with both IPv4 and IPv6.

IPv4

The Internet Protocol was officially standardized in September 1981. Included in the standard was a requirement for each host connected to an IP-based network to be assigned a unique, 32-bit address value for each network connection. This requirement resulted in some networking devices, such as routers and gateways, that have interfaces to more than one network, as well as host computers with multiple connections to the same or different network being assigned a unique IP address for each network interface. Figure 2-7 illustrates two bus-based Ethernet LANs connected by using a pair of routers. Note that each router has two interfaces, one represented by a connection to a LAN and the second represented by a connection to a serial interface that provides router-to-router connectivity via a wide area network. Thus, each router will have two IP addresses, one assigned to its LAN interface and the other assigned to its serial interface. By assigning addresses to each specific device interface, this method of addressing enables packets to be correctly routed when a device has two or more network connections.

The Basic Addressing Scheme

When the IP was developed, it was recognized that hosts would be connected to different networks and that those networks would be interconnected to form an Internet. Thus, in developing the IP addressing scheme, it was also recognized that a mechanism would be required to identify a network as well as a host connected to a network. This recognition resulted in the development of a two-level addressing hierarchy, as illustrated in Figure 2-8.

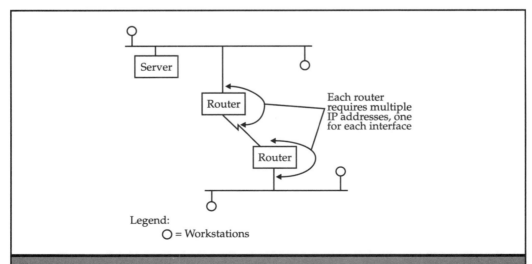

Figure 2-7. IP network addressing requires a unique 32-bit network number to be assigned to each device network interface.

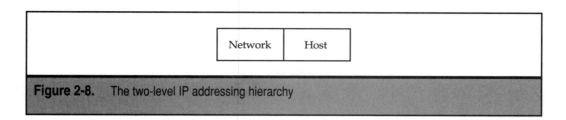

Figure 2-8.   The two-level IP addressing hierarchy

Under the two-level IP addressing hierarchy, the 32-bit IP address is subdivided into network and host portions. The composition of the first 4 bits of the 32-bit word specifies whether the network portion is 1, 2, or 3 bytes in length, resulting in the host portion being either 3, 2, or 1 bytes in length. In this addressing scheme, all hosts on the same network must be assigned the same network prefix, but they must have a unique host address to differentiate one host from another. Similarly, two hosts on different networks must be assigned different network prefixes; however, the hosts can have the same host address.

Address Classes

When IP was standardized, it was recognized that the use of a single method of subdivision of the 32-bit address into network and host portions would be wasteful with respect to the assignment of addresses. For example, if all addresses were split evenly, resulting in 16 bits for a network number and 16 bits for a host number, the result would allow a maximum of 65,534 ($2^{16} - 2$) networks with up to 65,534 hosts per network. In that case, the assignment of a network number to an organization that had only 100 computers would result in a waste of 65,434 host addresses, which could not then be assigned to another organization. Recognizing this problem, the designers of IP decided to subdivide the 32-bit address space into different address classes, resulting in five address classes being defined. Those classes are referred to as Class A through Class E.

Class A addresses are for very large networks, while Class B and Class C addresses are for medium-size and small networks, respectively. Class A, B, and C addresses incorporate the two-level IP addressing structure previously illustrated in Figure 2-8. Class D addresses are used for IP multicasting, where a single message is distributed to a group of hosts dispersed across a network. Class E addresses are reserved for experimental use. Both Class D and Class E addresses do not incorporate the two-level IP addressing structure used by Class A through Class C addresses.

Figure 2-9 illustrates the five IP address formats, including the bit allocation of each 32-bit address class. Note that the address class can be easily determined by examining the values of one or more of the first 4 bits in the 32-bit address. Once an address class is identified, the subdivision of the remainder of the address into the network and host address portions is automatically noted. Let's examine the composition of the network and host portion of each address when applicable, as doing so will provide some basic information that can be used to indicate how such addresses are used. Concerning the allocation of IP addresses, it should be noted that specific class addresses are assigned by the InterNIC.

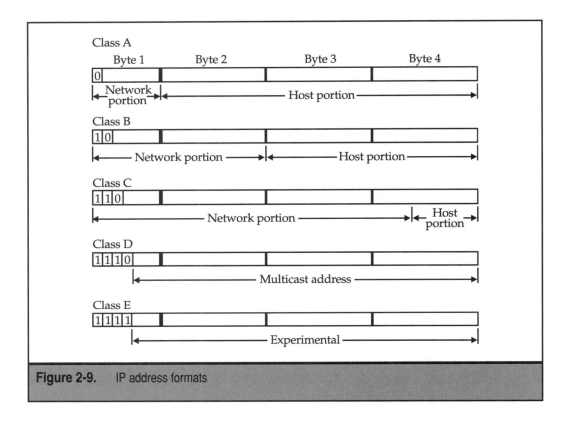

Figure 2-9. IP address formats

Class A A Class A IP address is defined by a 0-bit value in the high-order bit position of the address. This class of addresses uses 7 bits for the network portion and 24 bits for the host portion of the address. As a result of this subdivision, 128 networks can be defined, with approximately 16.78 million hosts capable of being addressed on each network. Due to the relatively small number of Class A networks that can be defined and the large number of hosts that can be supported per network, Class A addresses are primarily assigned to large organizations and countries that have national networks.

Loopback Address In the block of Class A addresses there is one network address that warrants attention. That address results from the setting of all bits in the network portion of the Class A address to 1 to represent 127 in decimal. The resulting sub-block of Class A addresses of the form 127.x.x.x is referred to as a *loopback address* and cannot be assigned as a unique IP address to a host. Instead, it is commonly used to test the operational status of the protocol stack on a host. That is, if you enter the command Ping 127.1.1.1, you would receive a response from your computer that informs you that your TCP/IP protocol stack is operational.

Class B   A Class B network is defined by the setting of the two high-order bits of an IP address to 10. The network portion of a Class B address is 14 bits in width, while the host portion is 16 bits wide. This results in the ability of Class B addresses to be assigned to 16,384 networks, with each network having the ability to support up to 65,536 hosts. Due to the manner by which Class B addresses are subdivided into network and host portions, such addresses are normally assigned to relatively large organizations with tens of thousands of employees.

Class C   A Class C address is identified by the first 3 bits in the IP address being set to the value 110. This results in the network portion of the address having 21 bits, while the host portion of the address is limited to 8-bit positions.

The use of 21 bits for a network address enables approximately 2 million distinct networks to be supported by the Class C address class. Since 8 bits are used for the host portion of a Class C address, this means that each Class C address can theoretically support up to 256 hosts. Due to the subdivision of network and host portions of Class C addresses, they are primarily assigned for use by relatively small networks, such as organizational LANs. Since it is quite common for many organizations to have multiple LANs, it is also quite common for multiple Class C addresses to be assigned to organizations that require more than 256 host addresses but are not large enough to justify a Class B address. Although Class A through Class C addresses are commonly assigned by the InterNIC to Internet service providers for distribution to their customers, Class D and Class E addresses represent special types of IP addresses.

Class D   A Class D IP address is defined by the assignment of the value 1110 to the first 4 bits in the address. The remaining bits are used to form what is referred to as a *multicast address.* Thus, the 28 bits used for that address enable approximately 268 million possible multicast addresses.

Multicast is an addressing technique that allows a source to send a single copy of a packet to a specific group through the use of a multicast address. Through a membership registration process, hosts can dynamically enroll in multicast groups. Thus, the use of a Class D address enables up to 268 million multicast sessions to simultaneously occur throughout the world.

Until recently, the use of multicast addresses was relatively limited; however, its use is increasing considerably, as it provides a mechanism to conserve bandwidth, which is becoming a precious commodity.

To understand how Class D addressing conserves bandwidth, consider a digitized audio or video presentation routed from the Internet onto a private network for which users working at ten hosts on the network wish to receive the presentation. Without a multicast transmission capability, ten separate audio or video streams containing audio would be transmitted onto the private network, with each stream consisting of packets containing ten distinct host-destination addresses. In comparison, through the use of a multicast address, one data stream would be routed to the private network.

Since an audio or video stream can require a relatively large amount of bandwidth in comparison to interactive query-response client-server communications, the ability to eliminate multiple data streams via multicast transmission can prevent networks from being saturated. This capability can also result in the avoidance of session timeouts when client-server sessions are delayed due to high-LAN-utilization levels, providing another reason for the use of multicast transmission.

Class E The fifth address class defined by the IP address specification is a reserved address class known as Class E. A Class E address is defined by the first 4 bits in the 32-bit IP address having the value of 1111. This results in the remaining 28 bits being capable of supporting approximately 268.4 million addresses. Class E addresses are restricted for experimentation.

Dotted-Decimal Notation

Recognizing that the direct use of 32-bit binary addresses is both cumbersome and unwieldy, a technique more acceptable for human use was developed. That technique is referred to as *dotted-decimal notation* in recognition of the fact that the technique developed to express IP addresses occurs via the use of four decimal numbers separated from one another by decimal points.

Dotted-decimal notation divides the 32-bit Internet Protocol address into four 8-bit (1-byte) fields, with the value of each field specified as a decimal number. That number can range from 0 to 255 in bytes 2, 3, and 4. In the first byte of an IP address, the setting of the first 4 bits in the byte that is used to denote the address class limits the range of decimal values that can be assigned to that byte. For example, from Figure 2-9, a Class A address is defined by the setting of the first bit position in the first byte to 0. Thus, the maximum value of the first byte in a Class A address is 127. Table 2-1 summarizes the numeric ranges for Class A through Class E IP addresses.

Class	Length of Network Address (Bits)	First Number Range (Decimal)
A	8	0–127
B	16	128–191
C	24	192–223
D	32	224–239
E	32	240–255

Table 2-1.   Class A Through Class E Address Characteristics

To illustrate the formation of a dotted-decimal number, let's first focus on the decimal relationship of the bit positions in a byte. Figure 2-10 indicates the decimal values of the bit positions within an 8-bit byte. Note that the decimal value of each bit position corresponds to 2^n, where n is the bit position in the byte. Using the decimal values of the bit positions shown in Figure 2-10, let's assume the first byte in an IP address has its bit positions set to 01100000. Then the value of that byte expressed as a decimal number becomes 64 + 32, or 96. Now let's assume that the second byte in the IP address has the bit values 01101000. From Figure 2-10, the decimal value of that binary byte is 64 + 32 + 8, or 104. Let's further assume that the last 2 bytes in the IP address have the bit values 00111110 and 10000011. Then the third byte would have the decimal value 32 + 16 + 8 + 4 + 2, or 62, while the last byte would have the decimal value 128 + 2 + 1, or 131.

Based on the preceding, the dotted-decimal number 96.104.62.131 is equivalent to the binary number 01100000011010000011111010000011. Obviously, it is easier to work with (and remember) four decimal numbers separated by dots than a string of 32 bits.

Reserved Addresses

There were three blocks of IP addresses originally reserved for networks that would not be connected to the Internet. Those address blocks were defined in RFC 1918, Address Allocation for Private Internets, summarized in Table 2-2.

Both security considerations as well as difficulty in obtaining large blocks of IP addresses resulted in many organizations using some of the addresses listed in Table 2-2 while connecting their networks to the Internet. Since the use of any private Internet address by two or more organizations connected to the Internet would result in addressing conflicts and the unreliable delivery of information, those addresses are not directly used. Instead, organizations commonly install a proxy firewall that provides address translation between a large number of private Internet addresses used on the internal network and a smaller number of assigned IP addresses. Not only does this technique allow organizations to connect large internal networks to the Internet without being able to obtain relatively scarce Class A or Class B addresses, but the proxy firewall hides internal addresses from the Internet community. This provides a degree of security, because any hacker who attempts to attack a host on your network actually has to attack your organization's proxy firewall.

Address Blocks

10.0.0.0–10.255.255.255

172.16.0.0–172.31.255.255

192.168.0.0–192.168.255.255

Table 2-2.   Reserved IP Addresses for Private Internet Use

128	64	32	16	8	4	2	1

Figure 2-10. The decimal value of the bit positions in a byte corresponds to 2^n where n is the bit position that ranges from 0 to 7.

APPLICATION NOTE: The translation of IP addresses by a router or firewall adds a slight delay to packets as they flow through the device. If your voice-over-IP application is stretched toward the maximum amount of tolerable delay, you may wish to consider placing stations that depend upon the application on their own network, which avoids the necessity of network address translation and its delay.

Protocol Spoofing

In the wonderful world of TCP/IP, networking routers only check the destination address during the routing process. This means you or a hacker can change the source address in your protocol stack before you initiate a behavior you do not wish to be traced to your IP address. Rather than using their IP address, hackers commonly use certain addresses as their source address. Those commonly used addresses include an address on the target network to be attacked, and RFC 1918 address, a broadcast address, or a loopback address. The use of any of these addresses or another address in place of a computer's actual IP address is referred to as *protocol spoofing*.

In an attempt to combat protocol spoofing many network administrators configure their router access lists with anti-spoofing statements at the beginning of the list. While this is indeed a good security measure, the extra router cycles required to check each packet occurring on an older router can result in a few milliseconds of delay that can make the difference between acceptable and non-acceptable reconstructed voice. Thus, you may wish to consider moving statements in your access control list that enable access to a voice gateway above the anti-spoofing statements.

APPLICATION NOTE: Anti-spoofing statements in a router access control list add a slight delay to the delivery of packets to their intended destination. By moving statements in the router access control list that provide access to a voice gateway above the anti-spoofing statements you can eliminate that delay.

Networking Basics

As previously noted, each network has a distinct network prefix, and each host on a network has a distinct host address. When two networks are interconnected by the use of a router, each router port that represents an interface is assigned an IP address that reflects the network to which it is connected. Figure 2-11 illustrates the connection of two networks via a router, indicating possible address assignments. On the left portion of Figure 2-11 note that the first decimal number (192) of the 4-byte dotted-decimal numbers associated with two hosts on the network denotes a Class C address. This is because 192 decimal is equivalent to 11000000 binary. Since the first 2 bits are set to the bit value 11, Figure 2-11 indicates a Class C address. Also note that the first 3 bytes of a Class C address indicate the network, while the fourth byte indicates the host address. Thus, the network shown in the left portion of Figure 2-11 is denoted as 192.78.46.0, with device addresses that can range from 192.78.46.1 through 192.78.46.254.

In the lower-right portion of Figure 2-11, two hosts are shown connected to another network. Note that the first byte for the 4-byte dotted-decimal number assigned to each host and the router port is decimal 222, which is equivalent to binary 11011110. Since the first 2 bits in the first byte are again set to 11, the second network also represents the use of a Class C address. Thus, the network address is 222.42.78.0, with device addresses on the network ranging from 222.42.78.01 to 222.42.78.254.

Although it would appear that 256 devices could be supported on a Class C network (0 through 255 used for the host address), in actuality the host-portion field of an IP address has two restrictions. First, the host-portion field cannot be set to all-0 bits. This is because an all-0 host number is used to identify a base network or subnetwork number. Concerning the latter, we will shortly discuss subnetworking. Second, an all-1 host number represents the broadcast address for a network or subnetwork. Due to these restrictions, a maximum of 254 devices can be defined for use on a Class C network. Similarly, other network classes have the previously discussed addressing restrictions, which reduces the number of distinct addressable devices that can be connected to each type of IP network by two. Since, as previously explained, an all-0 host number identifies a base network, the two networks shown in Figure 2-11 are shown as 192.78.46.0 and 222.42.78.0.

Subnetting   One of the problems associated with the use of IP addresses is the necessity to assign a distinct network address to each network. This can result in the waste of many addresses as well as a considerable expansion in the use of router tables. To appreciate these problems, let's return to Figure 2-11, which illustrates the connection of two Class C networks via a router.

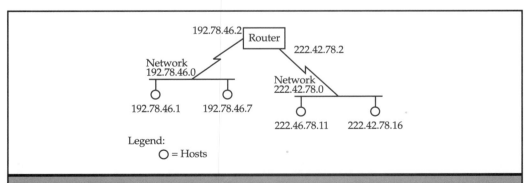

Figure 2-11.   Router connections to networks require an IP address for each connection.

Assume each Class C network supported 29 workstations and servers. Adding an address for the router port, each Class C network would use 30 out of 254 available addresses. Thus, the assignment of two Class C addresses to an organization that needs to support two networks with a total of 60 devices would result in 448 (254 × 2 – 60) available IP addresses in effect being wasted. In addition, routers would have to recognize two network addresses instead of one. When this situation is multiplied by numerous organizations requiring multiple networks, the effect on routing tables becomes more pronounced, resulting in extended search times as routers sort through their routing tables to determine an appropriate route to a network. Because of these problems, RFC 950 became a standard in 1985. That standard defines a procedure to subnet or divide a single Class A, B, or C network into subnetworks.

Through the process of subnetting, the two-level hierarchy of Class A, B, and C networks shown in Figure 2-9 is turned into a three-level hierarchy. In doing so, the host portion of an IP address is divided into a subnet portion and a host portion. Figure 2-12 provides a comparison between the two-level hierarchy initially defined for Class A, B, and C networks and the three-level subnet hierarchy.

Through the process of subnetting, a Class A, B, or C network address can be divided into different subnet numbers, with each subnet used to identify a different network internal to an organization. Since the network portion of the address remains the same, the route from the Internet to any subnet of a given IP network address is the same. This means that routers within the organization must be able to differentiate between different subnets, but routers outside the organization consider all subnets as one network.

The subnet process facilitates the use of IP addresses by reducing waste and decreasing routing table entries. Let's examine the process. In doing so, we will discuss the concept of masking and the use of the subnet mask, both of which are essential to the extension of the network portion of an IP address.

To illustrate the concept of subnetting, let's return to the two networks illustrated in Figure 2-11: networks 192.78.46.0 and 222.42.78.0. Let's assume that instead of two networks geographically separated from one another at two distinct locations, we require the establishment of five networks at one location. Let's further assume that each of the five networks will support a maximum of 15 stations. Although your organization could apply for four additional Class C addresses, doing so would waste precious IP address space since each Class C address supports a maximum of 254 devices. In addition, if your internal network were connected to the Internet, entries for four additional networks would be required in a number of routers in the Internet in addition to your organization's internal routers. Instead of requesting four additional Class C addresses, let's use subnetting, dividing the host portion of the IP address into a subnet number and a host number. Since we need to support five networks at one location, we must use a minimum of 3 bits from the host portion of the IP address as the subnet number. Since a Class C address uses one 8-bit byte for the host identification, this means that a maximum of five bit positions can be used (8 – 3) for the host number. Assuming we intend to use the 192.78.46.0 network address for our subnetting effort, we would construct an extended network prefix based on combining the network portion of the IP address with its subnet number.

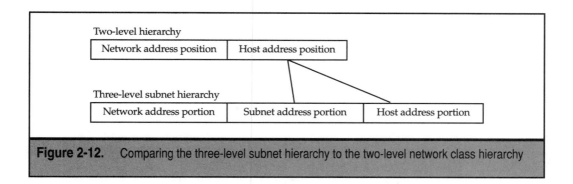

Figure 2-12.   Comparing the three-level subnet hierarchy to the two-level network class hierarchy

Figure 2-13 illustrates the creation of five subnets from the 192.78.46.0 network address. The top entry in Figure 2-13, labeled "Base network," represents the Class C network address with a host address byte field set to all 0s. Since we previously decided to use 3 bits from the host portion of the Class C IP address to develop an extended network prefix, the five entries in Figure 2-13 below the base network entry indicate the use of 3 bits from the host position in the address to create extended prefixes that identify five distinct subnets created from one IP Class C address. To the Internet, all five networks appear as the network address 192.78.46.0, with the router at an organization responsible for directing traffic to the appropriate subnet. It is important to note that externally (that is, to the Internet) there is no knowledge that the dotted-decimal numbers shown in the right column represent distinct subnets. This is because the Internet views the first byte of each dotted-decimal number and notes that the first 2 bits are set. Doing so tells routers on the Internet that the address is a Class C address for which the first 3 bytes represent the network portion of the IP address and the fourth byte represents the host address. Thus, to the outside world, address 192.78.46.32 would not be recognized as subnet 1. Instead, a router would interpret the address as network 192.78.46.0, with host address 32. Similarly, subnet 4 would appear as network address 192.78.46.0, with host address 128. However, within an organization, each of the addresses listed in the right column in Figure 2-13 would be recognized as a subnet. To visualize this dual interpretation of network addresses, consider Figure 2-14, which illustrates the Internet versus the private network view of subnets.

As we might logically assume from our prior discussion of Class C addresses, any address with the network prefix 192.78.46.0 will be routed to the corporate router. However, although we noted how subnet addresses are formed, we have yet to discuss how we assign host addresses to devices connected to different subnets or how the router can break down a subnet address so it can correctly route traffic to an appropriate subnet. Thus, we need to expand our discussion of host addressing on subnets to include the role of the subnet mask.

```
Base network:   11000000.01010000.00101110.00000000 = 192.78.46.0
Subnet #0:      11000000.01010000.00101110.00000000 = 192.78.46.0
Subnet #1:      11000000.01010000.00101110.00100000 = 192.78.46.32
Subnet #2:      11000000.01010000.00101110.01000000 = 192.78.46.64
Subnet #3:      11000000.01010000.00101110.01100000 = 192.78.46.96
Subnet #4:      11000000.01010000.00101110.10000000 = 192.78.46.128
```

Figure 2-13.   Creating extended network prefixes via subnetting

Host Addresses on Subnets   We previously subdivided the host portion of a Class C address into a 3-bit subnet field and a 5-bit host field. Since the host field of an IP address cannot contain all 0 bits or all 1 bits, the use of 5 bits in the host portion of each subnet address means that each subnet can support a maximum of $2^5 - 2$, or 30 addresses. Thus, we could use host addresses 1 through 30 on each subnet. Figure 2-15 illustrates the assignment of host addresses for subnet 3, whose creation was previously indicated in Figure 2-13. In examining Figure 2-15, note that we start with the subnet address 192.78.46.96, for which the first 3 bits in the fourth byte of the address are used to indicate the subnet. We use the remaining 5 bits to define the host address on each subnet. Thus, the address 192.78.46.96 represents the third subnet, while addresses 192.78.46.97 through 192.78.46.126 represent hosts 1 through 30 that can reside on subnet 3.

Although we now have an appreciation for creating subnets and host addresses on subnets, we haven't yet discussed how devices on a private network recognize subnet addressing. For example, if a packet arrives at an organizational router with the destination address 192.78.46.97, how does the router know to route that packet onto subnet 3? The answer to this question involves what is known as the *subnet mask.*

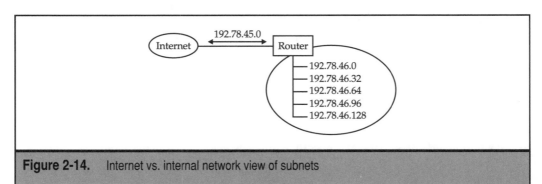

Figure 2-14.   Internet vs. internal network view of subnets

Subnet #3:   11000000.01010000.00101110.01100000 = 192.78.45.96

Host #1:   11000000.01010000.00101110.01100001 = 192.78.46.97

Host #2:   11000000.01010000.00101110.01100010 = 192.78.46.98

Host #3:   11000000.01010000.00101110.01100011 = 192.78.46.99

•••   •••

Host #30:   11000000.01010000.00101110.01111110 = 192.78.46.126

Figure 2-15.   Assigning host addresses by subnet

The Subnet Mask

The *subnet mask* represents a mechanism that enables devices on a network to determine the separation of an IP address into its network, subnet, and host portions. To accomplish this, the subnet mask consists of a sequence set to 1 bit that denotes the length of the network and subnet portions of the IP network address associated with a network. For example, let's assume our network address is 192.78.46.96, and we want to develop a subnet mask that can be used to identify the extended network. Since we previously used 3 bits from the host portion of the IP address, the subnet mask would become 11111111.11111111.11111111.11100000.

Similar to the manner in which IP addresses can be expressed using dotted-decimal notation, we can also express subnet masks using that notation. Doing so, we can express the subnet mask as 255.255.255.224.

The subnet mask tells the device examining an IP address which bits in the address should be treated as the extended network address consisting of network and subnet addresses. Then the remaining bits that are not set in the mask indicate the host on the extended network address. However, how does a device determine the subnet of the destination address? Since the subnet mask indicates the length of the extended network, including the network and subnet fields, knowing the length of the network portion of the address provides a device with the ability to determine the number of bits in the subnet field. Once this is accomplished, the device can determine the value of those bits, which indicates the subnet. To illustrate this concept, let's use the IP address 192.78.46.97 and the subnet mask 255.255.255.224, with the latter used to define a 27-bit extended network. The relationship between the IP address and the subnet mask is shown in Figure 2-16.

IP address:192.78.46.97   11000000.01010000.00101110.01100001
Subnet mask: 255.255.255.244 ⎸11111111.11111111.11111111.111⎹ 00000

Extended network address

Figure 2-16.   The relationship between the IP address and the subnet address

Since the first 2 bits in the IP address are set, this indicates a Class C address. Since a Class C address consists of 3 bytes used for the network address and 1 byte for the host address, this means the subnet must be 3 bits in length (27 – 24). Thus, bits 25 through 27, which are set to 011 in the IP address, identify the subnet as subnet 3. Since the last 5 bits in the subnet mask are set to 0, this means that those bit positions in the IP address identify the host on subnet 3. Since those bits have the value 00001, this means the IP address references host 1 on subnet 3 on network 192.78.46.0.

APPLICATION NOTE   Although modern routers are based on relatively fast microprocessor technology, it is important to remember that most networks include routers manufactured three, four, or even five years ago. Although they may provide a high level of support for traditional data transfer operations, if you are using subnetting, the extra cycles may add several milliseconds of delay that could represent the figurative straw that breaks the back of a real-time voice transport application. Instead of replacing the router, you might want to consider placing stations requiring real-time voice transport on their own network, thus avoiding the extra cycles associated with subnet processing.

Configuration Examples

When configuring a workstation or server to operate on a TCP/IP network, most network operating systems require you to enter a minimum of three IP addresses and an optional subnet mask or mask bit setting. The three IP addresses are as follows: (1) the IP address assigned to the workstation or server, (2) the IP address of the gateway or router responsible for relaying packets with a destination that is not on the local network to a different network, and (3) a name resolver that is referred to as the Domain Name Service (DNS). The latter is a computer responsible for translating near-English mnemonic names assigned to computers into IP addresses.

Figure 2-17 illustrates the first configuration screen in a series of screens displayed by the NetManage Chameleon Custom program. The IP Configuration screen illustrated in Figure 2-17, which is displayed by selecting an appropriate entry from the Setup menu, provides you with the ability to enter an IP address that is assigned to the workstation or server running the Chameleon TCP/IP protocol stack. The configuration screen also provides you with the ability to enter the number of subnet mask bits, which the program then converts into an appropriate decimal number. Table 2-3 compares the number of subnet bits to host bits and indicates the resulting decimal mask.

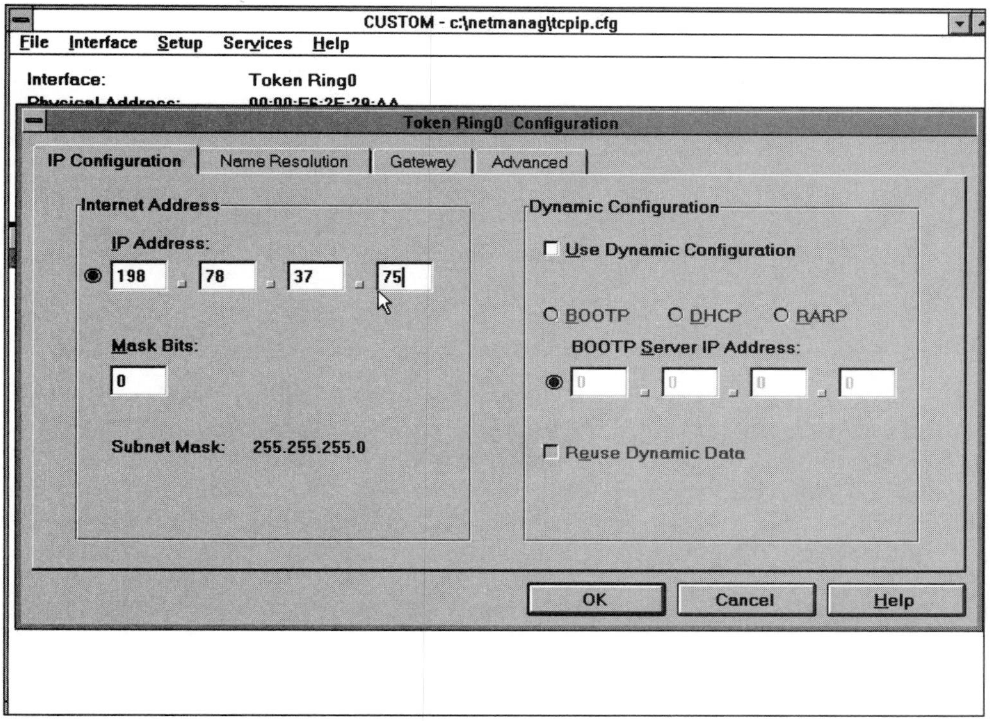

Figure 2-17. The NetManage Chameleon IP Configuration screen enables you to set the IP address of the host running the program's TCP/IP protocol stack.

Subnet Bits	Host Bits	Decimal Mask
0	8	0
1	7	128
2	6	192
3	5	224
4	4	240

Table 2-3. Subnet Masks

Subnet Bits	Host Bits	Decimal Mask
5	3	248
6	2	252
7	1	254
8	0	255

Table 2-3. Subnet Masks (continued)

In examining the screen displayed in Figure 2-17, note that simply clicking on different tabs results in the display of new configuration screens. For example, Figure 2-18 illustrates the Name Resolution Configuration screen. Note that you would enter the address of one or more domain servers as well as the name assigned to your host and its DNS domain name. In this example, the host name entered was "gil," while the DNS domain name entered was "feds.gov." This informs the domain server at the indicated address that requests to access the host with the near-English mnemonic gil.feds.gov should be routed to the IP address previously entered into the IP Configuration screen. Thus, this display screen provides network users with the ability to have their computers identified by a name rather than by a more cumbersome IP address. The specification of the IP address of at least one domain server also enables the use of near-English mnemonic names to access other computers. This is because the computer now knows to send the name-to-IP address resolution requests to the indicated domain server IP address.

Classless Networking

As previously noted, the use of individual Class A, B, and C addresses can result in a significant amount of unused address space, which makes them very inefficient to use. Recognizing the inefficiency associated with Class A, B, and C addressing, another method was developed to assign IP addresses to organizations. This method results in a more efficient assignment of IP addresses, because the number of distinct IP addresses is more closely tied to the requirements of an organization. Since the technique does away with network classes, it is commonly referred to as *classless networking*.

Under classless networking, an organization is assigned a number of bits to use as the local part of its addresses that best correspond to the number of addresses it needs. For example, if an organization requires 4000 IP addresses, it would be given 12 bits (4096 distinct addresses) to use as the local part of its address. The remaining 20 bits in the 32-bit address space are then used as a prefix to denote what is referred to as a *supernetwork*. To denote the network part of a classless network, the forward slash (/) is used, followed by the number of bits in the prefix. Thus, the previously mentioned classless network would be denoted as /20.

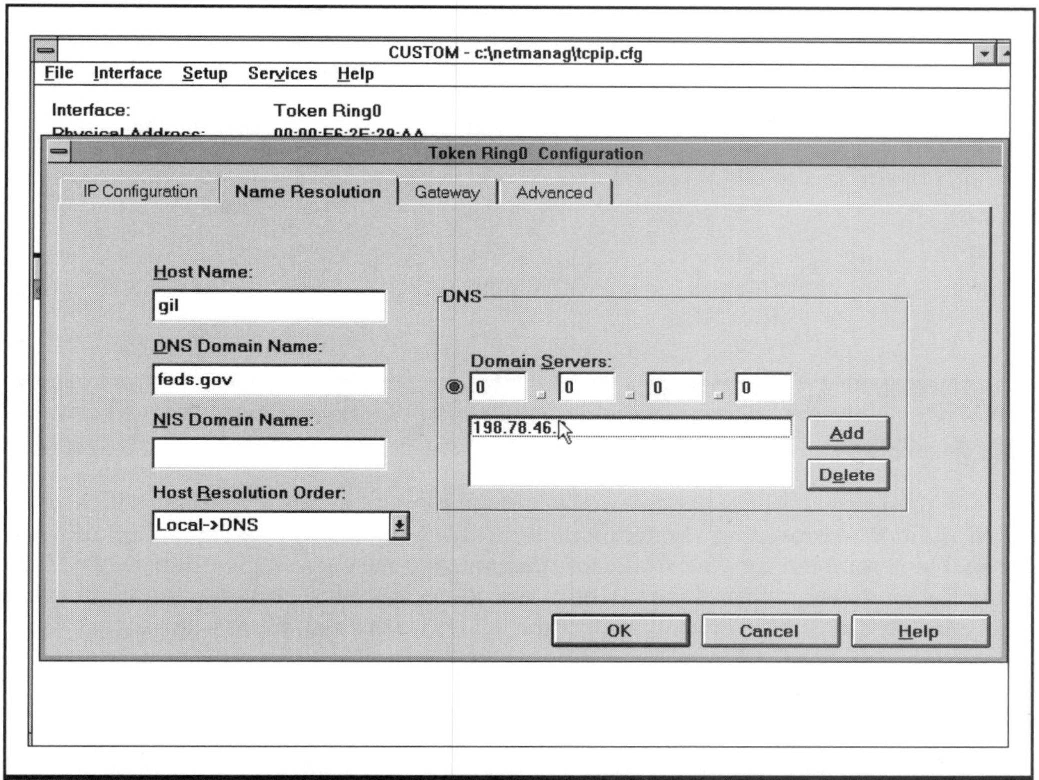

Figure 2-18. The Chameleon Name Resolution screen enables a host to be configured so that it can be identified by its near-English mnemonic name.

Currently, address allocations used for classless networking are taken from available Class C addresses. Thus, obtaining a 20-bit prefix is equivalent to obtaining 16 continuous Class C addresses. Table 2-4 lists the classless address blocks that can be assigned from available Class C address space.

Network Part	Local Bits	Equivalent Number of Class C Addresses	Distinct Addresses
124	8	1	256
123	9	2	512
122	10	4	1024

Table 2-4. Classless Network Address Assignments

Network Part	Local Bits	Equivalent Number of Class C Addresses	Distinct Addresses
121	11	8	2048
120	12	16	4096
119	13	32	8192
118	14	64	16384
117	15	128	32768

Table 2-4.   Classless Network Address Assignments *(continued)*

In addition to providing a better method for allocating IP addresses, classless addressing enables a router to forward traffic to an organization using a single routing entry. Due to the tremendous growth of the Internet, classless addressing provides a more efficient mechanism for locating entries in router tables. This is because one classless entry can replace up to 129 Class C addresses, enabling a router to locate entries faster as it searches its routing tables. Thus, you can expect the use of classless addressing to increase as a mechanism to both extend the availability of IP addresses and enable routers to operate more efficiently as we wait for IPv6 to be deployed.

IPv6

IPv6 was developed as a mechanism to simplify the operation of the Internet Protocol, provide a mechanism for adding new operations as they are developed through a header daisy chain capability, add built-in security and authentication, and extend source and destination addresses to an address space that could conceivably meet every possible addressing requirement for generations. The latter is accomplished through an expansion of source and destination addresses to 128 bits and is the focus of this section.

Address Architecture

IPv6 is based on the same architecture used in IPv4, resulting in each network interface requiring a distinct IP address. The key differences between IPv6 and IPv4 with respect to addresses are the manner in which an interface can be identified and the size and composition of the address. Under IPv6, an interface can be identified by several addresses to facilitate routing and management. In comparison, under IPv4, an interface can be assigned only one address. Concerning address size, IPv6 uses 128 bits, or 96 more bits than an IPv4 address.

Address Types

IPv6 addresses include unicast and multicast, which were also included in IPv4. In addition, IPv6 adds a new address category known as *anycast*. Although an anycast address identifies a group of stations similarly to a multicast address, a packet with an anycast address is delivered to only one station, the nearest member of the group. The use of anycast addressing can be expected to facilitate network restructuring while minimizing the amount of configuration changes required to support a new network structure. This is because you can use an anycast address to reference a group of routers, and the alteration of a network by stations using anycast addressing would enable them to continue to access the nearest router without a user having to change the address configuration of his or her workstation.

Address Notation

Since IPv6 addresses consist of 128 bits, a mechanism is required to facilitate their entry as configuration data. The mechanism used is to replace those bits with eight 16-bit integers separated by colons, each integer being represented by four hexadecimal digits. For example,

6ACD:00001:00FC:B10C:0001:0000:0000:001A

To facilitate the entry of IPV6 addresses, you can skip the leading 0s in each hexadecimal component. That is, you can write 1 instead of 0001 and 0 instead of 0000. Thus, this ability to suppress 0s in each hexadecimal component would reduce the previous network address to the following:

6ACD:1:FC:B10C:1:0:0:1A

Under IPv6, a second method of address simplification was introduced: the double colon (::). Inside an address a set of consecutive null 16-bit numbers can be replaced by two colons. Thus, the previously reduced IP address could be further reduced as follows:

6ACD:1:FC:B10C:1::1A

It is important to note that the double colon can be used only once inside an address. This is because the reconstruction of the address requires the number of integer fields in the address to be subtracted from 8 to determine the number of consecutive fields of zero value the double colon represents. The use of two or more double colons would create ambiguity that would not allow the address to be correctly reconstructed.

Address Allocation

The use of a 128-bit address space provides a high degree of address-assignment flexibility beyond that available under IPv4. IPv6 addressing enables Internet service providers to be identified and has the ability to identify local and global multicast addresses, private-site addresses for use within an organization, hierarchical geographical global unicast addresses, and other types of addresses. Table 2-5 lists the initial allocation of address space under IPv6.

Allocation	Prefix (Binary)	Fraction of Address Space
Reserved	0000 0000	1/256
Unassigned	0000 0001	1/256
Reserved for NSAP allocation	0000 001	1/128
Reserved for IPX allocation	0000 010	1/128
Unassigned	0000 011	1/128
Unassigned	0000 1	1/32
Unassigned	0001	1/16
Unassigned	001	1/8
Provider-based unicast address	010	1/8
Unassigned	011	1/8
Reserved for geographic-based unicast address	100	1/8
Unassigned	101	1/8
Unassigned	110	1/8
Unassigned	1110	1/16
Unassigned	1111 0	1/32
Unassigned	1111 10	1/64
Unassigned	1111 110	1/128
Unassigned	1111 1110 0	1/512
Link-local use addresses	1111 1110 10	1/1024
Site-local use addresses	1111 1110 11	1/1024
Multicast addresses	1111 1111	1/256

Table 2-5. IPv6 Address Space Allocation

The Internet Assigned Numbers Authority (IANA) was given the task of distributing portions of IPv6 address space to regional registries around the world, such as the InterNIC in North America, Reseaux IP Europeans (RIPE) in Europe, and Asian Pacific

Network Information Center (APNIC) in Asia. To illustrate the planned use of IPv6 addresses, let's turn our attention to what will probably be the most common type of IPv6 address—the provider-based address.

Provider-Based Addresses   The first official distribution of IPv6 addresses will be accomplished through the use of provider-based addresses. Based on the initial allocation of IPv6 addresses as shown in Table 2-5, each provider-based address will have the 3-bit prefix 010. That prefix will be followed by fields identifying the registry that allocated the address, the service provider, and the subscriber. The latter field actually consists of three subfields: a subscriber ID that can represent an organization and variable network and interface identification fields used in a similar manner to IPv4 network and host fields. Figure 2-19 illustrates the initial structure for a provider-based address.

Special Addresses   Under IPv6, there are five special types of unicast addresses that were defined, of which one deserves special attention. That address is the version 4 address, which was developed to provide a migration capability from IPv4 to IPv6.

In a mixed IPv4 and IPv6 environment, devices that do not support IPv6 will be mapped to version 6 addresses using the following form:

▼   0:0:0:0:0:FFFF:w.x.y.z

Here w.x.y.z represents the original IPv4 address. Thus, IPv4 addresses will be transported as IPv6 addresses through the use of the IPv6 version 4 address format. This means that an organization with a large number of workstations and servers connected to the Internet has only to upgrade its router to support IPv6 addressing when IPv6 is deployed. Then it can gradually upgrade its network on a device-by-device basis to obtain an orderly migration to IPv6. Now that we have an appreciation for IPv4 and IPv6 addressing, let's turn our attention to the address resolution process prior to exploring ICMP and the TCP and UDP headers.

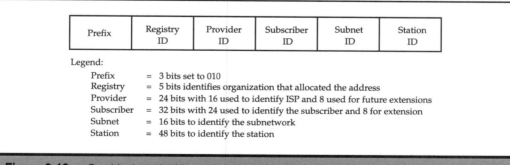

Figure 2-19.   Provider-based address structure

Address Resolution

The physical address associated with a local area network workstation is often referred to as its *hardware* or *media access control* (MAC) *address*. In actuality, that address can be formed via software to override the burned-in address on the network adapter card, a technique referred to as *locally administrated addressing*. When the built-in hardware address is used, this addressing technique is referred to as *universally administrated* addressing, as it represents a universally unique address whose creation we will shortly discuss. For both techniques, frames that flow at the data link layer use 6-byte source and destination addresses formed either via software or obtained from the network adapter.

Figure 2-20 illustrates the formats for both Ethernet and Token Ring frames. Both networks were standardized by the IEEE and use 6-byte source and destination addresses. The IEEE assigns blocks of addresses six hex characters in length to vendors that represent the first 24 bits of the 48-bit field used to uniquely identify a network adapter card. The vendor then encodes the remaining 24 bits, or six hex character positions, to identify the adapter manufactured by the vendor. Thus, each Ethernet and Token Ring adapter has a unique hardware of burned-in identifiers that denote the manufacturer and the adapter number. If an organization decides to override the hardware address, it can do so via software; however, a 48-bit address must still be specified for each station address.

When an Ethernet or Token Ring station has data to transmit, it encodes the destination address and source address fields with 48-bit numbers that identify the layer 2 locations on the network to receive the frame and the layer 2 device that is transmitting the frame. In comparison, at the network layer, IP uses a 32-bit address that has no relation to the MAC or layer 2 address. Thus, a common problem associated with the routing of an IP datagram to a particular workstation on a local area network involves the delivery of the datagram to its correct destination. This delivery process requires an IP device that needs to transmit a packet via a layer 2 delivery service to obtain the correct MAC or layer 2 address so it can take a packet and convert it into a frame for delivery. In the opposite direction, a workstation must be able to convert a MAC address into an IP address. Both of these address-translation problems are handled by protocols developed to provide an address resolution capability. One protocol, known as the Address Resolution Protocol, translates an IP address into a hardware address. The Reverse Address Resolution Protocol (RARP), as its name implies, performs a reverse translation, or mapping, converting a hardware layer 2 address into an IP address.

Operation   To obtain a general appreciation for the operation of ARP, let's assume one computer user located on an Ethernet network wants to transmit a datagram to another computer located on the same network. The first computer would transmit an ARP packet that would be carried as an Ethernet broadcast frame to all stations on the net-

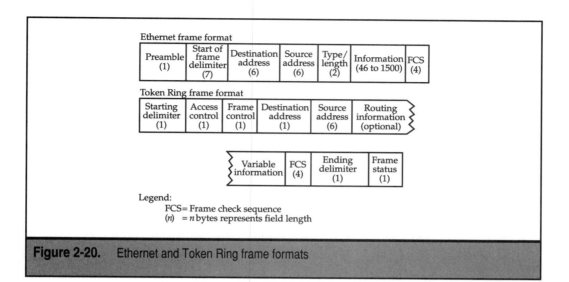

Figure 2-20.   Ethernet and Token Ring frame formats

work. Thus the packet would be transported to all devices on the Ethernet LAN. The packet would contain the destination IP address, which is known because the computer is transmitting the IP address to a known location. Another field in the ARP packet used for the hardware address would be set to all 0s, as the transmitting station does not know the destination hardware address. Each device on the Ethernet LAN will read the ARP packet as it is transmitted as a broadcast frame. However, only the station that recognizes that it has the destination field's IP address will copy the frame off the network and respond to the ARP request. When it does, it will transmit an ARP reply in which its physical address is inserted in the ARP address field that was previously set to 0.

To illustrate the necessity to constantly transmit ARP packets as well as to lower the utilization level of the LAN, the originator will record received information in a table known as an ARP *cache,* allowing subsequent datagrams with previously learned correspondences between IP addresses and MAC addresses to be quickly transmitted to the appropriate hardware address on the network. Thus, ARP provides a well-thought-out methodology for equating physical hardware addresses to IP's logical addresses and allows IP addressing at layer 3 to occur independently from LAN addressing at layer 2.

APPLICATION NOTE: Many routers and workstations run operating systems that dynamically update the ARP cache. This means that old entries are purged to make space available for new entries. This also means that if an entry for a voice gateway is purged, and then the layer 2 address requires resolution for an inbound packet transporting digitized voice, there will be a delay as a router attempts to resolve the layer 3 address to a layer 2 address so it can transmit the packet to the gateway.

To avoid ARP delays to devices that operate on digitized voice packets, consider configuring permanent ARP entries in your router that support communications to such devices. Doing so will eliminate the delay associated with the address resolution protocol and may shave a few additional milliseconds off end-to-end communications.

Now that we have an appreciation for IP addressing and the method by which IP addresses are equated to hardware layer 2 addresses, let's turn our attention to the Internet Control Message Protocol (ICMP) and how it provides both a messaging capability to convey information about abnormal conditions as well as a testing capability.

ICMP

Although the Internet Protocol (IP) represents a host-to-host datagram delivery service, on occasion a router or destination host will require a mechanism to inform the host of an abnormal condition. To provide this error-reporting capability, the Internet Control Message Protocol (ICMP) is used. As we will soon note, ICMP also provides a mechanism for determining the latency or delay through a network. ICMP used IP for delivery; however, it is considered an integral part of IP and by default must be supported in each IP module.

Overview

ICMP messages are transmitted within an IP datagram, resulting in an IP header prefixing the ICMP message. Figure 2-21 illustrates the transmission format of an ICMP message. Note that although each ICMP message has its own format, they all commence with the same three fields. Those fields are an 8-bit Type field, an 8-bit Code field, and a 16-bit Checksum field.

When IP transports ICMP, the Protocol field in the IP header has a value of 1. This indicates to the receiver that an ICMP message follows. To obtain an appreciation for the type of ICMP messages, let's turn our attention to the ICMP Type and Code fields.

ICMP Type Field

The ICMP Type field defines the meaning of the message. In addition, the Type code will indirectly provide information about the format of the message since certain message

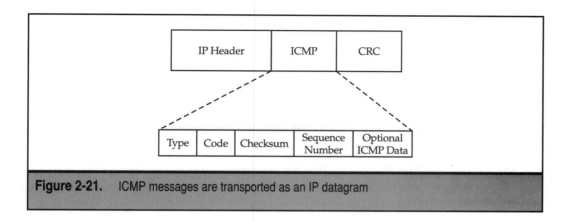

Figure 2-21. ICMP messages are transported as an IP datagram

types contain field values that other message types lack. Two of the more popularly used ICMP messages use type values 0 and 8. A Type field value of 8 represents an Echo Request, while a Type field value of 0 represents an ICMP Echo Reply. Although their official names are Echo Request and Echo Reply, most persons are more familiar with the term Ping, which is used to reference an application built into the TCP/IP protocol suite that uses both the request and reply.

The use of Ping provides an indication of the round-trip delay from the station issuing the Ping to the destination address used by the application. Thus, half of the round-trip delay indicates the one-way latency through a network to a destination address and can provide an indication of whether or not a voice-over-IP application can be expected to operate such that reconstructed voice does not sound abnormal.

APPLICATION NOTE: You can use Ping to estimate the delay from source to destination to determine ahead of time if a voice-over-IP application that provides a decent quality of reconstructed voice can be implemented.

Although the use of Ping provides an important tool for determining whether or not a voice-over-IP application can be expected to operate decently, it is not the only tool we have. Other tools include the use of Traceroute as well as various network utilization programs. In Chapter 11 we will focus our attention on management issues, including examining the operation of Ping and Traceroute. Table 2-6 lists ICMP Type field values that identify specific types of ICMP messages. Note that an ICMP Type field value of 30 is the Traceroute application.

Type	Name
0	Echo Reply
1	Unassigned
2	Unassigned
3	Destination Unreachable
4	Source Quench
5	Redirect
6	Alternate Host Address
7	Unassigned
8	Echo Request
9	Router Advertisement
10	Router Selection
11	Time Exceeded
12	Parameter Problem
13	Timestamp
14	Timestamp Reply
15	Information Request
16	Information Reply
17	Address Mask Request
18	Address Mask Reply
19	Reserved (for Security)
20–29	Reserved (for Robustness Experiment)
30	Traceroute
31	Datagram Conversion Error
32	Mobile Host Redirect
33	IPv6 Where-Are-You
34	IPv6 I-Am-Here
35	Mobile Registration Request
36	Mobile Registration Reply
37	Domain Name Request
38	Domain Name Reply

Table 2-6. ICMP Type Field Values table

Type	Name
39	SKIP
40	Photuris
41–255	Reserved

Table 2-6.   ICMP Type Field Values table *(continued)*

ICMP Code Field

The function of the ICMP Code field is to provide additional information about certain messages defined in the Type field. For example, an ICMP Type field value of 3 indicates that the destination host was not reachable. The Code field value indicates the possible reason for the Type field value, which further defines the scope of the problem. Here a Code field value of 0 indicates the network was unreachable, a code value of 1 indicates the host was unreachable, and so on. Table 2-7 lists Code field values for ICMP Type field values. Note that many Type field values to include Echo Reply (Type field value of 0) and Echo Request (Type field value of 8) have only a Code field value of 0 and thus do not have Code field options. The code fields for such Type fields are not included in Table 2-7.

Type Field	Code Field
3	0 Net reachable
	1 Host unreachable
	2 Protocol unreachable
	3 Port unreachable
	4 Fragmentation needed and DF set
	5 Source route failed
5	0 Redirect datagrams for the network
	1 Redirect datagrams for the host
	2 Redirect datagrams for the Type of Service and Network
	3 Redirect datagrams for the Type of Service and Host
11	0 Time-to-live exceeded in transit
	1 Fragment reassembly time exceeded
12	0 Pointer indicates the error

Table 2-7.   ICMP Code Field Values table

TCP AND UDP HEADERS

Both TCP and UDP represent layer 4 transport protocols. As discussed in the first section in this chapter, there are significant differences between the functionality of each transport protocol. TCP is a connection-oriented, reliable transport protocol that creates a virtual circuit for the transfer of information. In comparison, UDP is a connectionless, unreliable transport protocol that results in routers forwarding datagrams without requiring setup of a session between originator and recipient. This method of transmission represents a best-effort forwarding method and does not require the handshaking process used by TCP to establish and maintain a communications session. This means that you can consider TCP versus UDP as a trade-off between reliability and performance. Now that we've reviewed the general differences between TCP and UDP, let's turn our attention to the format of their headers to include their port number field, which is used in conjunction with IP address fields by routers and firewalls as a mechanism to filter packets.

The TCP Header

At the transport layer, TCP accepts application data in chunks of up to 64 Kbytes in length. Those chunks are fragmented into a series of smaller pieces that are transmitted as separate IP datagrams, typically 512 or 1024 bytes in length. Since IP provides no mechanism that guarantees datagrams will be correctly received as to both content and sequence, it is up to the TCP header to provide the mechanism for reliable and orderly delivery of data. To do so, the TCP header includes a field that is used for the sequencing of datagrams and a Checksum field for reliability. Because traffic from different applications, such as FTP and HTTP, can flow from or to a common host, a mechanism is required to differentiate the type of data carried by each datagram. This data differentiation is accomplished by the use of a Destination Port field containing a numeric that identifies the process or application in the datagram. In actuality, the TCP header plus data is referred to as a *segment*, so the port number identifies the type of data in the segment, and the IP header is added to the TCP header to form the datagram that will contain the source and destination IP address. Now that we have a general appreciation for the TCP header and its relationship to the application process and IP header, let's turn our attention to the fields in the TCP header whose structure is illustrated in Figure 2-22.

Source and Destination Port Fields

The Source and Destination Port fields are each 16 bits in length. Each field identifies a user process or application, with the first 1024 out of 65,536 available port numbers standardized with respect to the type of traffic transported via the use of a specific numeric value. The source port field is optional and, when not used, is set to a value of 0. The term *well-known port*, which is commonly used to denote an application layer protocol or process, actually refers to a port address at or below 1023. Both TCP and UDP headers contain fields for identifying source and destination ports. For example, Telnet, which is transported by TCP, uses the well-known port number 23, while SNMP, which is transported by UDP, uses the well-known port number 161.

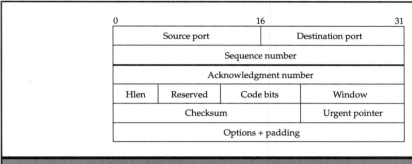

Figure 2-22.   The TCP header

Sequence and Acknowledgment Number Fields

The Sequence Number field is 32 bits in length and provides the mechanism for ensuring the sequentiality of the data stream. The Acknowledgment Number field, which is also 32 bits in length, is used to verify the receipt of data.

Hlen Field

The Hlen field is 4 bits in length. This field contains a value that indicates where the TCP header ends and the data field starts. This field is required because the inclusion of options can result in a variable-length header.

Code Bits Field

The Code Bits field is also referred to as a *Flags field,* as it contains 6 bits, each of which is used as a flag to indicate whether a function is enabled or disabled. Two bit positions indicate whether or not the acknowledgment and urgent pointer fields are significant. The purpose of the urgent bit or flag is to recognize an urgent or a priority activity, such as when a user presses the CTRL-BREAK key combination. Then the application will set the Urgent flag, which results in TCP immediately transmitting everything it has for the connection. The setting of the urgent bit or flag also indicates that the Urgent Pointer field is in use. Here, the Urgent Pointer field indicates the offset in bytes from the current sequence number where the urgent data is located. Other bits or flags include a PSH (push) bit, which requests the receiver to immediately deliver data to the application and forego any buffering, an RST (reset) bit to reset a connection, a SYN (synchronization) bit used to establish connections, and a FIN (finish) bit, which signifies the sender has no more data and the connection should be released.

Window Field

The Window field is 2 octets in length. This field is used to indicate the maximum number of blocks of data the receiving device can accept. A large value can significantly improve

TCP performance, as it permits the originator to transmit a number of blocks without having to wait for an acknowledgment and permits the receiver to acknowledge the receipt of multiple blocks with one acknowledgment. Although each field in the TCP header is important, the goal of this chapter is to provide an understanding of the operation of voice over IP and the configuration of equipment required to support it, so we will not probe deeper into the TCP header. Instead, we will conclude this chapter with an examination of the UDP header that will provide us with a background for understanding QoS and IP telephony-related protocols, which are the focus of Chapter 3.

The UDP Header

Through the use of UDP, an application can transport data in the form of IP datagrams without having to first establish a connection to the destination. This also means that when transmission occurs via UDP, there is no need to release a connection, which simplifies the communications process. This in turn results in a header that is greatly simplified and much smaller than TCP's header.

Figure 2-23 illustrates the composition of the UDP header, which consists of 16 bytes followed by actual user data. Similarly to TCP, an IP header will prefix the UDP header. The resulting message, consisting of the IP header, the UDP header, and user data, is referred to as a UDP *datagram.*

Source and Destination Port Fields

The Source and Destination Port fields are each 2 octets in length and function in a similar manner to their counterparts in the TCP header. That is, the Source Port field is optional and filled with 0s when not in use, while the destination port contains a numeric that identifies the application or process. Since UDP is commonly used by several Internet telephony products, you must determine the port a specific product uses. Then you will probably have to reprogram your organization's router access list and modify the configuration of your organization's firewalls to enable UDP datagrams using ports previously blocked to transport Internet telephony data onto your private network via the Internet.

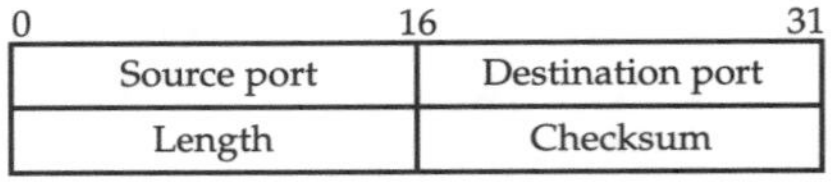

0	16	31
Source port	Destination port	
Length	Checksum	

Figure 2-23. The UDP header

APPLICATION NOTE: There are currently no standards concerning the use of different UDP port numbers for the transmission of digitized voice. This means that there exists a high degree of probability that different applications will use different UDP ports. This also means that if you use a router access list or firewall to enable certain applications to flow into and out of your private network, you will have multiple statements that require checking, adding a delay to delay-sensitive digitized voice packets. The best way to minimize this delay is to standardize on one or two products instead of having your router or firewall administrator enter literally dozens of statements to support a large number of products. Another technique is to move your UDP checking statements toward the top of your access list statements, positioning them directly below any anti-spoofing address statements to minimize delay.

Length Field

The Length field indicates the length of the UDP datagram to include header and user data. This 2-octet field has a minimum value of 8, which represents a UDP header without data.

Checksum Field

The Checksum field is 2 octets in length. The use of this field is optional and is filled with 0s if the application does not require a checksum. If a checksum is required, it is calculated on what is referred to as a *pseudo header*. This new logically formed header consists of the source and destination addresses and the Protocol field from the IP header. By verifying the contents of the two address fields through its checksum computation, the pseudo header ensures that the UDP datagram is delivered to the correct destination network and host. However, it does not verify the contents of the datagram.

Firewall and Router Considerations

Since an IP header will prefix TCP and UDP headers, there are four addresses that can be used for enabling or disabling the flow of datagrams. Those addresses are the source and destination IP addresses contained in the IP header and the source and destination port numbers contained in the TCP and UDP headers. Both firewalls and routers include a packet-filtering capability that enables users to program access lists to permit or deny the flow of packets from the Internet onto a private network or the reverse. Although most firewalls are very flexible and permit a high degree of user configuration capability, some firewalls are limited to supporting only a subset of all possible source and destination port values. This can create problems when running certain voice processes over IP if the application uses a high-value port number not supported by the firewall.

Now that we have an appreciation for the basics of the TCP/IP protocol suite, it is time to move on. In Chapter 3 we will use our knowledge obtained in this chapter to understand how we can obtain a Quality of Service capability in a TCP/IP environment as well as focus our attention upon IP telephony-related protocols. So, take a break, grab a Coke and some munchies, and let's continue our investigation of how to integrate voice over networks designed to transport data.

CHAPTER 3

QoS and IP Telephony-Related Protocols

One of the problems associated with the use of the TCP/IP protocol stack is the fact that in a normal operating environment the ability to prioritize one type of traffic over another is difficult, if not impossible. In a voice-over-IP environment we must minimize the latency experienced by packets transporting digitized voice as they flow across the packet network. To do so requires the use of more modern and evolving standards that provide what is referred to as a Quality of Service (QoS), which is the subject of the first portion of this chapter.

Quality of Service represents a broad term that in a voice-over-IP environment can be considered to represent the ability to expedite traffic through a network. Because packets can commence their journey on a LAN we will begin our examination of QoS by turning our attention to the IEEE 802.1p standard. Once packets flow off a LAN into a corporate TCP/IP intranet or into the Internet, there are several standards that govern the ability to expedite traffic into and through a TCP/IP network. Those standards include Differentiated Services (DiffServ), the ReSerVation Protocol (RSVP), and Multi-Protocol Label Switching (MPLS), each of which will be described as we turn our attention to QoS. In addition, because we can configure router queues to expedite traffic into a network, we will also examine this topic as we obtain an understanding of QoS.

While the ability to expedite traffic through a network is extremely important, so also is the ability to communicate signaling information. In the second portion of this chapter we will turn our attention to this topic, examining several more modern telephony-related protocols and standards whose utilization makes voice over IP a reality. Those protocols and standards include the Real Time Protocol (RTP), the H.323 standard, the Session Initiation Protocol (SIP) and the Media Gateway Control Protocol (MGCP).

QUALITY OF SERVICE

In this section we will turn our attention to different standards that permit traffic to be expedited through a packet network. Those standards, which define Quality of Service, affect traffic flow on a switched LAN as well as into and through a TCP/IP network. Because any discussion of QoS in a packet network environment requires a frame of reference, we will first review the manner by which the telephone company provides this capability.

Telephone Operation

When we pick up a telephone and call a distant party, we obtain a quality of service that makes a voice conversation both possible and practical. The practicality of the call results from the basic design of the telephone company network infrastructure. That infrastructure digitizes voice conversations into a 64-Kbps data stream and routes the digitized conversation through a fixed path established over the network infrastructure. For the entire path 64 Kbps of bandwidth is allocated on an end-to-end basis to the call. The fixed path is established through the process referred to as circuit switching.

Under circuit switching, a 64 Kbps time slot is allocated from the entry (ingress) point into the telephone network through the network to an exit (egress) point. The 64-Kbps time slot is commonly referred to as a Digital Signal (DS) level 0 (DS0), and the path through which the DS0 signal is allocated occurs by switches, reserving 64-Kbps slices of bandwidth.

An example of this allocation of bandwidth is shown in Figure 3-1. In examining Figure 3-1 note that the subscriber loop routed from an office or residence to a telephone company central office normally represents an analog transmission facility. At the central office a path is established between the central office serving the call originator and the central office serving the called party. That path represents a 64-Kbps segment of bandwidth that is allocated for the duration of the call. The rationale for reserving 64 Kbps of bandwidth is based upon the use of Pulse Code Modulation (PCM) by telephone companies as the preferred method of voice digitization. Regardless of the state of the conversation, 64 Kbps remain allocated until a party terminates the call.

As voice is digitized at the ingress point into the telephone company network, a slight delay of a few milliseconds occurs. As each switch performs a cross-connection operation, permitting digitized voice to flow from a DS0 contained in one circuit connected to the switch onto a DS0 channel on another circuit connected to the switch, another delay occurs. Although each cross-connection introduces a slight delay to the flow of digitized voice, the switch delay is minimal, typically a fraction of a millisecond or less. Thus, the total end-to-end delay experienced by digitized voice as it flows through a telephone network can be considered as minimal.

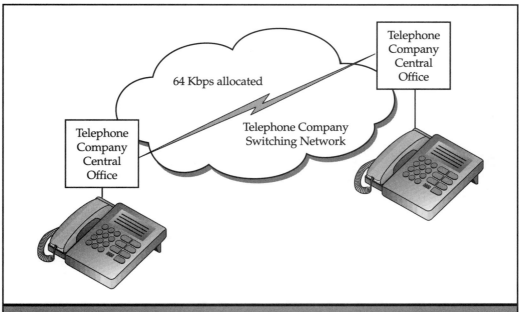

Figure 3-1.   As a voice conversation occurs through a telephone company network, 64 Kbps of bandwidth is reserved regardless of speech activity.

Another characteristic of the flow of digitized voice through the telephone network infrastructure concerns the variability or latency differences between each digitized voice sample. Although voice digitization and circuit switching processes add latency to each voice sample, that delay is uniform. Thus, we can characterize the telephone network as either a low-delay or a uniform or near-uniform delay transmission system. Those two qualities—low-delay and uniform or near-uniform delay—represent two key Quality of Service metrics. The two metrics are important considerations for obtaining the ability to transmit real-time data, such as voice and video. However, the telephone company infrastructure also provides a third key QoS metric, equally important. That metric is a uniform, dedicated, 64-Kbps bandwidth allocated to each voice conversation. Because that bandwidth is dedicated on an end-to-end basis you can view it as being similar to providing an expressway that allows a stream of cars to travel from one location to another while prohibiting other cars destined to other locations to share the highway.

A fourth QoS characteristic provided by the telephone company infrastructure is the fact that digitized voice flows end-to-end essentially lossless. That is, there is no planned dropping of voice samples during periods of traffic congestion. Instead, when the volume of calls exceeds the capacity of the network, such as on Mother's Day or Christmas Eve, new calls are temporarily blocked and the subscriber encounters a "fast" busy signal when dialing. Table 3-1 provides a summary of commonly used QoS metrics and their normal method of representation.

Although the telephone company network infrastructure provides the QoS necessary to support real-time communications, its design is relatively inefficient. This inefficiency results from the fact that unless humans shout at one another, a conversation is normally half-duplex; this results in half of the bandwidth utilization being wasted. In addition, unless we talk like the man in the Federal Express commercial that was popular a few years ago, we will periodically pause as we converse. Because 64 Kbps of bandwidth is allocated for the duration of the call, this means that the utilization of bandwidth is far from being optimized.

Metric	Normal Representation
Dedicated bandwidth	bps, Kbps, or Mbps
Latency (delay)	msec
Variation (jitter)	msec
Data Loss	Percent of frames or packets transmitted

Table 3-1. Common QoS Metrics

Packet Network

In comparison to a circuit switched network where the use of bandwidth is dedicated to a user, packet networks allow multiple users to share network bandwidth. While this increases the efficiency level of network utilization, it introduces several new problems. To obtain an appreciation for those problems, let us review the operation of a generic packet network; this could be a TCP/IP network such as the Internet, a corporate intranet, or even a frame relay network.

Figure 3-2 illustrates the flow of data from two different locations over a common backbone packet network infrastructure. In this example, two organizations—labeled 1 and 2—share access via packet network node A to the packet network. Let us assume that packets destined from organization 1 flow to the network address Z connected to mode E, while packets from organization 2 flow to location Y, also connected to packet network mode E.

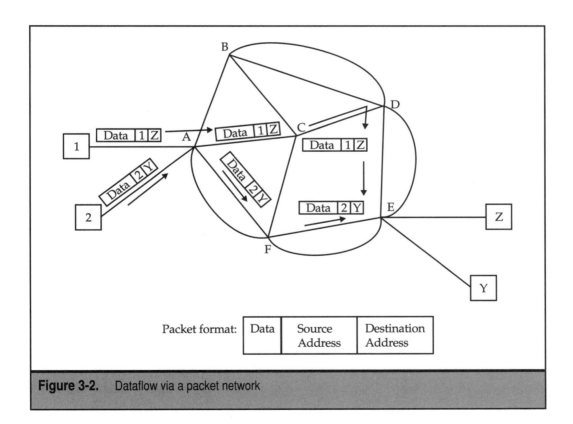

Figure 3-2. Dataflow via a packet network

In Figure 3-2, packets could flow over different backbone routes; however, their ingress and egress locations are shown to be in common. If we assume that location 2 is transmitting real-time information to location Y, what happens when periodically a packet from location 1 arrives at node A ahead of the packet from location 2? When this situation occurs, the packet from location 1 delays the processing of the packet arriving from location 2.

Suppose data sources connected to nodes B, C, and D all require access to devices connected to node A. When this situation arises, the device at node A may be literally swamped with packets beyond its processing capability. In this situation, the network device at node A may be forced to drop packets. While applications such as a file-transfer could simply retransmit a dropped packet without a user being aware of this situation, if real-time data such as voice or video was being transmitted, too many packet drops would become noticeable; they cannot be compensated for by retransmission that further delays real-time information.

Consider what happens when the packet from location 2 is serviced at node A. If other packets require routing to node F, packets from location 2 could be further delayed. After packets from location 2 are forwarded onto the circuit between nodes A and F they will be processed by node F. At this location packets arriving from nodes C and E could delay the ability of node F to forward packets destined to node E. Next, packets are forwarded on towards node E for delivery to address Y. For the previously described data flow several variable delays will be introduced, each adversely affecting the flow of packets from location 2 to address Y. In addition, once a packet reaches node E, it could be delayed by the need to process other packets. Examples include the one arriving from location 1 and destined to address Z.

Another characteristic of a packet network is the fact that when a node becomes overloaded it will send some packets to the great bit bucket in the sky. This is a normal characteristic of packet networks and, in fact, a frame relay performance metric involves the packet discard rate. Based upon the preceding examination of data-flow in a packet network, we note there is no guarantee that packets will arrive at all or arrive with minimal delay nor with a set amount of variation between packets. Because this situation makes it difficult if not impossible to transport real-time data over a packet network, various techniques were developed in an attempt to provide a QoS capability to packet networks. Those techniques can be categorized into three general areas. Those areas include expediting traffic at the ingress point into the network, expediting traffic through the network, and expediting delivery of traffic at the destination or egress point in the network. For each area, there are several techniques being supported by different hardware and software vendors to provide QoS capability. Some techniques are standardized, while others will be standardized in the near future.

The manner by which an organization connects equipment to the ingress and egress points on a packet network will have a bearing upon whether or not additional QoS tools and techniques are required; they would provide an end-to-end transmission capability within certain limits for delay, jitter, and obtainable bandwidth. Keeping this in mind, attention is turned to the ingress point of a packet network; let us examine the flow of traffic from a LAN to the ingress point. For lack of a better term, we will call this the LAN egress location.

LAN EGRESS AND THE IEEE 802.1P STANDARD

There are several methods by which a local area network can be connected to a packet network. Although it is quite common to connect a LAN to a packet network via the use of a router, there are numerous network configurations that can reside behind the router that represent the structure of the corporate LAN or even an intranet.

Because we are concerned with QoS, let us turn attention to a network configuration that allows traffic from several LAN and non-LAN based sources to be differentiated from one another as the data is passed to a router. The key to this capability is the IEEE 802.1p standard.

In the International Standards Organization (ISO) Open System Interconnection (OSI) Reference Model layer 2 represents the data link layer. That layer is responsible for the creation of frames to include applicable addressing and the computation of a cyclic redundancy check as well as the transmission of such frames. Other functions performed at layer 2 include error detection and correction as well as the use of positive and negative acknowledgments to indicate whether or not the destination received frames error free. One function omitted from the original OSI Reference Model for layer 2 operations is the topic of this section—Quality of Service.

Although the OSI Reference Model does not define QoS as a layer 2 function, the efforts of the Institute of Electrical and Electronic Engineers (IEEE) resulted in the development of a traffic expediting standard for layer 2 operations.

Overview

Many years ago the IEEE was tasked by the American National Standards Institute (ANSI) with developing LAN-related standards. While the series of Ethernet (IEEE 802.3) and Token Ring (IEEE 802.5) standards are well known, the IEEE also developed many additional standards since its original tasking that are equally important. Some of those standards include the use of the spanning tree algorithm for bridging, flow control as a mechanism to regulate the flow of data devices, virtual LANs, and traffic expediting.

Conventional LAN Dataflow

We can obtain an appreciation for the problems associated with obtaining QoS on a LAN by examining the dataflow on a conventional, shared-media LAN. Figure 3-3 illustrates a shared-media Ethernet LAN that we will use to examine dataflow.

Let's assume station 1 is performing a file transfer to the local server while station 2 is in the process of conducting a voice-over-IP (VoIP) call via the router to a destination on the Internet or on a corporate intranet. Because the LAN is a shared media network, this means that each station must contend for access to the media. If station 1 just began transmitting when station 2 "listens" to the LAN, the latter station will note the network is busy and will wait. In this example the wait will more than likely represent the time required to transmit a frame with a maximum length Information field of 1,500 bytes since we assumed station 1 was performing a file transfer operation. That type of operation uses a maximum length Information field for all frames but the last. Note that there is no

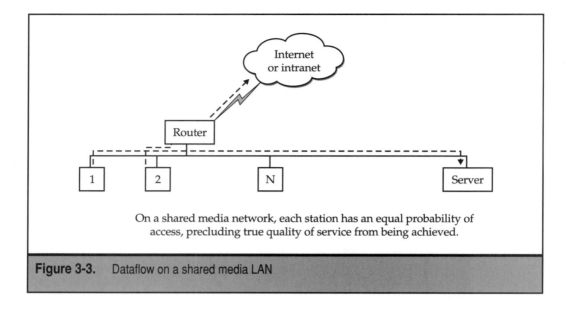

On a shared media network, each station has an equal probability of access, precluding true quality of service from being achieved.

Figure 3-3. Dataflow on a shared media LAN

method to prioritize gaining access to the media among stations in this example. Thus, this precludes the ability to obtain a true Quality of Service that would require bandwidth to be allocated on a shared-media network. In actuality, the preceding is true for shared-media Ethernet networks. In a Token-Ring networking environment the Control field of a frame contains three priority bits that enable certain types of applications to gain access to the media ahead of other types of applications. While the priority scheme expedites traffic and forms the basis for subsequent efforts by the IEEE, the priority scheme does not reserve bandwidth and represents a limitation carried over into the LAN switch environment. However, while the Token-Ring standard does not define the mapping of frames with different priority values into queues, the IEEE 802.1p standard does and can be considered as a more sophisticated evolution resulting from the Token Ring effort.

LAN Switching

In an Ethernet shared-media environment it is not possible to favor access to the media. Recognizing this limitation, the IEEE focused its traffic expediting efforts upon the LAN switch environment.

A LAN switch can be considered to represent a contention device similar to a telephone switch. When two persons dial the same number one will receive a busy signal while the other party will make a connection through the switch, allowing the dialed phone to ring. If we view a LAN switch as making connections on a frame by frame basis a similar analogy can be made, with frames routed towards the same destination causing blockage. However, unlike a telephone company switch that does not hold calls, a LAN switch includes buffer storage that can be used to temporarily hold frames when two or more input ports have data destined to the same output port.

Figure 3-4 illustrates the general flow of data through a layer 2 LAN switch when two input port data sources contend for access to the same destination port. In this example two clients are shown attempting to access a switch port connected to a server. Because many layer 2 switches allocate buffer storage for queuing frames destined to a common output port the switch contention example shown in Figure 3-4 shows two frames in a queue in memory, with one frame in the process of exiting the queue while new frames are shown flowing towards the queue.

The queue shown in Figure 3-4 is technically referred to as a first-in, first-out (FIFO) queue. Thus, frames that reach the queue are processed in the order in which they arrive. Also note that if the queue becomes full succeeding frames that reach the queue are dropped. Thus, a single FIFO queue has no mechanism to distinguish the needs of different applications concerning their ability to flow through a switch with minimal delay or delay variation.

Although FIFO queuing was simple to implement and equitable in the allocation of switch resources, it failed the differentiation test, being incapable of distinguishing different application requirements. While some switch vendors began to develop proprietary priority queuing techniques, the IEEE was in the process of developing virtual LAN standards and recognized the need to provide a common traffic expediting method. The initial IEEE effort, referred to as 802.1Q, was oriented towards developing a standard for virtual LANs. Here the term "virtual LAN" references a broadcast domain. The goal behind vLANs was to enable network administrators to position switches based upon organization, application, network protocol, or another criteria that enabled users to be grouped dynamically into a broadcast domain. Since one domain does not "hear" the broadcasts associated with other domains, the use of vLANs can boost switch performance while providing administrative flexibility.

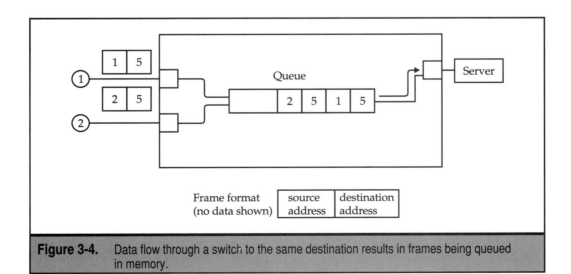

Figure 3-4.   Data flow through a switch to the same destination results in frames being queued in memory.

As part of the IEEE 802.1Q effort a tag was added to the layer 2 MAC frame as it enters a vLAN compliant switch. The 32-bit tag is inserted after the frame's normal Destination and Source Address fields as illustrated in Figure 3-5. In examining Figure 3-5 note that the IEEE "killed two birds with one frame modification" by defining 3 bits within the 32-bit header as a priority field. As indicated in Figure 3-5, the three priority field bits provide the ability to specify eight levels of priority per frame, with the default priority set to a value of 000. The 2-byte Tag Protocol Identifier (TPID) identifies the frame as a tagged frame. The 2-byte Tag Control Information (TCI) field contains three subfields to include the previously mentioned 3-bit user priority field, a 1-bit Token Ring encapsulation flag, and a vLAN Identifier (VID). The Token Ring encapsulation flag indicates if the encapsulated frame is in a native Token Ring (802.5) format. The VID uniquely identifies the vLAN to which the frame belongs and is also referred to as the vLAN tag.

Priority Assignments

Turning our attention to the priority subfield shown in Figure 3-5, you will note that priority levels range from 7 (highest priority) to 0 (lowest priority). As part of a revised 802.1D bridging standard the IEEE ratified its 802.1p specification, which provides a mechanism for layer 2 switches to prioritize traffic. Although network managers must determine actual mappings, the IEEE has made broad recommendations concerning priority settings. Table 3-2 indicates those broad recommendations.

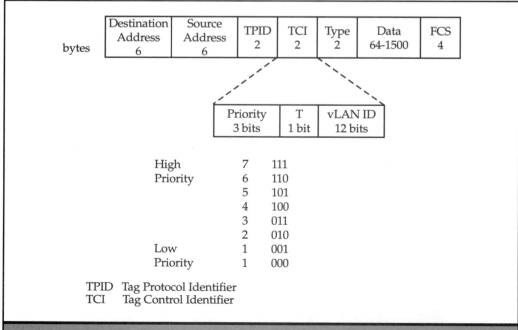

Figure 3-5.   IEEE Traffic prioritization uses 3 bits in the 2-byte Tag Control Information field, which represents half of the 4-byte virtual LAN frame extension.

111 (7)	Network Critical
110 (6)	Interactive Voice
101 (5)	Interactive Multimedia
100 (4)	Streaming Multimedia
011 (3)	Business Critical
010 (2)	Standard
001 (1)	Background
000 (0)	Best Effort (Default)

Table 3-2.   IEEE 802.1p Priority Setting Recommendations

In examining the entries in Table 3-2, network-critical traffic could represent Routing Information Protocol (RIP) and Open Shortest Path First (OSPF) table updates. Priority 6 could be used for VoIP applications, while priorities 5 and 4 would represent different multimedia applications. Business-critical traffic could include Service Advertising Protocol (SAP) traffic, while a zero value is used as a best-effort default. That is, if an application does not set the priority subfield its default value is set to zero.

The priority bits are within the 802.1Q frame. Those bits can be set by desktop clients, servers, routers, and switches. In desktops clients and servers 802.1p-compliant network adapter cards can form 802.1Q frames as well as set the bits in the priority subfield. The actual setting of the priority bits is commonly invoked by monitoring the network socket. Here the term "socket" references the IP address and upper layer port that identifies the application.

Based upon the preceding, we can note that the IEEE 802.1p standard represents a layer 2 (Media Access Control (MAC) layer) signaling technique that permits network traffic to be prioritized. This standard is implemented by relatively recently manufactured layer 2-compliant switches and routers, which can classify traffic into eight levels of priority. This means that through the use of IEEE 802.1p-compliant equipment, it becomes possible for time-critical applications on a LAN to receive preferential treatment over non-time-critical applications. It is important to note that the IEEE 802.1p standard is a layer 2 standard. This means that a priority tag that is added to LAN frames to differentiate traffic is removed at the layer 2 to layer 3 conversion point, when LAN frames are converted to packets for transmission over a WAN. Thus, the IEEE 802.1p standard is only applicable to expedite traffic on a LAN.

Operation

Figure 3-6 illustrates an example of the use of the IEEE 802.1p standard. In this example, a PBX is shown connected to a voice gateway, which in turn is connected to a port on a

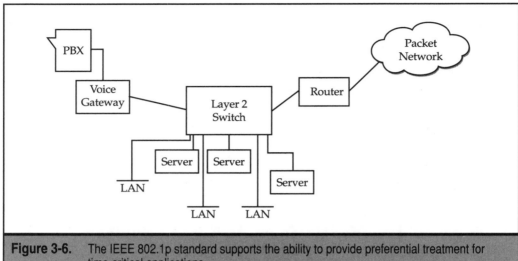

Figure 3-6.   The IEEE 802.1p standard supports the ability to provide preferential treatment for time-critical applications.

layer 2 LAN switch. Other switch connections include support for several LAN and server connections as well as a connection to a router, with the latter providing connectivity to a packet network.

A non-IEEE 802.1p-compliant switch would treat all flow requests equally. If a user on one LAN required access to the router at approximately the same time as a user on another LAN, the second request received would be queued behind the first. This is the classic first-in, first-out queuing method.

Under the IEEE 802.1p standard, traffic can be placed into different queues, based upon their level of priority. Thus, voice calls digitized by the voice gateway could be assigned a high level of priority, enabling the switch to service frames generated by the voice gateway and destined to the router prior to frames originated from other connections destined to the router. To obtain an appreciation for the operation of the standard, let's turn our attention to the use of the 802.1 priority tag.

Using the Priority Tag Values

Through the 3-bit priority tag IEEE 802.1p frames are assigned a priority value in the range 0 through 8. As frames flow through a hierarchy of 802.1p-compliant switches, each switch port that receives a frame will regenerate it based upon the priority assigned to the frame and a User Priority Regeneration table associated with the reception port.

The User Priority Regeneration table contains eight entries, one for each possible user priority value that can occur in a received frame. Based upon the user priority in the frame, the receiving port can be configured to map the user priority to any value in the range 0 to 7; however, by default the regenerated user priority value equals the value in the received frame. This is indicated in Table 3-3, which indicates the received switch port user priority regeneration options.

User Priority Value in Received Frame	Default Regeneration User Priority	Range of Regenerated User Priority Values
0	0	0-7
1	1	0-7
2	2	0-7
3	3	0-7
4	4	0-7
5	5	0-7
6	6	0-7
7	7	0-7

Table 3-3.   Receive Switch Port User Priority Regeneration Options

Frame Forwarding

In an IEEE 802.1p LAN switch environment frames can be temporarily stored in different queues prior to being forwarded out of the switch. Each switch port follows an order when deciding upon which frames to forward. That transmission order results in unicast frames with a given user priority value being processed prior to multicast frames. Within each class or category of frame (unicast or multicast) frames are first placed into an applicable storage queue based upon the user priority value in the frame in conjunction with a traffic class table.

User Priority to Traffic Class Mappings

The actual placement of a frame into a queue is anything but "cut and dried." This is because the IEEE recognized that the number of queues available per port on a switch would, from a practical standpoint, vary in conjunction with the processing power and buffer memory of a switch. To provide the ability to support a variable number of queues per switch port, the developers of the IEEE 802.1p standard defined a series of user priority to traffic class mappings, with different traffic class mapping used when a different number of queues were available for each switch port.

Table 3-4 illustrates the IEEE-recommended user priority to traffic class mappings. The default assumes a queue is available for each traffic class or eight queues per switch port to support the recommended mappings. For each switch port frames are placed into a queue based upon the user priority to traffic class mappings. Frames are then selected for extraction only if higher order queues are empty. For each queue unicast frames are processed prior to multicast frames.

		Number of Available Traffic Classes							
		1	2	3	4	5	6	7	8
U s e r P r i o r i t y	0	0	0	0	1	1	1	1	2
	1	0	0	0	0	0	0	0	0
	2	0	0	0	0	0	0	0	1
	3	0	0	0	1	1	2	2	3
	4	0	1	1	2	2	3	3	4
	5	0	1	1	2	3	4	4	5
	6	0	1	2	3	4	5	5	6
	7	0	1	2	3	4	5	6	7

Table 3-4.   Recommended User Priority to Traffic Class Mappings

Under the IEEE 802.1p standard traffic assignments that represent user priority occur based upon the number of queues supported by a switch port. For example, if there is only one queue all eight user priority values would fall into that queue and be extracted on a first-in, first-out basis with the exception that unicast frames would precede multicast frames. If there are two queues per switch port, frames will flow into each queue based upon the value of the user priority tag in the frame. For example, frames with priority values from 0 through 3 would flow into one queue while frames with priority values 4 through 7 would be placed into the second queue associated with each port. Similarly, when there are three queues associated with a switch port frames would flow into each of three queues based upon the user priority value in the frame.

Operational Example

To illustrate an example of the use of the 802.1p traffic-expediting standard, let's assume your organization uses 802.1p-compliant switches in a tier or hierarchical structure as illustrated in Figure 3-7. In this example let's assume client 1 is performing a file transfer to the server attached to switch 3 (shown as a sequence of x's) while client 2 is in the process of transferring a file via switches 1 and 2 to a distant location on the Internet or a corporate intranet (shown as a series of dots (.)). At the same time the preceding activity is occurring, client 3 is conducting a videoconference via switches 3 and 2 to a distant party on the

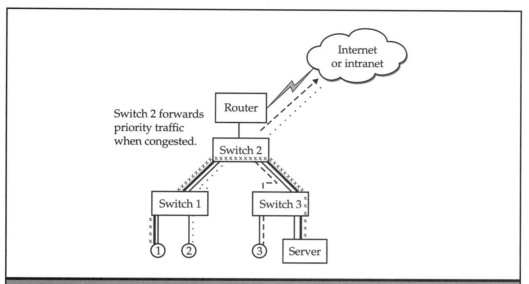

Figure 3-7.   Priority traffic control example. Assuming switch 2 is 802.1p-compliant, it places traffic from client 2 (dots) and client 3 (dashes) into different queues and extracts frames from client 3 for transmission to the router prior to extracting frames from the client 2 queue.

Internet or corporate intranet (shown as a sequence of dashes (-)). Although the flow of data from clients 2 and 3 flow over a common path from switch 2 to the router and then to the Internet or corporate intranet, the actual exiting of frames out of switch 2 to the router occurs based upon the priority bit settings of the applications forming the data flow from clients 2 and 3. That is, during periods of congestion when frames from both sources arrive at the switch destined for the same output port, the priority queuing scheme governs which frames gain access to exiting the port. For example, assume frames from client 2 have a default priority setting of 0 while frames from client 3 have a priority setting of 6. When frames from clients 2 and 3 contend for output via switch 2 to the router, they would flow into different queues. The switch would empty frames from the high priority queue prior to extracting frames from the lower priority queue, providing a traffic-expediting capability to the VoIP application being conducted by client 3.

Limitations

While the 802.1p specification provides a method to expedite traffic, it has several key limitations. Those limitations include its layer 2 operation, method of tagging, the need for mapping to obtain an end-to-end QoS capability, lack of admission control, and inability to limit the amount of resources an application can use. As we will shortly note, by itself 802.1p represents one piece of the QoS puzzle instead of a complete solution.

The first limitation of the 802.1p specification is the fact that it is restricted to layer 2 operations. This means that when frames cross a router boundary and are converted into network packets they lose their priority tag. Thus, the 802.1p specification does not address true end-to-end network performance.

A second limitation associated with the 802.1p specification concerns the fact that the 802.1Q vLAN tag extends the MAC frame. When placed on the media the maximum frame length is four bytes more than the 1526 associated with a maximum length Ethernet frame. While 802.1Q-compliant equipment has no problem recognizing the extended frame, it is quite possible that "legacy" Ethernet equipment could treat such frames as jumbo frames, sending them to the great bit bucket in the sky. Thus, you need to examine the compatibility of equipment with respect to extended frames and may need to replace some equipment if such equipment is not compatible with extended frames.

Another problem related to vLAN tags is the manner by which tagged and non-tagged frames are handled. Under the IEEE 802.1Q and 802.1p specifications non-tagged frames are treated the same as frames tagged with a 0 user priority level. That is, the non-tagged frame is treated on a best-effort basis, which may not be your intention.

A third limitation associated with the 802.1p specification is related to the first limitation. Because the 802.1p specification is limited to layer 2 operations one or more additional QoS mechanisms are required to provide QoS on an end-to-end basis. Those mechanisms can include the use of Differential Service if communications via the router is over a TCP/IP-based network or mapping applicable frames into constant Bit Rate (CBR) cells if the router is connected to an ATM-based network. In the event DiffServ is not supported by the TCP/IP network and it is impractical or not cost-effective to use an ATM infrastructure, it may be able to prioritize traffic through an existing router-based TCP/IP network infrastructure.

Two additional limitations associated with the 802.1p standard are potentially related to one another. Those limitations are the lack of any admission control and the inability to limit the amount of resources an application can use. Both of these limitations can be tackled by a Common Open Policy Service (COPS) server that can reside on a network and be configured to provide both admission control and allocation of resources according to a predefined criteria. Of course, this adds an additional layer of complexity to your network as well as an additional flow of traffic since stations must first access the COPS server.

COPS

The Common Open Policy Service represents a TCP/IP query/response protocol that was being evaluated by the Internet Engineering Task Force (IETF) when this book revision was prepared. Under COPS, messages governing admission control to network resources beyond a LAN are transported between a server and LAN clients.

The operation of COPS begins with a user (client) requesting a network connection from a Policy Enforcement Point (PEP) switch. COPS transports the request to a policy server where the request is authenticated and processed. During processing the server examines its state table to determine if sufficient bandwidth is available to satisfy the client request. The policy server then issues its response to the PEP. If bandwidth is available the

PEP switch will enable the application to access resources beyond the LAN. Otherwise, the application will have to wait for bandwidth to become available.

Now that we have an appreciation for the manner by which traffic can be expedited through LAN switches, let's turn our attention to the flow of traffic into the WAN.

EXPEDITING TRAFFIC INTO THE WAN

There are several methods you can consider by themselves or in conjunction with one or more other techniques to expedite the flow of traffic into the WAN. Those methods include the use of different types of router queues, the use of the Service Type byte in the IP header, and a revision to the use of that byte that is referred to as DiffServ. The last two methods can be loosely categorized as traffic-expediting methods occurring at the edge of a WAN. They supplement another QoS traffic-expediting method referred to as Multi-Protocol Label Switching that operates inside the WAN.

Router Queuing

Although only one connection from the layer 2 switch to the router is shown in Figure 3-7, in actuality a router can have several LAN-side and WAN-side connections. Because a heavily utilized router can add a significant delay to real-time traffic, Cisco added several techniques to expedite the flow of traffic through their products. Some of those techniques represent compliance with industry standards, while other techniques are proprietary features incorporated into their products.

There are several queuing methods beyond FIFO queuing supported by router manufacturers. Some of those additional queuing methods include priority queuing, custom queuing, and weighted fair queuing.

Priority queuing permits users to prioritize traffic based upon network protocol, incoming interface, packet size, and source and destination address. Because digitized voice traffic is transported in relatively short packets, you could use priority queuing to expedite digitized voice ahead of file transfers and other types of traffic carried in relatively long packets.

Under custom queuing, you can share bandwidth among applications. Thus, through its use you could ensure a voice or video application obtains a guaranteed portion of bandwidth at an entry point into a WAN. Under weighted fair-queuing, interactive traffic is assigned to the front of a queue to reduce response time; the remaining bandwidth is shared among high-bandwidth traffic flow. In Chapter 7 when we turn our attention to voice-over-IP implementation, we will note the Cisco IOS statements applicable to configuring different queuing methods.

Traffic Expediting

In this section we will turn our attention to the manner by which traffic can be expedited through a network.

Overview

As previously noted at the beginning of this section, there are two methods that can be used to expedite traffic into a WAN while a third method is used to expedite traffic through the WAN. In this section we will first obtain an overview of each method and then focus our attention upon the two methods used to expedite traffic into a WAN.

Within the IP header is a field labeled "Type of Service," which is also referred to as the Service Type field. This 8-bit field was intended to allow applications to indicate the type of routing path they would like, such as low delay, high throughput, and high reliability for a real-time application. Although a great idea, this field is rarely used directly within an IP network and is usually set to a value of 0. However, because it is easy for an application to set the ToS byte, it is possible to use it as a mapping mechanism, as we will shortly note.

A second traffic-expediting method recognizes the limited use of the "Service Type field" and both renamed and reassigned values to the field. This method reuses the "Service Type field" as a DiffServ field. Under DiffServ, traffic definitions are assigned to denote the manner by which data flows are handled by routers. At the present time, Assured Service and Preferred Service (each with slightly different definitions of service) have been defined by the IETF.

A third traffic-expediting method involves mapping a flow of traffic between two locations and adding a label to packets to expedite their routing. The goal behind this label method is to avoid searching through each router's address table to find the relevant port to output a packet. Because a router could have thousands of entries in its address tables the ability to bypass the address search expedites the flow of traffic through the router. This technique was originally referred to as *tag-switching* by Cisco. It was standardized by the IETF as Multi-protocol Label Switching (MPLS). Since this section is focused upon traffic expediting into the WAN, we will defer a discussion of MPLS until the next section in this chapter.

The ToS Byte

The function of the ToS byte is to denote the importance of the datagram as well as the datagram's requirements with respect to delay, throughput, and reliability.

Figure 3-8 again illustrates the format of the ToS byte so you won't have to refer back to Chapter 2. Note that the first three bit positions can be used to assign a precedence to a datagram. Although eight levels of precedence are available, the highest two values are normally reserved for internal network utilization, allowing the ToS byte to provide a mechanism for partitioning traffic in up to six classes of service.

The setting of a precedence value is normally a function of an application. Although the original intention of the ToS byte was to provide a mechanism for both specifying precedence and the reliability, throughput, delay, and cost requirements of a datagram, like many good intentions its use never approached its goal. In fact, a popular joke is that by default IP traffic is unreliable, slow, and costly, which perhaps explains why applications did not turn on any of the relevant ToS bits. However, if you obtain an application that can be configured to turn on an applicable precedence bit you can use a Cisco extended IP

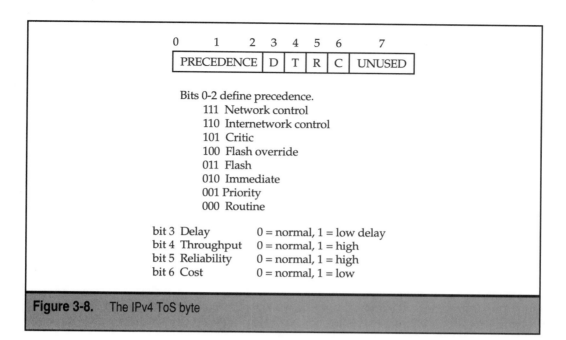

```
0     1    2   3   4   5   6      7
┌──────────────┬───┬───┬───┬───┬─────────┐
│  PRECEDENCE  │ D │ T │ R │ C │ UNUSED  │
└──────────────┴───┴───┴───┴───┴─────────┘
```

Bits 0-2 define precedence.
    111 Network control
    110 Internetwork control
    101 Critic
    100 Flash override
    011 Flash
    010 Immediate
    001 Priority
    000 Routine

bit 3	Delay	0 = normal, 1 = low delay
bit 4	Throughput	0 = normal, 1 = high
bit 5	Reliability	0 = normal, 1 = high
bit 6	Cost	0 = normal, 1 = low

Figure 3-8.   The IPv4 ToS byte

access list in conjunction with a priority-list command to associate traffic with an applicable precedence bit setting into a predefined queue. A second possibility that exists concerning the use of the ToS byte is to map its setting to another transport protocol. In particular, the ability to map IP into ATM affords you the possibility to assign voice over IP to transport via ATM's constant bit rate (CBR), which guarantees a QoS over an ATM backbone that could be used to interconnect geographically separated WANs. In Chapters 7 and 9 we will examine the manner by which the ToS byte can be used.

While Cisco and other router manufacturers provide a mechanism for traffic differentiation based upon the setting of the ToS precedence bits, it will more than likely be a warm day in Antarctica before its use within an IP network, if ever, becomes popular. Recognizing the low level of utilization of the ToS byte, the Internet Engineering Task Force worked on a series of RFCs that redefined the use of that byte. Under the revision, the byte is used to provide DiffServ information to routers . Thus, let's turn our attention to this topic.

Differentiated Services

Differentiated Services can be considered to represent a set of technologies that permits network service providers to offer different categories of QoS to different customer traffic streams. In actuality, DiffServ does not provide a true QoS capability. Instead, it provides a mechanism where more bandwidth can be assigned to higher priority traffic flows than that of lower priority flows during a period of time. Due to this priority- or traffic-expediting scheme, higher-priority traffic receives better transmission performance than lower-priority flows. The use of Differentiated Services requires the establishment of a Service Level Agreement (SLA). The SLA is established between the service provider and the customer prior to the use of Differentiated Services. The SLA can specify the manner by which packets are handled based upon one or more metrics. Those metrics can include the expected throughput obtained, the probability of a packet being dropped, the delay or latency resulting from packets flowing through a network, and the variation in the flow of packets received when a stream is presented to the network. Here the latter metric is commonly referred to as *jitter*.

The SLA is used by the network operator to define the manner by which packets will be handled based upon the composition of the DiffServ byte. Here "handling" is a loose term that references the forwarding service customers will receive for a particular packet class or set of classes. Once the SLA is established, the customer submits packets with applicable DiffServ bit settings to indicate the service desired. The service provider then becomes responsible for assuring that their routers are configured with applicable forwarding policies to provide the QoS specified in the SLA for each packet class.

Instead of taking the traditional approach to QoS in which routers along the path from source to destination reserve bandwidth for a traffic stream, DiffServ simply assigns a value to a byte in the IP header that allows all routers on a path to forward traffic according to the byte setting. This action eliminates the necessity to convey separate signaling information between routers. By eliminating the need for separate signaling information, DiffServ avoids the high overhead associated with RSVP (discussed later in this chapter), which made it impractical for use on the Internet. Another advantage associated with DiffServ is the fact that many traffic streams can be aggregated to one of a small number of behavior aggregates (BAs). That is, all traffic with the same DiffServ byte settings are treated in the same manner by each DiffServ-compatible router, regardless of source, destination, or composition of a packet. This means that it is possible for a voice conversation and a near-real-time application between pairs of dissimilar addresses to be treated in the same manner as their packets flow through a network. The BA traffic streams are forwarded by each DiffServ-compatible router based upon one of a limited number of per-hop behaviors (PHBs) further simplifying data flow as well as minimizing the processing effort of routers.

Operation

Differentiated Services represents a service provided by network service providers to customers. This service requires negotiation ahead of time between the two parties so that customer traffic can be processed through the network based upon the manner by which the DiffServ byte is marked or classified in each packet at the edge of the network.

The DiffServ byte is based upon the use of the ToS field in the IPv4 header and the Traffic Class field in the IPv6 header; however, it maintains backward compatibility with the RFC 791 IPv4 specification that defined the original use of the bit positions in the ToS field. To obtain backward compatibility, the DiffServ byte uses bit positions 0, 1, and 2 for priority settings that provide a general mapping to the Precedence field in the ToS byte. Thus, routers that use the Precedence field in the ToS byte will not choke on the DiffServ byte. Instead, such routers will provide a priority flow similar to that provided to a packet where the IPv4 header uses a ToS byte.

The DiffServ byte both reorganized and renamed the first three bit positions of the IPv4 ToS byte. Table 3-5 lists the eight precedence levels now defined under DiffServ.

In examining Table 3-5, note that precedence levels 6 and 7 remain the same as they are still assigned to network operations, such as routing protocols and keep alive signals. Collectively, Classes 1 through 4 are referred to as Assured Forwarding (AF) and are used in conjunction with bit positions 3 and 4 of the DiffServ byte. By setting bits 3 and 4 it becomes possible to specify the forward percentage of packets for each class of traffic.

The bit settings shown in Table 3-5 result in three types of PHBs, including best effort forwarding, expedited forwarding, and assured forwarding, which we can categorize as follows.

Best effort forwarding (BE) represents the default forwarding method. Most Internet traffic flows use BE, resulting in packets marked BE being transmitted only when network bandwidth is available. Thus, the transmission performance of BE marked packets will be good when the network is not congested and will deteriorate as traffic carried by the network increases.

Expedited forwarding (EF) represents a higher priority of transmission than BE. A network can reserve a portion of its resources and bandwidth for packets marked EF. It is possible to preclude BE marked packets from using resources and bandwidth allocated for EF, guaranteeing EF flows a level of performance and better service than BE marked packets.

The third type of PHB is Assured forwarding (AF). AF results in a minimum bandwidth being allocated to each AF flow, with the potential to subdivide flows into several priority classes. This permits each AF marked packet to be assured a service quality and provides more granularity than EF.

Precedence Level	Bit Settings	Meaning
7	111	Network
6	110	Network
5	101	Expedited forwarding
4	100	Class 4
3	011	Class 3
2	010	Class 2
1	001	Class 1
0	000	Best effort

Table 3-5.   DiffServ Byte Precedence Levels

Collectively, bits 3 and 4 of the DiffServ byte are referred to as the DiffServ Code Point (DSCP). Table 3-6 illustrates the use of the DSCP codes that enables traffic to be differentiated within a class. Note that the reason six bit positions are shown for each drop percentage for each class is due to the fact that bit 6 is always assigned a value of zero.

Although bits 3 and 4 represent the DSCP bits, in some literature the first six bits in the DiffServ byte are referred to as the DS codepoint. That codepoint takes the following form to indicate a drop percentage for each of the four traffic classes:

xxxyy0

where xxx represents a zero or 1 and is the revised ToS precedence level, while yy represents the DSCP bit values, with a trailing 0 in the sixth position of the byte.

Two other codepoint bit compositions included in RFC 2474 that define the DiffServ byte are xxxx11 and xxxx01, with x again representing either a binary 0 or 1. The first format is currently reserved for experimentation. The second format is also reserved for experimentation; however, it can be allocated for future use by a subsequent DiffServ specification. Last but not least, a DSCP value of all zeros (000000) represents a default packet class best effort forwarding request. This means that packets marked with this value are forwarded in the order they are received whenever bandwidth is available.

Also note that under this router forwarding mechanism the forwarding action that occurs based upon the setting of bits 3 and 4 is not standardized. That is, there is no precise definition of "low," "medium," and "high" drop percentages. Because not every router will initially be DiffServ-compliant and recognize bit positions 3 and 4, its previously mentioned backward compatibility with the precedence field in the ToS byte permits routers to forward packets based upon a lower level of granularity by only considering a packet precedence value. Also note that instead of including a granular capability for a precedence value of 5, the use of expedited forwarding provides a higher level of service that enables service providers to guarantee a minimum service level. To do so, a service provider could use a queuing method that results in packets with a DiffServ precedence of 5 receiving a minimum amount of bandwidth. This would compensate for the situation where a severely congested circuit could result in poor performance for classes 1 through 4.

	Traffic Class			
Drop Percentage	Class 1	Class 2	Class 3	Class 4
Low	001010	010010	011010	100010
Medium	001100	010100	011100	100100
High	001110	010110	011110	100110

Table 3-6.   DSCP Bits Enable Packets Within a Class to Be Differentiated.

From a simplistic point of view a router might treat packets by policing the input interface to segregate traffic into two queues based upon whether the DiffServ byte indicates expedited forwarding or best effort. Because the specification of expedited forwarding behavior dictates that packets so marked should be given priority at the output interface over best effort marked packets, packets could be placed into two queues. An example of this queuing action is illustrated in Figure 3-9.

The router would then be configured to extract packets from its high order queue prior to extracting packets from its low order queue. The key problem with this simplistic example is the fact that if too much traffic flowing through the router is marked for expedited forwarding the ability to extract best effort marked packets would be similar to waiting for snow in Miami. That is, it may not happen. With this in mind let's probe deeper into the manner by which DiffServ processing occurs.

DSCP Processing

Once a packet's DSCP bits are set at the entry to a network, they are examined by all routers within the network. Because the Internet represents a network of interconnected networks, the term "domain" is used to reference an administrative entity, such as an Internet Service Provider (ISP). Within a domain packets receive a consistent level of service based upon the setting of the DSCPs, which enables the service provider to negotiate SLAs with their customers. A customer can be a user organization or another domain. Concerning the latter, when a customer is located within another domain, the home domain will attempt to forward packets to the foreign domain requesting appropriate service to match the original requested level of service. Unfortunately, there is no standard billing mechanism or available metering software that examines DiffServ traffic classes, which means that the practicality of interdomain DiffServ communications may take some time to occur.

Routers within a domain can be classified as either boundary nodes or interior nodes. Interior node routers perform forwarding based upon the composition of the bits in the DiffServ byte. This forwarding treatment is referred to as per-hop behavior. In comparison,

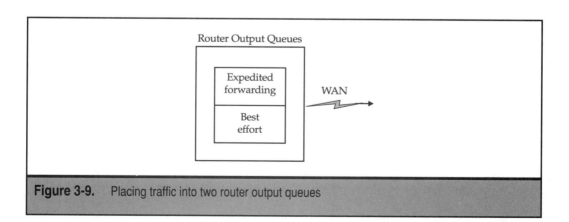

Figure 3-9.   Placing traffic into two router output queues

boundary routers are responsible for the classification of packets. With this in mind, let's turn our attention to the functions performed at the edge of the network, which provides the information used by the interior nodes within a domain.

Ingress Functions As data from a host enters a DiffServ-compatible router at the edge of a network, a traffic-conditioning process will occur. That process is based upon four QoS control functions. First, each message will be classified based upon a set of rules, resulting in the name "classifier" associated with this function. A second function involves measuring submitted traffic to determine if it conforms to the negotiated SLA. This function is referred to as *metering*. A third function involves performing one or more actions applicable to the classification of the packet. One action could be applying a policy or set of policies to traffic by changing the assignment of codepoints, a technique referred to as *marking*. Another action can result in traffic being delayed to ensure it does not exceed the traffic rate specified for a particular class. Because this action alters the flow rate it is referred to as *traffic shaping*. A third action results in the discard of packets when their rate exceeds that of the rate specified for their class and the flow cannot be delayed via traffic shaping. This action is referred to as *dropping*. Finally, the fourth QoS control function involves the queuing of traffic for output in an appropriate queue based upon its DSCP bit settings.

The relationship of the four previously described DiffServ control functions are illustrated in Figure 3-10. In examining Figure 3-10 note that all packets are first classified. Once classified, the resource utilization of a sequence of packets that represent a flow must be measured. This measurement occurs by the metering function measuring the volume of packets in the flow over a predefined time interval. This measurement permits the compliance of the flow with the SLA to be determined and permits an applicable action to occur. As previously noted, available actions include marking, traffic shaping, and dropping. However, there is a fourth action that is not shown in Figure 3-10. That action, or perhaps a better term would be inaction, is to do nothing and simply pass the classified packet to the queuing control function.

Classifier Operation A classifier can be considered to represent a logical 1:N fan-out device. Although it forwards packets in a serial stream to the marker, it generates N logically separate traffic streams as output. The simplest type of classifier is a behavior aggregate (BA) classifier. The BA classifier uses only the DSCP in the IP packet header to determine an appropriate output stream to which the packet should be directed. A more

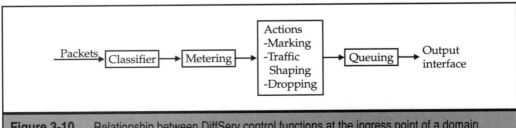

Figure 3-10. Relationship between DiffServ control functions at the ingress point of a domain

complex classifier would be a multifield classifier. This type of classifier is based on the use of one or more fields plus the DSCP field. One common multifield classifier is represented by the use of the destination address, source address, and IP Protocol Type fields in the IP header, and the source and destination ports in the UDP or TCP header following an IP header. Because the preceding field values are used in conjunction with the DSCP, this multifield classifier is also referred to as a *sextuple classifier*.

Metering   Metering also represents a logical 1:N fan-out operation, with the rate of traffic flow compared to SLA thresholds to determine the flow's conformance to a particular SLA. There are three levels of conformance that can be referred to as "colorful" due to the use of different color terms. A green level of conformance indicates that a flow conforms to its associated SLA. In comparison, yellow indicates particular conformance, while red indicates non-conformance. These three levels or conformance can trigger an applicable action, such as marking or dropping, or can be used to trigger a particular type of queuing.

Action   In Figure 3-10 we noted three action elements and previously described a fourth, which is to do nothing. This is a so-called null action. Because there is probably no free lunch, we can expect service providers to charge different fees or surcharges for the flow of packets within different traffic classes. Due to this, another potential action is counting. Counting would represent a passive action with respect to traffic that could be used for billing. In addition, this action could provide a mechanism for service verification as well as provide the service provider with qualitative information that could be useful for network engineering purposes.

Because DiffServ was only recently standardized via a series of RFCs, a large amount of effort needs to occur to both implement the technology as well as make it effective. Concerning the latter, suppose your organization entered into an SLA with a service provider and wanted to verify the level of service obtained. While you might trust your service provider to furnish a counting of packets in different classes, the ability to use SNMP to directly query counters might be a more suitable method. However, it may be quite some time until customers obtain this capability as it cannot be expected to represent a vendor implementation priority.

Routing   As briefly mentioned earlier in this section, the routing or forwarding of packets within a domain is referred to as per-hop behavior. Because an edge router is responsible for classifying, metering, and several action elements, interior routers are normally relieved of those functions. Thus, an interior router will normally have its DiffServ operations focused upon examining the DSCP bit settings in inbound packets and transferring the packet to an appropriate queue based on the previously mentioned bit settings. PHB is expected to be supported over various underlying technologies, such as the mapping of the DSCP byte to a Multi-Protocol Label Switch (MPLS) label value, or via ATM. In fact, the ATM Forum is working on defining a new VC call setup information element to transport PHB ID codes between ATM switches. Another area being considered by various committees is the development of an optional QoS agent that could provide either active or passive support for the Resource ReSerVation Protocol (RSVP). Here possible support would result in the agent snooping through RSVP messages to learn how to classify traffic without having to participate as an RSVP protocol peer. In comparison, under active support the QoS agent could participate in a per-flow-aggregate signaling of QoS requirements. A DiffServ-compliant router could then accept or reject RSVP admission requests to provide a mechanism of admission control to DiffServ-based services.

Operational Problems

When we examined DiffServ we noted that this traffic-expediting technique divides transmission into a small number of classes. Routers then apply a standardized set of behaviors to each class of traffic, which is referred to as a per-hop behavior. Because DiffServ-aware routers perform their operation without the need to know the path of traffic nor information about other routers in the network, the technology avoids the need for conveying signaling information other than the bit composition of the revised ToS byte. While this is a key advantage of DiffServ, it also results in the need to carefully consider as well as control the arrival rate of one or more types of traffic classes. For example, consider the use of expedited forwarding (EF), which represents a simple per-hop behavior that informs routers that packets marked for EF status should be forwarded with minimal delay and loss. The only way that a network operator can guarantee minimal delay and loss at each hop to all packets marked for expedited forwarding is to limit the arrival rate of such packets to less than the rate at which routers can forward EF packets. While this task may appear simple, in actuality this method of flow control can be quite complex, as we will note from an examination of the data flow in Figure 3-11.

In examining Figure 3-11 let's assume routers 1 and 2 are connected via a T1 circuit in the middle of a network while router 1 has connections via circuits labeled A, B, and C to three other devices. This means that the network operator needs to limit the aggregate arrival rate of EF marked packets via circuits A, B, and C destined to router 2 via router 1 to less than 1.544 Mbps. In actuality, the network operator must also consider other traffic classes and control the aggregate arrival rate of EF marked packets to a rate significantly

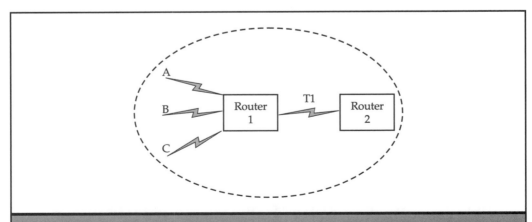

Figure 3-11.   The arrival rate for all ingress ports at a router must be kept under the egress rate of the traffic class for each route outbound from the router.

under 1.544 Mbps to enable router 1 to transfer other traffic to router 2. Because routers connected to circuits A, B, and C may have one or more inputs that will flow from router 1 to router 2, EF marked packets at another hierarchy in the network must also be considered. As a network increases in complexity it becomes harder for the network operator to configure devices at the edge of the network to mark packets that allow routers within the network to guarantee a consistent handling of the packets as they flow through the network. For this reason an alternate approach to provide a QoS capability through a network may be desirable.

In actuality there are several standards that involve providing a QoS capability through a network. One standard, referred to as Integrated Services, represents a suite of evolving QoS transport capabilities over an IP network. A second standard referred to as Resource ReSerVation Protocol provides the ability to guarantee bandwidth, while a third standard, referred to as Multi-Protocol Label Switching, expedites traffic through a network. Collectively, these techniques will be referred to as "expediting traffic through the WAN."

EXPEDITING TRAFFIC THROUGH THE WAN

As previously discussed, in this section we will turn our attention to three standards that govern the manner by which traffic can be expedited through a wide area network. Those standards include Integrated Services, RSVP, and MPLS.

Integrated Services

Integrated Services (IntServ) references a suite of evolving standards intended to provide a QoS transport capability over IP-based networks. IntServ dates to the work of the IEFT during the mid-1990s and is defined in RFC 1633.

The basic design issues covered by IntServ concern sharing available bandwidth during times of congestion. Under IntServ, congestion management consists of the following:

▼ Admission control, which is invoked by a reservation protocol to determine if resources are available for the flow at the requested QoS.

■ A routing algorithm, which maintains a routing database that provides next-hop information for each destination address.

■ A queuing algorithm, which controls the flow of traffic through a router.

▲ A discard policy, which provides a uniform mechanism that denotes the conditions under which packets are sent to the great bit bucket in the sky.

Although IntServ includes an admission control function, it does not define the method of control to be used. However, when discussing admission control many persons incorrectly associate the ReSerVation Protocol as part of IntServ. In actuality, RSVP can be used under IntServ but is not a required member of the architecture. Because RSVP is currently the only mechanism that can be employed to deliver guaranteed bandwidth within an IP network, let's turn our attention to the basics of this signaling protocol.

RSVP

RSVP represents a signaling protocol that enables applications that require guaranteed bandwidth to request such bandwidth from a network. In doing so RSVP provides an admission control capability on an end-to-end basis since the availability of required bandwidth determines whether or not the application gains access to the network. Through the ability to guarantee bandwidth it becomes possible for real-time audio and video to flow through a network with an assurance of delivery that matches their QoS requirements. That is, bandwidth will be allocated by application to ensure the application receives a predefined portion of the communications circuit linking two routers.

Unlike most protocols that are sender-driven, RSVP is receiver-driven. While at first glance this may appear a bit awkward, there is a valid reason for this approach. The rationale is based upon the need to accommodate different members of a multicast group that can have different resource requirements. For example, one member of a multicast group could be connected to an IP network via a 56-Kbps digital circuit while a second member is connected via a corporate LAN connected using a T1 line operating at 1.544 Mbps.

Operation

Under RSVP a sender will transmit a Path message downstream towards all receivers. The Path message includes information on the traffic characteristics of the data stream that will be generated. Each receiver requests a specific QoS from the network by responding to the Path message with a reservation (Resv) message. As a Resv message flows towards the originator, each router in the path checks to determine if sufficient resources are available. If so, a reservation is established and the Resv message is forwarded to the next upstream router. Otherwise, the reservation fails and an error message is returned to the receiver.

Flowspec and Filterspec

To reduce signaling requirements, a sequence of packets from a common address with a common destination are treated as a flow specification (flowspec). The flowspec describes the service requirements in terms of a desired QoS or reservation request and can include such information as a service class defined by the application and a reservation specification (Rspec), which defines the bandwidth required.

Another integral part of each RSVP-compliant router is a filter specification (filterspec). The filterspec specifies packets that will be serviced and the manner in which they will be serviced based upon the flowspec. An example of the relationship between the filterspec and flowspec is shown in Figure 3-12.

Classifiers and Schedulers

Each RSVP-compliant router includes a packet classifier and packet scheduler. The classifier determines the route of the packet, while the scheduler is responsible for servicing and forwarding decisions required to achieve the requested QoS. Note that the packet classifier aggregates a series of packets that flow to a common destination. The filterspec

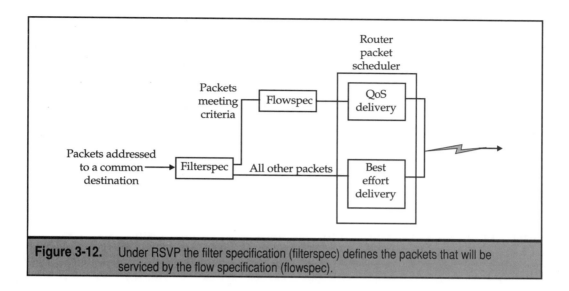

Figure 3-12.   Under RSVP the filter specification (filterspec) defines the packets that will be serviced by the flow specification (flowspec).

specifies packets that will be serviced by the flowspec, while the flowspec parameters are used by the router to place packets into applicable QoS delivery queues.

Returning to the flowspec, it can consist of up to three components—a service class, an Rspec, and a Tspec. The service class is defined by the application. The QoS is defined by the Rspec where "R" represents "reserve." The Tspec describes the traffic flow. The Tspec can be considered to form one side of a "contract" between the data flow and the service. That is, once a router accepts a specified QoS, it must continue to provide that level of service as long as the flow is within the Tspec. However, if traffic should exceed the expected level, the router can then drop packets, revert to servicing the flow on a best-effort basis, or employ another service mechanism.

Message Types

As noted earlier in this section, RSVP supports two basic message types: Resv and Path. The Path message is originated by the sender and includes information concerning the traffic characteristics of the data stream that will be generated. Path messages are forwarded through the network to provide downstream routing information. The Resv message is sent by each receiver in the opposite direction toward the sender by reversing the paths of the Path messages. Thus, the Path message will indirectly provide upstream routing information. To maintain an RSVP session, a sender will periodically issue Path messages.

Both Resv and Path messages have timeout values that are used by routers and switches to set their internal timers. If those timers expire, the reservation and routing information associated with the reservation are turned down. This limits the seizure of resources by the failure of a receiver to terminate an RSVP session.

Status

Although RSVP is several years young its implementation is commonly restricted to intranets whose sizes are but a small fraction of the Internet. In fact, during 1997 the IETF cautioned the Internet community concerning the implementation of RSVP.

The reason for RSVP's limited use in intranets is primarily due to the signaling overhead associated with the protocol, the need for all routers in a network to be RSVP-compliant, and the inability to negotiate the cost of bandwidth when a QoS traffic flow crosses an Internet Service Provider boundary. Concerning the latter, the simplicity of allocating bandwidth to an application does not mean that it is easy to bill for the bandwidth. This is because allocated bandwidth can be temporarily de-allocated when the application is quiescent, permitting a router to use previously allocated bandwidth until the application needs it. Thus, an interesting issue is how, for example, would one ISP bill another for an application that required 8 Kbps of bandwidth that was available for use by the host ISP 60 percent of the time? While billing issues between ISPs may take a while to resolve, many carriers are taking another look at RSVP for internal use to include its proposed extension to support label distribution and explicit routing, which is referred to as RSVP-TE (traffic engineering.) Since this requires knowledge of label distribution, let's turn our attention to this topic in the form of Internet draft documents that describe MPLS.

MPLS

Multi-Protocol Label Switching represents a switching architecture where packets entering a network are assigned a label. Then, instead of forwarding packets based upon searching their routing table, a router uses the label as a forwarding criteria. Because there can be tens of thousands of entries in a routing table while label entries are associated with device interfaces, it is much faster for the router to make its forwarding decision based upon the contents of the label.

The original goal of MPLS was to provide the efficiency of layer 2 switching to layer 3 networking. While the manufacture of application-specific integrated circuit (ASIC) based routers permits relatively fast router table lookup operations that negate the original goal of MPLS, there are other benefits associated with label switching. Those benefits include the ability to define paths for different types of traffic through a network that is referred to as traffic engineering and the creation of IP tunnels through a network that facilitates the creation of virtual private networks (VPNs).

Evolution

MPLS is actually a rather old idea first proposed as a tag-switching technique by Ipsilon, a manufacturer of switching equipment. IBM, Cisco Systems, and other vendors developed proprietary versions of tag switching at layers 2 and 3 of the ISO Open System Interconnection (OSI) Reference Model. At the time this book revision was prepared the IETF was in the process of standardizing the technology that would define IP over ATM, IP over optical networks, and of course, IP expedited through an IP network with label-aware routers referred to as label switch routers (LSR) at the edge and within the network.

The MPLS Label

The MPLS label is 32 bits in length and consists of four fields as indicated below:

|-20 bits Label-|-3 bits CoS-|-1 bit stack-|-8 bits TTL-|

The 20-bit label field contains the actual value of the MPLS label and has local significance in the same manner as a Frame Relay Data Link Control Identifier (DLCI) is used to convey path information. The Class of Service (CoS) field permits packets to be placed into one of eight classes that can affect queuing and discard algorithms applied to the packet as it flows through each router in a network. The third field in the label is the Stack (S) field. This 1-bit field is used to indicate a hierarchical label stack and is referred to as a label stack. This means that it becomes possible for a packet to have multiple paths with the value of the label stack identifying a particular path to be taken.

The fourth field, time to live (TTL), consists of 8 bits that provide the functionality of the conventional IP TTL field in the MPLS header. That is, it prevents a labeled packet from endlessly wandering through a network.

The MPLS label is inserted after the layer 2 header, resulting in it preceding the IP header. The specific path through an MPLS network is referred to as a label switch path (LSP). The LSP is provisioned via the use of a label distribution protocol (LDP), which both establishes a path through an MPLS network as well as reserves resources required to satisfy the predefined service requirements for the data path. One pending LDP is the previously mentioned extension to RSVP to support label distribution and explicit routing referred to as RSVP-TE. While one group of vendors to include Cisco and Juniper Networks supports the extension of RSVP, a second group of vendors to include GDC and Nortel are backing a second signaling mechanism referred to as CR-LDP. While both proposed signaling protocols have many similarities, they are not interoperable and the IETF's MPLS working group has its work cut out in attempting to select one. If you point your browser to **www.mplsrc.com**—the home page of the MPLS Resource Center, a Web site that functions as a clearinghouse for information on the IETF's Multi-Protocol Label Switching Standard—you can determine if a selection between the two signaling protocols was made, as well as access a wealth of MPLS-related information.

Operation

Since the best way to note the operation of a technology is by example, let's turn our attention to the operation of MPLS via a small network. In the example we will use only four routers, although an MPLS-compliant network can be expected to have many more routers.

Figure 3-13 illustrates an example of the use of labels within MPLS-compliant routers as well as the flow of traffic through an MPLS-compliant network. In this example, let's assume traffic exits the MPLS network to routers connected to the 198.78.46.0 and 205.131.175.0 networks. As data flows into the MPLS network the leftmost LSR applies an initial label to the packet after performing a conventional longest-match lookup based upon the destination IP address in the IP header. In this example a label with a value of 2 is assigned to packets destined to the 204.131.175.0 network. Once a packet is labeled,

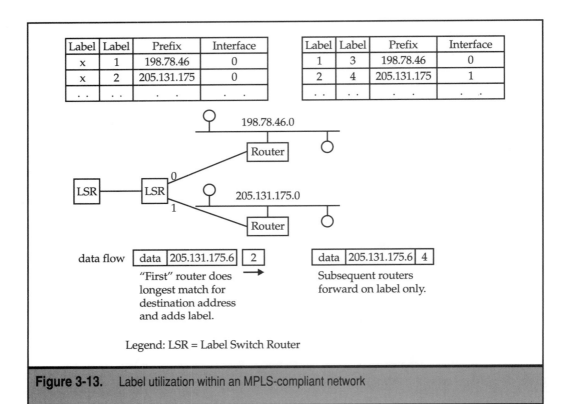

Label	Label	Prefix	Interface
x	1	198.78.46	0
x	2	205.131.175	0
. .	. .	. .	. .

Label	Label	Prefix	Interface
1	3	198.78.46	0
2	4	205.131.175	1
. .	. .	. .	. .

198.78.46.0

Router

LSR — LSR

205.131.175.0

Router

data flow | data | 205.131.175.6 | 2 | | data | 205.131.175.6 | 4 |

"First" router does longest match for destination address and adds label.

Subsequent routers forward on label only.

Legend: LSR = Label Switch Router

Figure 3-13.   Label utilization within an MPLS-compliant network

subsequent LSRs via the LSR advertising protocol obtain the capability of forwarding based upon the label value. Because most labels have local significance LSRs commonly replace the label on an incoming packet with a new value as they are forwarded through the network. This is shown in the lower-right portion of Figure 3-13 where the label value of 2 is replaced by a label value of 4. This practice is referred to as *label swapping*. The second LSR then forwards the packet to the router connected to the LAN whose network address is 205.131.175.0. At this time the LSR in the lower right removes the label as the packet is output, enabling "legacy" routers to operate transparently to the MPLS network.

Limitations

Similar to DiffServ and RSVP there are many limitations associated with MPLS. Those limitations include the selection of a label distribution protocol, the need for a network operator to upgrade all routers in their network, and problems that occur when data expedited through one MPLS-compliant network has to flow through a non-compliant MPLS network to reach its destination. In addition to these limitations it is also important to note that MPLS by itself does not provide a QoS capability but instead expedites traffic through a network. Thus, MPLS needs to be "married" to other DiffServ or to RSVP to both expedite and guarantee the flow of traffic through a network, which can represent a daunting task that may take years to occur.

IP TELEPHONY-RELATED PROTOCOLS

In this section we will focus our attention upon those protocols that make IP telephony a reality. Those protocols we will discuss include the Real Time Protocol (RTP), which provides the time stamping necessary to permit the removal of jitter from packets encountering variable delay as they flow through a network; the H323 standard, which provides a uniform mechanism for multimedia communications between LANs; the Session Initiation Protocol (SIP), which provides a signaling protocol for establishing calls via a TCP/IP network; and the Media Gateway Control Protocol (MGCP), which provides a mechanism to control telephone gateways.

RTP

RTP was approved as an Internet standard in late 1995 and is defined in RFCs 1889 and 1890. RFC 1889 is titled "RTP: A Transport Protocol for Real-Time Applications." RFC 1890 is titled "RTP Profile for Audio and Video Conferences with Minimal Control." RTP was developed to provide several desirable features for applications with real-time properties, including the ability to reconstruct timing, loss detection, security, and the identification of the content of packets. Several major vendors, including Intel, Microsoft, and Netscape, signaled their intention to construct their voice and video products by including RTP in existing standards. Although a few Internet audio applications use a protocol known as VAT, that protocol uses the audio-encoding method specified in RTP. Thus, we can expect a full migration to RTP to occur, which will facilitate the interoperability of products currently based on proprietary technology.

Overview

RTP was developed as an end-to-end delivery service for data with real-time characteristics, such as interactive audio and video. Those services include time stamping, sequence numbering, delivery monitoring, and identification of the type of data transported. The actual monitoring of the quality of service is performed by the RTP Control Protocol (RTCP).

RTP can be considered an application service. Applications such as Internet telephony will usually run RTP on top of UDP, with RTP and UDP forming distinct portions of the transport functionality required to support real-time data transfer. Instead of functioning as a distinct layer 4 protocol, RTP was designed to be embedded into an application process. Thus, the RTP specification provides a mechanism for denoting a common set of functions for applications that require the use of a Real-Time Transport Protocol. Although RTP provides a considerable degree of flexibility, readers should note that it does not contain any mechanism that guarantees the timely delivery of data, nor does it provide any other quality of service guarantees. Instead, RTP relies on lower-layer services, such as RSVP, to provide this capability.

When RTP is transported by UDP or similar protocols, the specification requires the use of an even port number, with a corresponding RTCP stream functioning as a control mechanism using the next higher odd port number. If an application provides an odd number for use as the RTP port, the specification requires its replacement with the next lower or even-numbered port.

The RTP Protocol

Similar to any data transfer protocol, RTP consists of a header followed by data to form a packet. Unlike other transport protocols that use the contents of fields within the protocol header for control purposes, RTP uses a separate control mechanism in the form of RTCP packets for control purposes.

The RTP Header

Figure 3-14 illustrates the format of the RTP header. This header contains 10 fields, of which the last field is optional and is included when audio packets are resynchronized to reconstruct a constant 20-ms spacing. The resynchronization is performed by an RTP-level relay, referred to as a *mixer*, that enables an audio stream to be varied based on the available bandwidth of different circuits. The mixer is an important part of the RTP protocol. In conjunction with translators, it enables stations with different capabilities to participate in real-time conferencing without requiring all stations to set themselves to operate at the lowest common denominator in terms of speech encoding and other characteristics that govern a real-time data stream. Mixers can be used to reconstruct audio and other media streams into lower-bandwidth, and usually lower-quality, data streams. Thus, they enable other stations to receive a degraded version of a multicast transmission instead of forcing all stations to receive the degraded version or excluding some stations from receiving the transmission due to their inability to service a higher-quality data stream.

A mixer receives a sequence of RTP packets from one or more sources and combines them into a new data stream. That data stream can be directed to a single or to multiple destinations, with the format of data either left as is or changed. A translator represents a simpler device, as it operates on one packet at a time, generating one outbound packet for

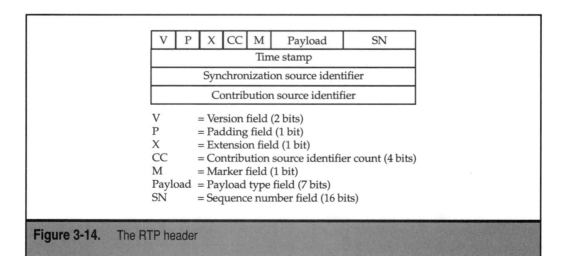

Figure 3-14. The RTP header

each inbound packet received. The translator can change the format of data in the packet as well as initiate the use of a different protocol for the transfer of data. Together, mixers and translators provide a mechanism that enables stations with different capabilities to independently receive real-time data streams without adversely affecting the capability of other stations to do so. To obtain an appreciation for the capability of RTP, let's turn our attention to the fields in its header.

Version Field

The Version field is 2 bits in length and identifies the version of RTP. The current version of RTP is 2, with a value of 1 used by the first-draft version and the value of 0 used by RTP as initially implemented in the VAT audio protocol.

Padding Field

The Padding field is a 1-bit flag that indicates whether the packet contains padding octets that are not part of the actual payload but are appended into the Payload field. The setting of this bit flag is designed to accommodate the use of encryption that requires a fixed block length or for transporting several RTP packets in a lower-layer protocol data unit.

Extension Field

The Extension field is also 1 bit in length and can be considered as a flag that indicates whether or not the fixed header is followed by a header extension. The header extension provides a mechanism that enables developers to experiment by adding payload-format-independent functions that require additional header information while allowing other interoperating implementations to ignore the extension.

CSRC Count Field

The Contributing Source (CSRC) Count field is 4 bits in length. This field indicates the number of CSRC identifiers that follow the fixed header—*contributing source* being a term used for the source of a stream of RTP packets that contributed to the combined stream produced by an RTP mixer. The mixer will insert a list, called the CSRC list, of the synchronization source identifiers of the sources that contributed to the generation of the packet into the RTP header. An example of the generation of a CSRC list would be an audio conference where a mixer would indicate all the talkers whose speech was combined to generate the transmitted packet.

Marker Field

The 1-bit Marker field functions as a flag; however, its interpretation is governed by the payload type. When set, the marker bit will indicate that the Payload Type field carries specific information defined to suit different requirements. For example, when video is transported, the Marker field would indicate the end of a frame. In comparison, for audio it would mark the beginning of the speech occurring between two silent periods.

Payload Type Field

This 7-bit field identifies the format of the RTP payload and determines the interpretation of its contents by the application. The code for the payload type identifies both the audio and video encoding schemes, the clock or sampling rate, and, if appropriate, the number of audio channels carried. For example, a payload type value of 2 indicates the ITU G.721 audio coding technique in which a clock rate of 800 Hz is used to provide a single audio channel.

Table 3-7 contains a list of a few of the defined RTP payload types and RTCP control packet types. By examining the entries in Table 3-7, it becomes obvious that RTP was developed as an all-encompassing multimedia time-stamping mechanism that may not represent the best method for time stamping and sequencing digitized voice for two-way communications. This is because more than 99 percent of all telephone calls are nonconference calls. Thus, forcing such calls to use RTP where 32-bit fields are unnecessary adds additional delay that could be removed by a streamlined version of RTP for two-way voice. Perhaps developers reading this book will initiate action on an RFC to provide a streamlined version of RTP for two-way communications, shaving another millisecond or more when a shortened header flows over a low-speed link.

Payload Type	Encoding Name	Audio/Video	Clock Rate (Hz)
2	G.721	A	8000
4	G.723	A	8000
7	LPC	A	8000
9	G.722	A	8000
15	G.728	A	8000
26	JPEG	V	90000
31	H.261	V	90000
34	H.263	V	90000

RTCP Control Packet Types

Value	Report Type
200	Sender report
201	Receiver report
202	Source description
203	Goodbye
204	Application defined

Table 3-7. Representative RTP Payload Types and RTCP Control Packets

Sequence Number Field

The 16-bit Sequence Number field provides a mechanism that allows a receiver to detect the loss of a packet. However, the method used to compensate for the loss of a packet is up to the application.

Time-Stamp Field

The 32-bit Time-Stamp field notes the sampling instant of the first octet in the RTP data packet. This field enables a receiver to determine if the arrival of a packet was adversely affected by a delay known as *jitter*. However, the actual method by which an application compensates for jitter is up to the application.

Through the use of a uniform Time-Stamp field, developers can receive packets with random spacing and buffer them prior to reconstructing speech. Then each packet can be removed from the buffer based upon the value of the Time-Stamp field to provide a uniform method of reconstructed speech that eliminates what would otherwise appear as random delays that would make portions of speech sound awkward.

Synchronization Source Identifier Field

This 32-bit field identifies the synchronization source. In actuality, an algorithm generates a random identifier so that no two synchronization sources within the same RTP session will have the same identifier.

Contributing Source Identifier Field

As previously discussed, this 32-bit field identifies the contributing sources for the payload contained in the packet. Up to 15 sources can be inserted by mixers into this field.

In a voice-over-IP environment an RTP header consisting of 12 bytes is used to prefix an 8-byte UDP header. The RTP and UDP headers are in turn prefixed with a 20-byte IP header. If RTP is employed with an 8-Kbps voice digitization method that results in 20 ms portions of speech being encoded, then the payload of the UDP segment will be 160 bits or 20 bytes. This means that the total length of the IP datagram conveying the digitized voice sample is 60 bytes, of which only 20 represent the actual payload. Through the use of RTP header compression it becomes possible to reduce the IP/RTP/UDP header from 40 bytes to between 2 to 5 bytes with the actual reduction dependent upon the susceptibility of the header's contents to compression. RTP compression is currently defined in an IETF draft appropriately titled "Compressed RTP (CRTP)." Several vendors including Cisco Systems have implemented RTP header compression to enhance communications on serial lines using frame relay, HDLC, Point to Point Protocol (PPP) encapsulation, or via an ISDN interface. Because there is no free lunch in communications, it is important to note that there is a relationship between router processing time required to perform header compression and the speed at which compressed data is provided to a router's interface. In general, the higher the data rate the higher the processing time required. Thus, RTP header compression is commonly recommended for use on slower links operating up to 512 Kbps where RTP header compression can reduce overhead without adversely affecting the processing capability of the router.

Summary

Although the use of RTP provides a standardized mechanism to transport audio, the manner in which it is operated depends on the application. Although this fact will differentiate applications from one another, the support of the different RTP payloads will provide a high degree of interoperability between applications.

The H.323 Standard

The H.323 standard can be considered to represent an umbrella recommendation from the ITU for multimedia transmission on LANs and via packet networks that have nonguaranteed bandwidth, such as IP and frame relay. As an umbrella recommendation, this means that not all parts of the recommendation have to be implemented. In addition, the standard does not define how different parts should be implemented, resulting in interoperability problems between equipment from different vendors that are H.323-compliant. Due to this problem several vendors formed iNOW! to develop interoperability among H.323 products.

Overview

The H.323 standard was originally developed for use on a single LAN. A second version of the H.323 standard introduced the concept of "zones" that enables the standard to scale to geographically separated LANs. However, it is still difficult to envision H.323 to scale to anywhere near the size of the Internet, which means that while organizations can use this standard on a private network or on the Internet between a limited number of LANs, it will more than likely not be deployed as a global mechanism.

Components

The H.323 standard defines four major components: terminals, gateways, gatekeepers, and multipoint control units. It should be noted that not all components are required and that the functions of two or more components can be implemented on a common platform.

An H.323 terminal represents an endpoint on a LAN that supports real-time, two-way communications. Thus, a workstation running an Internet telephony operation that is H.323-compliant would represent an H.323 terminal.

A gateway provides a translation service between H.323 conference endpoints and other H.323-compliant terminals. The presence of a gateway is optional and, when required, is commonly implemented on the same platform as a gatekeeper.

The gatekeeper controls access to the network for H.323 endpoints. Thus, it provides a control mechanism that prevents a LAN from being overloaded with too many voice and video calls. For example, if you set a threshold for the maximum number of simultaneous conferences, the gatekeeper will refuse to allow more conversations once the threshold is reached.

Another key function performed by the gatekeeper is address translation. Here the gatekeeper translates addresses from LAN aliases for terminals and gateways to IP addresses. The actual translation process is governed by the Registration, Admission, and Status (RAS) protocol, which conveys the registration of terminals and gatekeepers, admissions, bandwidth changes, and status messages between IP telephony devices and gatekeepers.

The fourth component of the H.323 standard is the multipoint control unit (MCU). The MCU supports conferences between three or more endpoints by functioning as a bridge.

The H.323 Protocol Stack

To obtain an appreciation for the operation of the H.323 protocol, let's first turn our attention to its protocol stack. Figure 3-15 illustrates the H.323 protocol stack and indicates various components of the protocol and their relationship to the TCP/IP protocol stack.

In examining Figure 3-15, note that H.225 represents a call signaling protocol that is required for establishing and terminating calls and is based upon the ISDN Q.931 standard. The H.245 protocol, in comparison, provides the negotiation capability between endpoints, allowing, for example, two devices to select the use of a commonly supported voice-compression method.

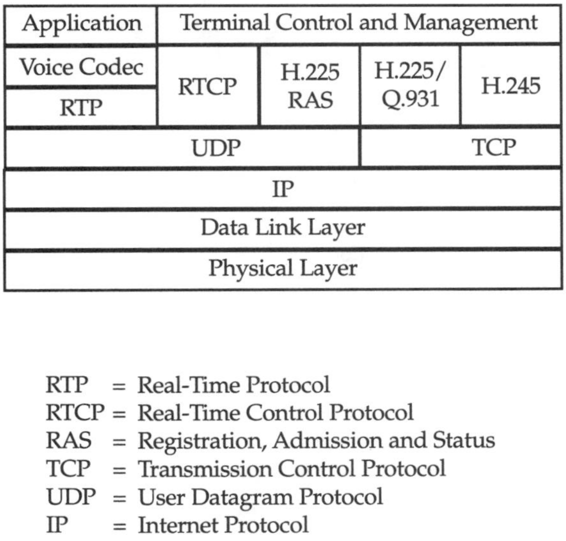

Application	Terminal Control and Management			
Voice Codec	RTCP	H.225 RAS	H.225/ Q.931	H.245
RTP				
UDP			TCP	
IP				
Data Link Layer				
Physical Layer				

RTP = Real-Time Protocol
RTCP = Real-Time Control Protocol
RAS = Registration, Admission and Status
TCP = Transmission Control Protocol
UDP = User Datagram Protocol
IP   = Internet Protocol

Figure 3-15.   The H.323 protocol stack

Operation

The H.323 standard defines several protocol exchanges between terminals, gateways, and gatekeepers that must occur prior to establishing an audio connection between two terminals. Table 3-8 summarizes the required protocol exchanges.

The explanation for the first exchange listed in Table 3-8 is as follows: When a gateway connects to a network it needs to register its presence with a gatekeeper and join the gatekeeper zone. To do so, it informs the gatekeeper of its IP address and related alias address, such as telephone number and host name.

Both a gateway and a conventional terminal device need to first locate the gatekeeper. To do so, each device will use the Gateway Discovery Protocol (GDP) to determine which gatekeeper they should register with.

Figure 3-16 illustrates the gatekeeper discovery procedure. Under this procedure an H.323 endpoint issues a Gatekeeper Request (GRQ), which is transmitted using port 1718 to all stations. Upon issuing the GRQ, the endpoint starts a timer. If the timer expires prior to receiving a response, the endpoint can send another GRQ. The gatekeeper will respond with either a Gatekeeper Confirmation (GCF) or a Gatekeeper Rejection (GRJ) message.

Once a gatekeeper is discovered the terminal registration process occurs. At this point in time a terminal uses UDP port 1719 to register its presence. An example of the terminal registration process is shown in Figure 3-17. Here the terminal issues a Gatekeeper Registration Request (RRQ) using UDP port 1719. The gatekeeper will respond with either a Registration Confirmation (RCF) or Registration Reject (RRJ) message. After a session is completed the terminal will issue an Unregister Request (URQ). The gatekeeper will respond with either a Gatekeeper Unregistration Confirm (UCF) or Gatekeeper Unregistration Reject (URJ) message. Assuming a confirmation occurs, one end of the session is now broken. Thus, the gatekeeper will then issue its own URQ, which the terminal responds to with an UCF message.

Protocol Exchange	Description
H.225–RAS	Gatekeeper discovery and terminal registration
H.225–RAS	Routed call setup between the terminals through the gatekeeper
H.225–Q.931	Can represent a terminal-to-gatekeeper-to-gateway-to-terminal channel or directly between gateways after gatekeeper allows admission
H.245	Initial communications and capability exchange
H.245	Establishes audio communication via opened logical channel
RTP/RTCP	Audio communications with packets time-stamped

Table 3-8. H.323 Protocol Exchanges

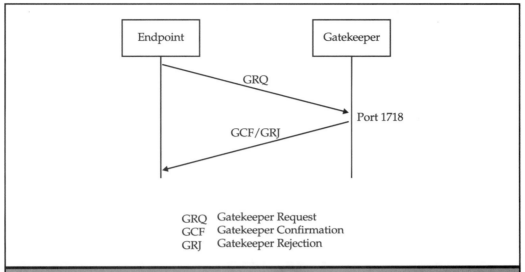

GRQ   Gatekeeper Request
GCF   Gatekeeper Confirmation
GRJ   Gatekeeper Rejection

Figure 3-16.   The gatekeeper discovery procedure is used by endpoints to determine which H.323 gatekeeper it should register with.

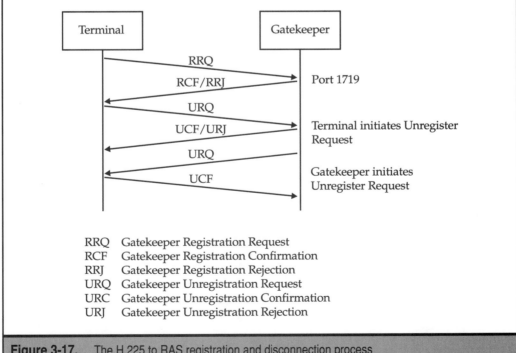

RRQ   Gatekeeper Registration Request
RCF   Gatekeeper Registration Confirmation
RRJ   Gatekeeper Registration Rejection
URQ   Gatekeeper Unregistration Request
URC   Gatekeeper Unregistration Confirmation
URJ   Gatekeeper Unregistration Rejection

Figure 3-17.   The H.225 to RAS registration and disconnection process

After the H.225-RAS protocol exchange that results in a call setup between the terminals through the gatekeeper, the H.225-Q.931 protocol exchange can take several forms. It can be a terminal-to-terminal session via a gatekeeper, a gatekeeper-to-gateway or gateway-to-gateway-to-gateway session. Figure 3-18 illustrates the establishment of a terminal-to-terminal session. Note that a terminal first requests admission from a gatekeeper via an Admission Request (ARQ). Assuming the gatekeeper responds with an Admission Confirm (ACF) message, a call setup process occurs, resulting in one terminal sending a call setup to another terminal. The gatekeeper responds to the call setup from the first terminal while the called terminal responds to the setup related through the gatekeeper to the gatekeeper. The called terminal will next issue a gatekeeper ARQ message, which the gatekeeper will respond to with either an Admission Confirm or Admission Reject (ARJ) message. Assuming admission is granted, the called terminal will first alert the calling terminal prior to connecting to that terminal.

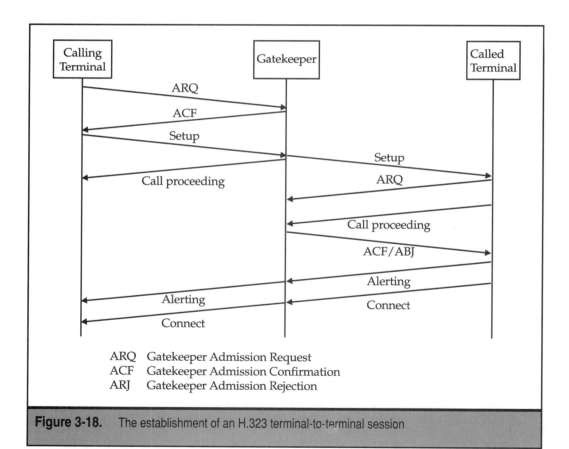

ARQ   Gatekeeper Admission Request
ACF   Gatekeeper Admission Confirmation
ARJ   Gatekeeper Admission Rejection

Figure 3-18.   The establishment of an H.323 terminal-to-terminal session

The fourth exchange in the table represents end-to-end control messages exchanged after a connection is established, which occurs at the bottom of Figure 3-18. For each call using H.225, one H.245 control channel is used. Once initial communications result in a capability exchange and a codec is selected, audio communications are established through the use of a logical channel. This is then followed by the use of RTP and TRCP to ensure packets are time-stamped.

Because the H.323 standard includes the ability to negotiate the capabilities of different terminals, it functions as a mechanism to provide common denominator support. In addition, because it also provides a mechanism to control voice and video activity, it represents a standard being supported by many vendors developing voice-over-IP products.

While H.323 provides different vendors with a mechanism to develop interoperable products, several caveats deserve mention. First, it is important to note that H.323 compliance is not sufficient for compatibility. As previously mentioned, H.323 represents an umbrella standard, which means that a vendor does not have to implement the full standard. In addition, the manner of implementation is not standardized. In fact, by the time the new edition of this book is published, there will be three versions of H.323: the original version introduced a few years ago; version 2, which added options for encryption; and version 3. With three versions, and no standard for implementation, it is highly doubtful that two H.323-compliant products from different vendors will interoperate. This is the reason for the iNOW! initiative being launched by vendors of Internet telephony products in an attempt to define options manufacturers may use with H.323 products and to designate how the use of such options is negotiated.

Another problem associated with the H.323 standard occurs when terminals in different zones need to communicate. In this situation it can take up to eight seconds for a call to be established, which is a significant delay in comparison to a conventional call setup via the PSTN.

APPLICATION NOTE: If using the H.323 standard with multiple zones, educating users concerning extended call setup time can prevent problems caused by employees believing something is amiss and killing an in-progress session setup.

Interoperability

In concluding this section on the H.323 standard we will turn our attention to three methods commonly used to obtain equipment interoperability. Those methods include gateway-to-gateway, combined gateway-to-gateway, and gatekeeper-to-gatekeeper interoperability.

Gateway-to-Gateway

Gateway-to-gateway interoperability provides a mechanism to enable different brands of gateway products to exchange calls. Because you must operate a matching gatekeeper for each brand of gateway, this method of interoperability can result in an extra degree of administrative burden on the network manager.

Gateway-to-Gateway Plus Gatekeeper-to-Gatekeeper

Although this method of interoperability requires a matching gatekeeper for each brand of gateway, each gatekeeper needs to know information only about its own brand of gateway. Thus, this method of interoperability is easier to administrate.

Gatekeeper-to-Gateway

This third method of H.323 interoperability enables gatekeepers of one brand of products to directly communicate and control gateways developed by a different vendor. However, to obtain this method of interoperability, the gatekeeper must truly interoperate with gateways produced by other vendors.

SIP

The Session Initiation Protocol (SIP) represents a signaling protocol for establishing calls via a TCP/IP network. Such calls can include audio, video, and various types of conferences.

One of the key advantages of SIP over H.323 is its ability to scale. Its scalability results from the fact that SIP callers and receivers are identified via an addressing scheme that works with DNS. For example, SIP addresses can have either of the following forms:

SIP: user@host.domain

SIP: user@network_address

The addressing scheme used by SIP enables it to reuse DNS MX records or Simple Mail Transport Protocol (SMTP) addresses. This means that similar to e-mail, it becomes possible to redirect calls based upon the status of the recipient.

Basic Operation

SIP is defined in RFC 2543. A caller first locates a server, typically through the use of DNS, although it can operate over any packet protocol to include TCP, UDP, and even X.25. The called server can map the name to the format user@host.com or another top-level domain. The called party can accept, reject, or forward the call request. If the call is accepted the called party responds with a confirmation and the conversation begins. At any time either the caller or called party can send a BYE message to terminate the call. Although the preceding represents a simplified version of the operation of SIP, it is important to note that SIP uses a core set of simple messages to perform its operation. Five of those messages and their meanings are listed in Table 3-9.

Modes of Operation

SIP supports two modes of operation—proxy and redirect. When in a proxy mode of operation a server acts on behalf of itself. Here a caller issues an INVITE message to a location server to obtain a user address in the form of user@host.domain. The proxy server can be an X.500 server, a SQL database, DNS, or even a WHOIS server. The proxy server then forwards the INVITE to the destination server. In the redirect mode a caller again issues an

Message	Description
INVITE	Used to invite a user to a call as well as to establish a new connection.
ACK	Used to acknowledge and INVITE.
BYE	Used to terminate a request or a search for a user.
REGISTER	Used for conveying information about a user's location to a SIP server.
OPTIONS	Used for soliciting information about the capabilities of a SIP server.

Table 3-9.   Core SIP Messages

INVITE message to a location server. Assuming the called party moved, is on vacation, or is on temporary duty at another location, it is possible to provide new contact information to the caller. The caller would then issue an INVITE using the new contact information. Due to the redirect mode of operation it becomes possible to support call redirection to a Web page, time-of-day routing, a follow-me operation, and similar redirections.

Figure 3-19 provides a comparison between proxy and redirect modes of operation. Note that SIP codes are structured in increments of 100, starting at 100, with 1xx used for information purposes, 2xx for confirmations, and a 4xx range to indicate a request failure.

Comparison to H.323

One of the key advantages of SIP in comparison to H.323 is the simplicity of the former. H.323 is complex, with its set of specifications resembling the IRS tax code at over 700 pages. In comparison, the SIP standard is far simpler and is defined through a 200-page document. Another key difference concerns conference calls. H.323 requires a multicast controller or H.332, while SIP has built-in support for conference calls. Finally, H.323 is limited to TCP, while SIP can also operate under X.25.

MGCP

In concluding this section we will turn our attention to the Media Gateway Control Protocol (MGCP). The function of MGCP is to provide a mechanism to control telephone gateways from external call control elements referred to as call agents or media gateway controllers. MGCP is primarily based on the Simple Gateway Control Protocol (SGCP), an ASCII string-based signaling protocol. SGCP is noteworthy in that it provided the foundation for MGCP and a version used on cable TV for cable telephony. In a TCP/IP protocol suite environment MGCP is transported via UDP and represents an application

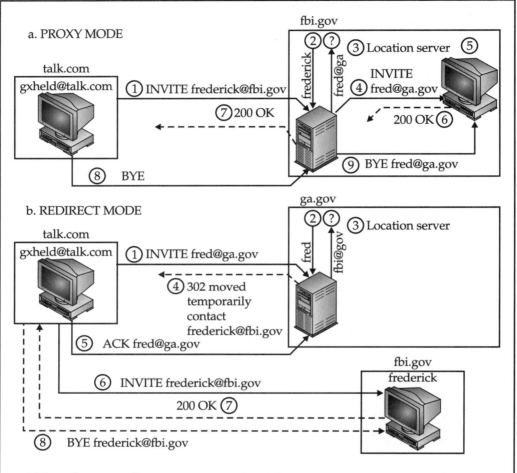

1. The caller uses a directory service, such as LDAP, to map a name to a user domain. This is accomplished through the use of an INVITE transmission method.
2. Using the caller domain, an SIP server is located via a DNS lookup.
3. The called server maps the name to the form user@host.
4. The user@host name is used to issue an INVITE to the callee. The callee can accept, reject, or forward the INVITE to a new address.
5. If the call is forwarded, the result is a return to step 2.
6. If the callee accepts the call, a confirmation is sent.
7. The conversation occurs.
8. The caller or callee terminates the session via a BYE

Figure 3-19. SIP modes of operation

residing at layer 5 in the TCP/IP protocol stack. To understand MGCP let's first briefly turn our attention to the basic operation of a telephone gateway, which we will cover in considerably more detail in Chapter 8.

Gateway Operation

The primary function of a telephone gateway is to provide a conversion operation between audio signals used on telephone circuits and data packets used on packet networks. When a telephone gateway operates in a business environment a common problem is the need to keep track of trunk groups and circuits in trunks that are connected to a gateway, which provides the connection between the packet network and the PSTN, a PBX, or even an ATM network. To accomplish this control call agents are used in conjunction with a configuration database to keep track of trunk groups and circuits in trunks.

MGCP Operation

MGCP performs its operation by issuing a sequence of ASCII commands to endpoints, where an endpoint represents a point of entry and exit of media flows. Each command consists of a verb that defines an action to be performed by the selected endpoint. The MGCP vocabulary consists of eight commands, each of which has a header and may contain other parameters to include a session description that sets up an endpoint to both recognize and generate an applicable media format. Table 3-10 lists the eight commands supported by MGCP.

Parameters supported by MGCP commands range in scope from a CallID that identifies a call to a SignalRequest that requests an endpoint to generate a specific signal, such as a dial tone. The support of specific parameters varies by MGCP command, with some parameters supported by only one command whereas other parameters are supported by multiple commands.

Verb	MGCP Code
CreateConnection	CRCX
ModifyConnection	MDCX
DeleteConnection	DLCX
NotificationRequest	RQNT
Notify	NTFY
AuditEndpoint	AUEP
AuditConnection	AUCX
ReStartInProgress	RSIP

Table 3-10.   MGCP Commands

Figure 3-20 illustrates the relationship between a gateway and trunks that can support multiple calls, such as a T1 link on which up to 24 calls can be placed. To provide information about the operation of a gateway, MGCP supports a variety of error codes. Examples of some MGCP error codes include digits that denote "phone is already off-hook," "endpoint is restarting," and "insufficient bandwidth."

Use with H.323

Although MGCP does not have to be part of H.323, if it is, an MGCP call agent will act as an H.323 gatekeeper. This enables an H.323-capable terminal to set up calls with MGCP endpoints. In this environment an H.323-compatible terminal would employ H.225/H.245 signaling to a call agent. The call agent would then use MGCP commands to control the endpoints in the gateway. Thus, MGCP should be viewed as a gateway control protocol that has a considerable degree of utilization flexibility.

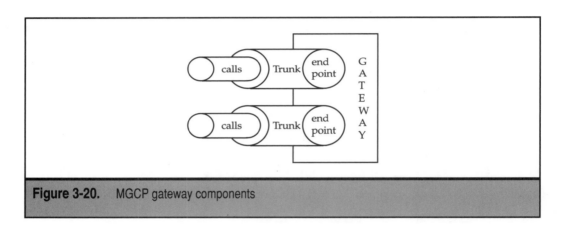

Figure 3-20.   MGCP gateway components

CHAPTER 4

Frame Relay

The ability to appreciate many of the issues associated with transporting voice over frame relay requires knowledge of the basic operation of the technology associated with this packet switching network. The purpose of this chapter is to provide readers with information concerning the development, operation, and utilization of frame relay as well as the basic cost components associated with the use of public frame relay networks. In doing so, we will examine the fields in the frame relay frame and their use, the flow of data through a frame relay network, and the use of parameters that define the operation and throughput associated with a connection to a frame relay network.

OVERVIEW

The rationale for the development of frame relay can be traced to the operation of its predecessor, X.25, as well as to advances in the acceptance of local area networks and the requirement of organizations for interconnecting LANs with a minimum amount of latency or transmission delay. X.25 was developed when most long-distance transmission occurred over relatively poor-quality microwave and copper transmission facilities. To compensate for the poor-quality transmission facilities, each switch at an X.25 network node performs an error check on each received packet. To do so, the switch must first buffer the packet in order to compute a cyclic redundancy check on the contents of the received packet, transmit a negative acknowledgment to the sender, and discard the packet if the locally generated CRC does not match the CRC appended to the end of the packet.

X.25 Network Delay Constraints

Although the delay associated with error checking at any individual X.25 node may be only between 100 and 200 ms, since error checking occurs at each X.25 node, the cumulative effect of error checking becomes significant. This significance results from the fact that an average session requires routing through three or four network nodes. This means that the typical latency or delay associated with the routing of data through an X.25 data network can range between 300 and 800 ms and commonly averages 400 ms. That makes this type of network unsuitable for low-delay data transfer applications such as transporting real-time audio and video data streams and supporting interactive query response between geographically separated local area networks interconnected via an X.25 network.

Fiber-optic-Based Backbones

The development of X.25 data networks primarily occurred during the 1970s and 1980s. During the latter part of the 1980s, communications carriers began to significantly improve the transmission capability of their networks via the installation of tens of thousands of miles of fiber-optic cable. In addition to greatly increasing the transmission capability in comparison to the existing microwave and copper infrastructure, the use of fiber-optic cable greatly lowered the error rate, in many cases by several orders of magnitude. This new communications-carrier infrastructure facilitated the development of a new type of transmission facility that placed responsibility for error detection and correction in the higher layers of the protocol instead of in each node and the endpoints in an X.25 network. Not only does this shift in the responsibility for error detection and correction provide for greater throughput, but it results in a lower amount of protocol overhead. This in turn allows frame relay to make more efficient use of transmission lines.

Comparison to X.25

Other significant differences between frame relay and X.25 include the layer in which they operate in the protocol stack, their network access rates, and how they handle network congestion.

Protocol Stack Operation

Frame relay is a layer 2 protocol that uses some of the core aspects of that layer. That is, a frame relay network will check the validity of a frame but will not request retransmission if an error is found. Instead, the frame is simply dropped, and the higher layers in the protocol become responsible for noting and correcting this situation. In comparison, X.25 is a fully featured protocol that operates at layer 3 in the OSI Reference Model. As a layer 3 protocol, X.25 includes responsibility for error detection and correction, flow control, and extensive support for controlling the delivery of packets and sending supervisory information between network nodes. These functions are, to a large degree, removed from frame relay, which reduces its overhead, enhances its throughput, and results in its name as frames are relayed from node to node through the network from source to destination, thus minimizing delay.

Network Access

Frame relay design was based on the use of more modern digital circuits as opposed to the analog infrastructure commonly available when X.25 networks were developed. This difference enabled frame relay designers to construct their networks to support higher data access rates, because the fiber backbone permits high-speed transmission between network nodes. Thus, while access to an X.25 network is limited to 56 Kbps, access to a frame relay network is supported at data rates of up to the 1.544-Mbps T1 and 2.048-Mbps E1 circuit operating rates, and T3 access offerings at approximately 45 Mbps recently became available and are offered by several vendors.

Network Congestion

The third area of difference between frame relay and X.25 networks lies in the manner in which they handle network congestion. In an X.25 network, congestion is handled by flow control, a process that regulates the flow of data through the network. Although flow control ensures that packets are not lost, this technique both delays the flow of packets through the network and adds a considerable amount of supervisory overhead that adversely affects the packet processing capability of switches in the network. In addition, since X.25 switches may be required to temporarily buffer a large number of packets during network flow control conditions, those switches include a relatively large amount of buffer memory. Since the processing of packets occurs on a first-in, first-out (FIFO) basis, flow control conditions can result in a considerable degree of packet latency. In comparison, frame relay uses a very simple procedure for dealing with periods of network congestion: packet dropping. That is, a frame relay network is constructed with sufficient capacity to service a baseline of traffic from each network subscriber that is technically referred to as the *committed information rate* (CIR), which we will describe and discuss in detail later in this chapter. When congestion occurs in a frame relay network, switches used by the frame relay communications carrier will simply drop or discard certain packets. Those packets, which are referred to as *frames* since frame relay is a layer 2 protocol, are examined for the setting of a discard eligibility (DE) bit that designates the frame can be dropped. Thus, a frame relay network compensates for network congestion by simply dropping certain frames, leaving it to higher layers at endpoints to recognize the fact that frames were dropped and to take corrective action by retransmitting those frames.

Feature Comparison

Table 4-1 provides a general comparison of frame relay and X.25 features. In examining the entries in Table 4-1, note that statistical multiplexing refers to the ability of each network to interleave by time packets and frames from different sources routed to the same or different destinations, enabling high-speed circuits linking network nodes to be used more efficiently. Figure 4-1 illustrates an example of statistical multiplexing. In this example, it is assumed that device X transmits a packet to device Z. Next, device Y transmits a

Feature	Frame Relay	X.25
Statistical multiplexing	Yes	Yes
Port sharing	Yes	Yes
OSI layer operation	2	3
Maximum access rate	1.544/45 Mbps	56 Kbps
Throughput delay	Low	High
Flow control	None	Built-in
Packet/frame discard	Yes	No
Service-level agreement	Yes	No
Priority queuing	Yes	No

Table 4-1.   Comparing Frame Relay and X.25

packet to Z, followed by device Y again transmitting a packet to Z. Although the path between nodes B and A are shared by time, the multiplexing is statistical since packets are transmitted only when a device has data to send.

Returning to Table 4-1, port sharing refers to the ability of both X.25 and frame relay to support the establishment of multiple virtual circuits via the connection of one port from an X.25 packet assembler/disassembler (PAD) or frame relay access device (FRAD) to an X.25 or a frame relay network. This concept is also illustrated in Figure 4-1 since device X, which is connected to node B on the packet network, is shown transmitting data to device Z on node A and device W on node C. Thus, two virtual circuits representing temporary connections from node B to nodes A and C are established to enable device X to communicate with devices W and Z.

For those not familiar with the term *virtual circuit*, it represents a path between endpoints in a network that is established based on information in a packet or frame header. There are two types of virtual circuits supported by most packet networks: permanent and switched. A permanent virtual circuit (PVC) is assigned by the network operator. Depending on the network operator, the actual path of a PVC can vary and PVCs can have alternate routing. In addition, as illustrated in Figure 4-1, multiple PVCs can be established over a single connection to a packet network.

In comparison to a PVC, which represents a fixed connection through a network set by the network operator, a switched virtual circuit (SVC) represents a temporary path through a network. The destination endpoint of an SVC call is set by the user and requires the use of a call setup procedure. Until recently, most frame relay networks were limited to supporting PVCs.

Returning to Table 4-1, the last two entries represent recent additions to offerings of many frame relay service providers that are related to one another and that can be essential for success when transmitting voice over frame relay. A service-level agreement

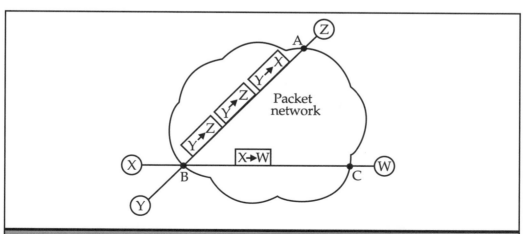

Figure 4-1.   Statistical multiplexing permits both X.25 and frame relay networks to share communications circuits among many users by time.

(SLA) is a contract between a customer and the frame relay operator that guarantees a defined level of performance. That level of performance can be expressed in a variety of ways, ranging from a mean time to repair a network failure to the percentage of frames that are delivered with and without their DE bit set. Perhaps the most important SLA with respect to voice over frame relay is an agreement that defines maximum latency. To provide a maximum latency guarantee, frame relay providers typically divide traffic into classes and assign classes to different queues in their network switches. By favoring the extraction of frames from high-priority queues over lower-priority queues, a maximum latency time can be guaranteed.

APPLICATION NOTE: When considering voice over frame relay, the ability to initiate a service-level agreement that guarantees a maximum end-to-end latency through the network enables you to develop a predictable voice transport mechanism.

Although frame relay builds on the port and bandwidth sharing of X.25 technology, its design was based on the use of a digital circuit infrastructure that permits higher-speed access and lower delays through the network. This in turn makes frame relay more suitable for transporting relatively bursty traffic associated with interconnecting local area networks, while the low latency makes it suitable for carrying digitized voice, even though the network was originally designed as a data transport facility. Although frame relay has many advantages over X.25, the use of the latter is probably several orders of magnitude higher than the former, even though frame relay utilization is increasing at a rate probably twice that of X.25. The key reason for this is the fact that a large percentage of credit card and other financial transactions are carried via X.25 networks, a trend expected to continue. Credit card and bank transactions are transported in relatively short packets, and the error checking performed at each node does not sufficiently delay the authorization in comparison to the willingness of customers at supermarket checkout counters, gas stations, or bank teller windows to wait a few seconds for the transaction to be completed.

Now that we have an appreciation for the general characteristics of frame relay in comparison to X.25, let's turn our attention to a few specifics and examine the technology in detail.

EVOLUTION AND STANDARDIZATION

Frame relay, while considered by many persons to represent a recently developed transmission method, actually has its roots in the efforts of the Consultative Committee for International Telephone and Telegraph (CCITT) during 1988. In that year, the CCITT, which is now known as the International Telecommunications Union (ITU), approved its Recommendation I.122, titled "Framework for Additional Packet Mode Bearer Series," which was a part of a series of ISDN-related specifications.

LAP-D

An important part of the I.122 recommendation is known as Link Access Protocol—D channel (LAP-D), which is used to transmit signaling information on the ISDN D channel. ISDN developers noted that LAP-D had several characteristics that made it useful for other applications. One such characteristic was the multiplexing of virtual circuits at layer 2 of the ISO Reference Model instead of at layer 3, which was used by X.25 networks. Recognizing the additional capabilities of LAP-D, the I.122 recommendation was written to provide a general framework to indicate how the protocol could be used to support applications other than ISDN signaling.

Bell Labs and ANSI Standards

Building on the CCITT Recommendation I.122, the research arm of AT&T, which at that time was Bell Laboratories, performed a significant amount of work to develop a fast-packet technology. The American National Standards Institute (ANSI) provided a significant focal point for the development of a series of frame relay standards that were approved beginning in 1990. Those standards, while forming an important foundation, were just a beginning. Recognizing the potential afforded by this evolving technology, a group of equipment manufacturers, including Cisco Systems, Digital Equipment Corporation, Stratacom, and Northern Telecom, formed a consortium known as the Frame Relay Forum during 1990.

The Frame Relay Forum

The goal of the Frame Relay Forum is to focus on the development of frame relay technology and to facilitate the interoperability of equipment by developing appropriate standards. In addition to developing a specification that conforms to the basic protocol standardized by ANSI and the CCITT, the Frame Relay Forum has led the way in developing *Implementation Agreements* (IAs), which define the methods for using frame relay with other technologies and for other applications. An example of the former is the IA for frame relay being carried over ATM, while an example of the latter is the IA for voice over frame relay. Table 4-2 lists the 14 Frame Relay Forum Implementation Agreements that were completed as of late 2000. As noted in the table, FRF.11 is the IA for Voice over Frame Relay. In Chapter 9, when we discuss voice over frame relay in detail, we will also examine the FRF.11 IA. Due to the effort of the Frame Relay Forum, developers have the ability to promote the interoperability of equipment and to standardize the use of new technology that enhances frame relay service.

Although FRF.11 IA is very important for obtaining network equipment interoperability, another important IA is FRF.13, which covers service-level agreements. The focus of FRF.13 is in the definition of parameters that address frame transfer rate, frame delivery ratio, data delivery ratio, and service availability, all of which make it easier to compare and contrast SLAs from different service providers.

FRF.1.1	User-to-Network Implementation Agreement
FRF.2.1	Network-to-Network Implementation Agreement
FRF.3.1	Multiprotocol Encapsulation Implementation Agreement
FRF.4	Switched Virtual Circuit Implementation Agreement
FRF.5	Frame Relay/ATM Network Internetworking Implementation Agreement
FRF.6	Frame Relay Customer Network Management Implementation Agreement
FRF.7	Frame Relay PVC Multicast Service and Protocol Description Implementation Agreement
FRF.8	Frame Relay ATM/PVC Service Interworking Implementation Agreement
FRF.9	Data Compression over Frame Relay Implementation Agreement
FRF.10	Frame Relay Network-to-Network Interface Switched Virtual Connections Implementation Agreement
FRF.11	Voice over Frame Relay
FRF.12	Frame Relay Fragmentation Implementation Agreement
FRF.13	Service-Level Definitions Implementation Agreement
FRF.14	Physical Layer Interface Implementation Agreement

Table 4-2. Frame Relay Forum Implementation Agreements

Under the FRF.13 IA, a representative network structure is defined for both intranetwork and internetwork communications, with the latter representing the flow of data on an end-to-end basis between two or more frame relay networks. Perhaps one of the more important aspects of the FRF.13 IA with respect to the transmission of voice over a frame relay network is a formal definition of frame transfer delay. Under the FRF.13 IA, the frame transfer delay is as follows:

$$FTD = t_2 - t_1$$

where t_1 is the time in milliseconds a frame leaves its source and t_2 is the time a frame arrives at its destination. It is important to note that this definition can be structured to produce end-to-end, edge-to-edge interface, or edge-to-edge egress parameters. Because it is important to understand the full effect of delay upon voice on an end-to-end basis, you should normally request any SLA delay to be specified on an end-to-end basis.

APPLICATION NOTE: When considering a service-level agreement covering delay, make sure the agreement covers latency on an end-to-end basis.

Currently the Frame Relay Forum has three working groups: (1) Market, Development, and Education; (2) Technical; and (3) Interoperability and Testing. In addition to facilitating the development of frame relay-related standards and promoting interoperability of vendor equipment, the Forum provides a conduit for end users to obtain information concerning the latest developments in frame relay technology as well as access to developing and finalized IAs. The Market, Development, and Education working group is similar to corporate PR and marketing departments.

The Technical working group is responsible for developing and modifying standards as well as providing guidance on technically related issues. One important example of the work of the Technical working group is an extension to the core frame relay specification known as the Local Management Interface (LMI). LMI addresses such issues as global addressing, multicasting, and the manner in which status updates can be transmitted as unsolicited messages by a frame relay network. Here the term "multicast" refers to the ability of a frame relay network to transmit messages for users that belong to a predefined group. Instead of transmitting one message per member of the group, only one message is transmitted, which significantly reduces the use of network bandwidth.

The third working group of the Frame Relay Forum is the Interoperability and Testing working group. As its name implies, this group is responsible for structuring tests to ensure interoperability issues can be resolved and equipment from different vendors can work with one another.

FRAME RELAY OPERATIONS

In this section, we will focus on the basic operation of frame relay. In doing so, we will first examine the format or composition of a frame relay frame. Using that information as a foundation will enable us to discuss the operation of frame relay to include the manner in which a connection is established through a frame relay network.

Frame Formats

Frame relay is a synchronous higher-level data link control (HDLC) type of protocol that traces its roots to LAP-D. The key difference between the frame relay protocol and LAP-D is the absence of a control field, since the passing of supervisory and control information, which is extensive and pervasive in X.25 and true HDLC, is conspicuous by its absence in frame relay. In actuality, the absence of control and supervisory information required through the use of a control field is by design, since frame relay does not directly support flow control and error detection and connection, and it uses higher-layer software on a separate channel for setting up and tearing down connections. In comparison, X.25 data networks use the control field to perform the preceding and more, placing a heavy administrative burden on the network.

Figure 4-2 illustrates the three basic formats of the frame relay frame. In actuality there is only one basic frame relay frame format, which has three address field variations, indicated in the lower portion of Figure 4-2. To help understand the operation of frame relay, let's focus on the fields in the frame.

Flag Fields

The flag fields are used to delimit the beginning and end of each frame. Each flag has the bit composition 01111110, and any sequence of five set bits in natural data is automatically modified by the insertion of a 0, a process referred to as zero insertion, to prevent the possibility of a false flag formed from natural data being misinterpreted as a flag. Thus, the flag fields function in the same manner in frame relay as their counterparts in HDLC.

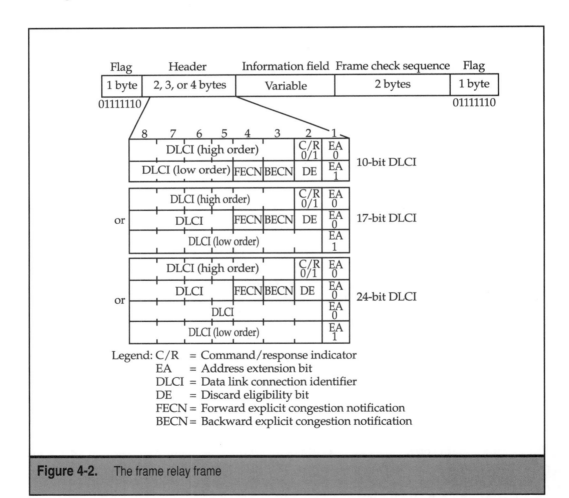

Figure 4-2. The frame relay frame

Header Field

The frame relay header provides addressing, congestion control, link management, and routing information through the use of six subfields:

▼ Data link control identifier (DLCI)

■ Command/response (C/R) bit

■ Address extension (EA) bits

■ Forward explicit congestion notification (FECN)

■ Backward explicit congestion notification (BECN)

▲ Discard eligibility (DE) bits

To obtain an appreciation for their roles, let's examine each of the six subfields in the header field.

DLCI Subfield

The DLCI is either 10, 17, or 24 bits in length, based on the setting of the extended address (EA) bits. The purpose of this field is to identify the logical connection that is multiplexed onto the physical channel. The DLCI identifies both directions of a virtual connection through a frame relay network and not an actual destination address.

To illustrate the use of DLCI addressing, consider Figure 4-3, which shows a four-node frame relay network with FRADs or routers supporting frame relays located in New York, Chicago, Atlanta, and Miami. If the router or FRAD in New York and Chicago establish a virtual circuit between them, they would both specify DLCIs that could be mapped to one another by the network. In this example DLCI = 20 in Chicago would be mapped to DLCI = 37 in New York. Thus, the DLCI can be considered to have local instead of network significance.

A second term that may require a degree of explanation for some readers is the FRAD. In addition, since many routers are frame relay-compliant, a brief discussion of the difference between the use of a FRAD and a router to access a frame relay network may be in order.

A frame relay access device (FRAD) directly converts a data source, such as IBM's SNA or an Ethernet LAN frame, into the frame relay frame illustrated in Figure 4-2. In examining Figure 4-2, you will note that the header is relatively short. When a router is used to provide a connection to a frame relay network, it uses IP encapsulation, resulting in the frame relay frame being transported within an IP packet. This action results in a much larger overhead; however, there are certain functions a router can perform that are not commonly included in a FRAD. Some of those functions include dial backup support and load balancing when operating dual access lines. Most frame relay network operators support both FRAD and router connectivity, leaving the choice of the access device to the subscriber.

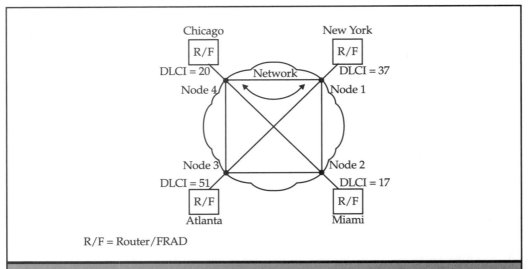

Figure 4-3. Using data link connection identifiers (DLCIs) to establish connections through a frame relay network

APPLICATION NOTE: If you have a choice between using a FRAD and a router to access a frame relay network and intend to transmit voice, use a FRAD. Doing so will eliminate IP encapsulation, which results in a significant degree of overhead that especially adds to latency when access to the network occurs over relatively low-speed fractional T1 lines.

To illustrate the mapping of DLCIs, consider node 1 in Figure 4-3. Let's assume it is a four-port router or FRAD, with one port connected to the circuit routed to the customer in New York. Let's further assume that the customer in New York is connected to the node via port 1, and ports 2, 3, and 4 are connected to nodes 2, 3, and 4 in the frame relay network. The routing table for node 1 showing the relationship between ports and DLCIs known to the network would be as follows:

Port	DLCI
1	37
2	20
3	51
4	17

There are certain DLCIs that have predefined significance. For example, DLCI = 0 and DLCI = 1023 are used for Local Management Interface (LMI) management control identification. Through the use of DLCI = 0, call control messages are placed in the information field and are used to request the establishment and clearing of a logical connection. Each side can request the establishment of a logical connection via the use of a SETUP message, with the other party responding with a CONNECT message if it accepts the connection, or a RELEASE COMPLETE message if it elects not to accept the connection request. The side sending the SETUP message can assign the DLCI by including it in its SETUP message; however, the other party can respond with a different DLCI in its CONNECT message. Once data is exchanged, either party can issue a RELEASE message to clear the logical connection, with the other party responding with a RELEASE COMPLETE message.

FECN, BECN, and DE Subfields

The forward explicit congestion notification (FECN), backward explicit congestion notification (BECN), and discard eligibility (DE) subfields each consist of a single bit. The setting of each bit to a value of 1 provides the network with the ability to perform congestion control.

When a frame relay switch begins to experience congestion, it can inform its upstream and downstream nodes of the problem by setting either or both FECN and BECN bits to a binary 1. The BECN bit is set in frames transmitted downstream to notify the source of the traffic that congestion exists at a switch in the virtual connection path. On receipt of this notification, the traffic source will ideally control the flow of its traffic until the congestion problem is cleared and the switch no longer sets the BECN bit in each frame transmitted downstream. When congestion is occurring downstream, a switch can also set the FECN bit in frames transmitted to upstream nodes that act as receivers in the virtual path.

Figure 4-4 illustrates the direction of FECN and BECN set bits in transmitted frames by a switch experiencing congestion—in this example, switch number 1. Although the setting of the FECN bit may appear illogical when congestion is caused from the other direction, it provides the possibility for the opposite party to take action to correct a congestion problem caused by the other party. For example, the status of the FECN bit could be passed to an upper-layer protocol that might slow down acknowledgments, which in turn reduces the flow of data to the destination via the congested switch.

In the event network congestion continues, the frame relay network will invoke a frame-discard strategy that will provide a degree of fairness among network users. To accomplish this, the network will use the discard eligibility (DE) bit in the frame header to determine which frames to send to the great bit bucket in the sky. This is because the DE bit is set to identify frames from users who exceed their negotiated throughput rates. The negotiated throughput rate represents a user's estimate of his or her normal traffic during a busy period and is technically referred to as a *committed information rate* (CIR).

The CIR   Each PVC has an associated CIR; however, the CIR represents a guaranteed transmission rate and not a cap on the transmission rate. That is, a PVC can burst transmission above its CIR, with the ability to do so and the data rate above the CIR that's achievable depends on several variables, including the *committed burst size* (B_c), the *excess burst size* (B_e), and the *network time interval* (T_c). In the next section we will examine those variables and their relationship to one another.

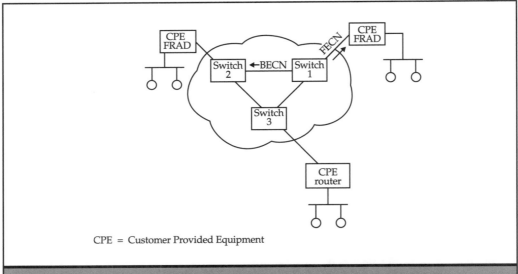

CPE = Customer Provided Equipment

Figure 4-4. The direction of FECN and BECN set bits generated by a congested switch

Committed Burst Size The committed burst size (B_c) represents the amount of data in bits that a frame relay network agrees to transfer under normal network conditions during a predefined time interval (T_c). Most frame relay network operators set T_c at 1 second. Thus, the CIR can be redefined as follows:

$$CIR = \frac{B_c}{T_c} = B_c$$

That is, the CIR represents the rate at which the network agrees to transfer B_c during normal network conditions on a PVC.

Excess Burst Size The excess burst size (B_e) represents the maximum amount of uncommitted data in bits above B_c that the network will attempt to deliver during the time interval T_c, with T_c normally set by most networks to 1 second. To illustrate the relationship between the CIR, B_c, B_e, and T_c and to show how the network alters the DE bit, let's assume you have a connection to a frame relay network via the use of a T1 line operating at 1.544 Mbps and have negotiated a 64-Kbps CIR. This means that when you are transmitting data at less than or equal to 64 Kbps, the network will not alter the DE bit. While a transmission rate of 64 Kbps may be sufficient for many networking activities, such as short query/responses, suppose you just initiated a file transfer. Since the operating rate of the frame relay access line connection exceeds the CIR, you can burst your transmission above the CIR. In this example, the B_e for a long file transfer would become

(1.536 Mbps–64 Kbps), or 1.472 Mbps, since the T1 line uses 8 Kbps for framing that cannot be used for information transfer. Thus, you obtain the ability to burst your transmission above the CIR; however, when this occurs, the network will set the DE bit to 1.

Figure 4-5 graphically illustrates the relationship between the line access rate, excess burst size, and committed burst size. Note that as user activity exceeds the CIR, the DE bit in frames entering the network are set.

Using the DE Bit   When the DE bit is set to 0, this tells each network switch that the frame should not be discarded unless there is absolutely no alternative. For most networks, the setting of the DE bit to a value of 0 essentially guarantees that the frame will reach its destination. In comparison, the setting of the DE bit to a value of 1 indicates to each switch in the frame relay network that the frame is eligible for the great bit bucket in the sky if the switch is experiencing network congestion. However, just because the DE bit is set to a value of 1 does not mean it will be discarded. In fact, the vast majority of frames with a DE bit set to a value of 1 reach their destination; however, there is no guarantee that they will do so, and they will be discarded much more frequently than frames with a DE bit set to a

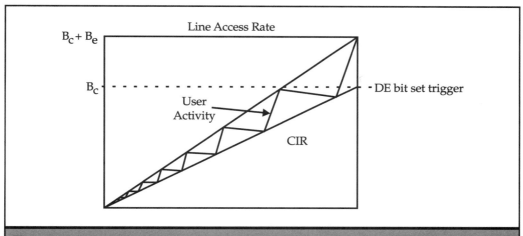

Figure 4-5.   The relationship between frame relay metrics

value of 0. In fact, many frame relay network operators guarantee that 99 percent of all frames with a DE bit set to a value of 1 will reach their destination. In addition to the examination of the DE bit as a criteria for discarding frames, there are several other criteria that will result in a frame being discarded. These criteria include a frame not being bound by opening and closing flags, the absence of an information field, an FCS error, and the use of an invalid DLCI value in a frame.

Command/Response Subfield

The command/response (C/R) subfield represents another 1-bit flag included in the header field. The C/R bit was included in the header to facilitate the support of polled protocols, such as IBM's SNA. Through the use of the C/R field, the direction of a poll can be designated. That is, a command frame would set the C/R bit to a value of 0, while a response frame would set the C/R bit to a value of 1. Thus, the C/R subfield can be considered to represent an application-specific subfield whose values are set and processed only by end stations.

Extended Address Subfield

The extended address (EA) subfield provides the mechanism by which the boundary of the address field is defined. This means that an additional header byte follows that contains additional address information. When the EA subfield is set to a value of 1, this means the header field has terminated and no additional addressing information follows.

When the first EA subfield has a value of 1, the DLCI is represented by 10 bits. This is the most common type of DLCI address and provides 1024 possible PVC numbers or identifiers. In actuality, 32 DLCIs are reserved for use by the frame relay network, resulting in 992 being available for assignment as PVC numbers.

Information Field

The third field in the frame relay frame is the information field. This field can theoretically be up to 8192 bytes in length and transports the payload or data being routed from source to destination. In actuality, there is no defined information field maximum length. However, most equipment is commonly set to transport frames using a 512- or 1024-byte information field.

Variable-Length Field Problems

One of the key problems associated with the use of a variable-length information field is the fact that a frame transporting a file transfer can adversely affect the transmission or reception of a frame carrying time-sensitive information such as audio or video data. To illustrate this concept, assume a FRAD has two internal private-network connections and one frame relay network connection as illustrated in Figure 4-6. Let's further assume that the frame relay network connection operates at 128 Kbps and that each internal network connection to the frame relay network has a PVC whose CIR is 64 Kbps, which enables a burst of up to 128 Kbps when the other PVC is inoperative. Assume at time $t = 0$ both internal private-network connections are not used. At $t = 1$, a LAN device initiates a file transfer and fills each frame so that the information field transports 8192 bytes of data, resulting in a frame length of 8198 when a 2-byte address field is used. If the file transfer bursts up to 128 Kbps, the delay introduced by the frame carrying the file transfer is 8198 bytes × 8 bits/byte/128 bits/second, or approximately 0.512 second. If the PBX sends a voice signal to the FRAD, which digitizes the signal and places it into a frame, each time the LAN station is serviced it will generate a frame that will delay the subsequent digitized voice frame by approximately 1/2 second without considering any network delay. This means that the ability to transport voice over a frame relay network along with data will require some type of priority scheme, an adjustment of the maximum frame length, or a quality of service mechanism. Otherwise, lengthy frames transporting data that are serviced ahead of digitized voice frames can induce an amount of delay or latency sufficient to produce awkward-sounding reconstructed voice.

Quality of Service

Until 1997, the only way a frame relay subscriber could enhance the ability of frames to flow end to end with a minimum amount of latency or delay was to increase their committed information rate. Although doing so can improve performance through a frame relay network, it adds to your monthly bill and still does not provide a quality of service (QoS). Recognizing the need for subscribers to obtain a better guarantee that frames will reach their destination with a minimal delay, frame relay network providers introduced several techniques to enhance the flow of data through their networks. Two of the more

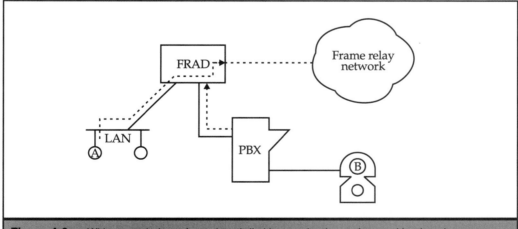

Figure 4-6. Without a priority or frame length-limiting mechanism, a frame with a lengthy information field transporting data can adversely affect the transmission or reception of frames carrying time-sensitive information.

interesting methods are MCI Communications Corporation's (now part of Worldcom) Priority Permanent Virtual Circuit (PPVC) and LCI International's guarantee concerning the delivery of frames.

The MCI Priority PVC enables subscribers to assign their PVCs high, medium, or low priority. The MCI network then uses a round-robin sampling scheme that samples high-priority PVCs more often than lower-priority PVCs. Although this technique does not provide subscribers with a mechanism to obtain priority over other MCI subscribers, it does enable subscribers to prioritize their traffic. Doing so can result in the partial avoidance of the previously described situation where a data packet can adversely delay the transmission of a frame carrying a digitized voice sample. Thus, although a priority mechanism is a significant improvement over nonprioritized traffic, it does not directly provide a QoS. In addition, MCI implements its Priority PVC through the use of Bay Network (now a subsidiary of Nortel) switches and software and can be considered to represent a proprietary technique. This means you cannot obtain a priority for your traffic if it is routed via MCI to a subscriber that uses a different frame relay network operator.

Another frame relay technique to prioritize the flow of frames, which provides a better approach to a true QoS, is a technique introduced by LCI International, Inc., in March 1997. LCI guarantees the arrival of frames at their destination based on their length. Frames of up to 1600 bytes in length are guaranteed to arrive in less than 0.25 seconds. Frames of less than 500 bytes are guaranteed to arrive in less than 95 milliseconds, or just under 1/10 second. Finally, frames of less than 100 bytes in length are guaranteed to arrive in less than 35 milliseconds. Although the LCI International frame arrival guarantee would minimize latency problems if subscribers could ensure that their voice-digitized frames were minimized with respect to their length, it is the function of the equipment and not the network that controls the length of the information field in the frame.

Since the introduction of MCI's Priority PVC and LCI International's latency delay guarantee, other service providers have added a variety of service-level agreement parameters to their frame relay offerings. Today you can readily select guaranteed latency based on frame length from several service providers. In addition, during the fall of 2000 Sprint introduced a managed voice-over-frame-relay service, an offering that not only provides customers with guaranteed service levels but also provides the potential to reduce an organization's communication cost. Under the Sprint managed voice-over-frame relay service that vendor will deploy Cisco 3810 routers at each custom site. The customer must obtain a separate PVC for voice services and will commonly request 10 Kbps per call. If a Sprint customer is on Sprint's Enhanced Broadband SONET network they can obtain a service-level agreement that guarantees 100 percent network availability. If the Sprint network is unavailable for up to one hour the customer will receive a credit for three days worth of PVC and port charges. For outages that exceed one hour Sprint provides customers with three days of PVC and port charges plus one day for each additional hour. Based upon the preceding, it is obviously also important to consider credits under any vendor SLA to obtain a valid comparison between vendor service-level agreements.

We can also obtain an appreciation for the direction of voice over frame relay based upon a Network World article published in September 2000. According to that article concerning the Sprint voice-over-frame-relay service, up to 10 percent of Sprint customer's were using voice over their frame relay networks prior to the vendor even announcing the availability of a managed service.

When we discuss voice over frame relay in Chapter 9, we will also discuss several methods used by equipment vendors to ensure that frames transporting digitized voice are provided to the frame relay network with an information field length that facilitates the transportation of the frame through the network.

Frame Check Sequence Field

The frame check sequence (FCS) field is used as a mechanism to detect bit errors in the frame header and information fields. However, unlike X.25 and other protocols that generate a negative acknowledgment and wait for a retransmission on detection of an FCS error, a frame relay network simply discards the frame. Thus, error recovery is left up to the higher layers at end-user devices.

Management

As briefly discussed earlier in this chapter, the Local Management Interface (LMI) specification provides a mechanism for network management functions to be performed. LMI is an extension to frame relay. Unfortunately, there are three LMI types: (1) ANSI, which is equivalent to the LMI extensions developed by the Frame Relay Forum; (2) an ITU version; and (3) a variation used by Cisco Systems. To further compound an interesting situation, some LMI extensions are referred to as *common* and are expected to be adopted by vendors that comply with a specific specification, while other LMI functions are referred to as *optional* and may or may not be supported by a vendor that supports a particular specification. Some FRADs and frame relay-compliant routers support all specifications using an automatic sensing feature to determine the specific LMI version used by the network. To do so, the device transmits a status request in each LMI version, listing on DLCI 0 and DLCI 1023 to the network response to configure the hardware of the version of LMI supported by the network. The device listens on both DLCI 0 and DLCI 1023 because the LMI protocol can operate on either data link control identifier.

Frame Types

The use of LMI is optional and requires the subscriber to originate all exchanges of information. By forcing subscribers to initiate the exchange of LMI frames, the network is precluded from transmitting unwanted information to subscribers whose FRADs or routers do not support the LMI protocol. Thus, LMI represents a polling scheme, with subscriber status requests answered by network reports. Once a device indicates it supports LMI, a third type of LMI frame can be transmitted. That frame is an unsolicited status update generated by the network for a subscriber.

LMI frames are identified by their DLCI values. In the original LMI specification, it was defined to be DLCI 1023; however, it can now be either DLCI 0 or DLCI 1023. The basic LMI protocol supports three types of information elements (IE): report type, keep-alive, and PVC status. The keep-alive frame is used to establish an interval at which the far-end FRAD or router will be polled by the network. Its use enables the network and distant FRAD or router to note that the interface and the circuit used to connect the subscriber to the network are alive. The default value for the keep-alive interval is 10 seconds, and this value must be set to a value less than a polling verification timer value, which is used by LMI to denote the maximum number of seconds that can transpire before a polling error is logged. Each LMI message contains one report type element and one keep-alive element. A full status message from the network to the user also contains one PVC status element for each PVC on the circuit connecting the subscriber to the network.

Figure 4-7 shows the general format for each type of LMI message. In examining the three formats shown in Figure 4-7, note that the LMI header is 6 bytes in length. The first 2 bytes contain the DLCI as well as the C/R, EA, FECN, BECN, and DE frame relay control bits. With the exception of the EA bit, the other control bits are not used. The first 2 bytes then define the DLCI. The third byte identifies each LMI frame as an unnumbered information frame, as specified by the LAP-D standard. The fourth byte contains a value that identifies the frame as one containing LMI information and is technically referred to as a *protocol discriminator*. Although it may appear redundant to use a DLCI value and protocol discriminator to define an LMI frame, the intention of the protocol discriminator is to allow other signaling protocols to be transported instead of, or along with, LMI. The fifth byte contains an LAP-D parameter referred to as a *call reference*, which is a dummy field in LMI frames and is always set to 0. The last byte in the header identifies the LMI message type as either a status enquiry transmitted from the subscriber or status transmitted from the network or an update status transmitted from the network.

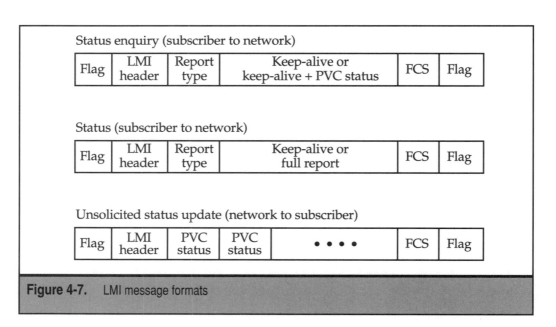

Figure 4-7. LMI message formats

One of the more important functions of common LMIs is the exchange of PVC status information between a subscriber and the network. Through the exchange of PVC status information, PVCs can be synchronized and can prevent the situation in which a subscriber transmits data into a nonexistent PVC, a situation referred to as a *black hole*. Thus, PVC status exchanges can be used to report the existence of new PVCs and the deletion of old PVCs.

Optional LMI Functions

Although common LMI functions are important, some of the optional LMI functions provide the capability to significantly enhance the utility of frame relay. Three very interesting optional LMI functions are global addressing, multicasting, and flow control.

Global Addressing   Global addressing provides a mechanism for each DLCI to uniquely identify a subscriber. To provide this capability, both the network and subscriber equipment must support the extension of the DLCI field, and the network operator must support unique DLCIs. Through the use of global addressing, subscribers can enhance the interconnection of LANs, because the process requiring the observation of DLCIs in use and an appropriate selection of an available DLCI is eliminated.

Multicasting   Multicasting enables frame relay to function similarly to LANs. That is, multicasting allows a subscriber to transmit a single frame and have it delivered to all members of the multicast group. LMI multicasting is accomplished by the use of DLCIs between 1019 and 1022. A subscriber who wants to transmit a message to all members of a predefined group has only to transmit the message once on a multicast DLCI. This feature can significantly reduce the utilization of bandwidth through a frame relay network and provides a mechanism to support conference calls via a frame relay network.

Flow Control   A third optional LMI function is flow control. Unlike X.25, which can perform flow control between network nodes, LMI flow control is used as a mechanism for the network to report congestion to subscribers. Although an LMI-compliant device supporting optional LMI functions will receive notification of flow control, the device does not have to act on this information.

COST OF USE

In concluding this chapter on frame relay, we will turn our attention to a topic near and dear to network managers and administrators as well as to the bean counters located in many organizations. That topic is the cost associated with the use of a frame relay network. Since there are several components associated with the use of a commercial frame relay network, perhaps the best place to commence an examination of the cost of using this type of packet network is by reviewing how we normally connect an organization's data facility to a frame relay network.

Cost Components

Figure 4-8 illustrates the connection of an organization's FRAD or router to a public frame relay network. From a physical perspective, you will need (1) an access line routed from your organization or customer site to the frame relay network operator's point of presence (POP), and (2) a dedicated port on the frame relay network operator's frame relay switch. Let's first concentrate on the cost of these two physical entities; then we'll turn to other cost elements that can be associated with the use of a public frame relay network.

Local Access Line

In examining Figure 4-8, note that unless a bypass network operator is used, the local access line that connects the subscriber's site to the frame relay network is furnished by the local telephone company. Most frame relay network providers will acquire the local access line and bundle the cost of the circuit into the price of their service. Most frame relay network providers simply pass the cost of the access line to the subscriber, which precludes excess charges, but it also precludes the potential savings of using a bypass operator to obtain a connection to the frame relay provider's point of presence (POP).

Switch Port Cost

Each access line routed from a customer site to a frame relay network operator will terminate into a serial port on a frame relay switch located at the network operator's node. As you might expect, many frame relay network operators charge their subscribers for the use of the port on their switch, basing the monthly fee on the operating rate of the line connected to the port.

PVC Charge

Most frame relay network operators base PVC prices on their CIR in a manner analogous to the cost of digital leased lines. That is, the cost of the PVC depends on the CIR. The higher the CIR, the higher the monthly cost. However, unlike leased lines, where the monthly cost depends on both the operating rate of the line and interexchange mileage, frame relay is not distance-sensitive. This means you can access Tulsa or Tampa for the same fee!

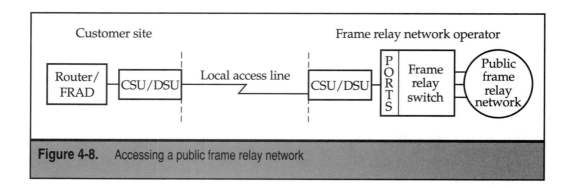

Figure 4-8.   Accessing a public frame relay network

Other Cost Considerations

The actual manner in which frame relay network providers bill subscribers can be quite complex. Although some frame relay providers may simply bill subscribers based on the local access, port charge, and CIR, there are additional charges you may have to consider. Those charges can include the number of PVCs used on a frame relay connection and a monthly fee for the use of a vendor's switch port. In addition, some frame relay network operators include a usage fee per PVC, which is based on the quantity of data transmitted on each PVC during the billing period.

Table 4-3 lists the common cost components that can be associated with the use of a public frame relay network. In examining the entries in Table 4-3, it is important to note that there can be significant differences in the methods by which different network operators bill subscribers. In addition, it should also be noted that it is important to perform a complete analysis of the cost components associated with the use of a frame relay network to be able to compare the estimated cost associated with the use of different vendor networks. This means you should not jump to conclusions about the economic merits associated with the use of one network over another just because one network operator appears to have a simplified billing structure in comparison to the billing structure of other frame relay network operators.

Returning to Table 4-3, note that it is structured to enable readers to enter their specific requirements, which can then be used as a mechanism to compare pricing from two or more vendors. Although Table 4-3 only lists two vendors, you can easily extend the columns to include a cost comparison for additional vendors. For example, to compare the cost of an access line, you would first enter the operating rate of the circuit under the requirements column. Then you would use that operating rate to compare the cost of this frame relay network cost component among different network operators.

Although the costs associated with access lines, network ports, and the number of PVCs and their CIRs are relatively straightforward and easy to determine, if a vendor also charges a fee based on PVC usage, this fee can only be estimated. Thus, it becomes important to obtain a reasonable estimate of the potential use of a frame relay network

Cost Component	Requirement	Vendor A	Vendor B
Access line	__Kbps/Mbps	____	____
Network port	__Kbps/Mbps	____	____
Network connection	__PVC/CIR	____	____
	__PVC usage	____	____

Table 4-3.   Common Frame Relay Network Usage Cost Components

due to the addition of the transmission of digitized voice over the network when a billing component includes a charge based on Mbytes of data. Concerning that charge, it should be noted that many "efficient" voice digitization methods result in the transmission of a voice conversation at a data rate of 8 Kbps. At that data rate, one minute of voice results in a data transfer of 8000 bps × 60 sec/min/8 bit/byte, or 60,000 bytes. Thus, a full hour of voice activity (ignoring the fact that some schemes do not transmit data during periods of silence) would result in the transmission of 3.6 Mbytes of data. If the vendor charges $.10 per Mbyte, the transmission of each hour of digitized voice over frame relay network would add $.36 to your organization's monthly bill and be equivalent to running voice at a cost of $.36/60 seconds, or $.006 per minute, not including the cost associated with establishing a PVC and associated CIR for the PVC that is used to transport voice. Since an 8-Kbps PVC might cost $10 per month, you can easily see that the use of digitized voice over a common local access line also used for data can provide a voice transmission facility for under a penny a minute. Thus, this little exercise illustrates that voice transported via frame relay can result in long-distance calls costing a penny a minute for the use of the network. Now that we understand the basic cost components of using a public frame relay network, let's look at an example of pricing so we can appreciate the cost of adding a PVC and its associated CIR to support digitized voice over an existing frame relay network connection. We will defer until Chapter 9 a more detailed economic analysis. In that chapter, we will turn our attention to voice over frame relay.

Pricing Example

To illustrate the potential cost of adding voice to an existing frame relay network connection requires a series of reasonable assumptions to be made. The first assumption is that the organization is using a T1 line operating at 1.544 Mbps for its connection to the frame relay network provider, and that sufficient bandwidth is available to add a PVC with an 8-Kbps CIR to the T1 line to support one voice conversation between two locations. Since the access line and switch port remain the same, there is no additional cost associated with those two frame relay pricing components. Thus, the additional cost associated with transporting voice will depend on the vendor pricing for a PVC at a given CIR. Based on a series of telephone calls made during the fall of 2000 to several frame relay network operators, this author determined that the average monthly cost of a 16-Kbps PVC/CIR is $10. Since two locations must set up PVCs to communicate with one another, this would result in an additional monthly cost of $20 to the frame relay network operator to support one voice conversation between two locations, assuming sufficient bandwidth was available on each access line. If a provider does not add a PVC usage component to its price, this is probably a bargain, as it allows you to make as many calls as you wish for one low monthly fee—regardless of usage. If you connect a PBX at each location to a FRAD or router and encourage personnel to use this new interoffice communications method, you might be able to replace several hours of long-distance PSTN traffic each day with the use of frame relay. Assuming just two hours per day can be moved onto a frame relay network that costs $.15 per minute on the PSTN, your savings can rapidly build up. For example, 120 minutes per day at $.15 per minute would eliminate $18 per day of public telephone usage. On a monthly basis, assuming 22 working days per month, this would eliminate $396 ($18/day × 22 working days) of toll charges in exchange for paying your frame relay network operator an additional $20 per month. Although this example does not consider the cost of equipment necessary to obtain the capability to transmit voice over frame relay, by comparing PSTN and frame relay network charges, it clearly shows that you can obtain a reasonable level of savings by moving toll calls onto an existing frame relay network. In fact, when you consider the cost of international calls, the savings become even more pronounced. In Chapter 9, we'll examine voice over frame relay in detail to *include* the cost associated with equipment that permits this capability, and we'll also discuss its use as a mechanism to interconnect national and international organizational locations for both voice and data transmission.

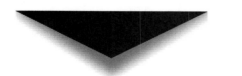

CHAPTER 5

Understanding Voice

You must first understand voice to select appropriate equipment for transmitting voice over data networks by weighing the advantages and disadvantages of different transmission methods. You'll need a knowledge of the basic properties of speech, including its production, how we comprehend it, and different techniques used to digitize a voice conversation—all topics covered in this chapter.

First, we'll focus on the basic properties of speech to learn how sounds are produced. Next, we will examine three distinct coding categories used to digitize voice and the different techniques within each category. From information presented in this chapter, we will understand the advantages and disadvantages associated with different voice-digitization techniques, which will in turn enable us to understand the advantages and disadvantages of using different products, or options supported by different products, to transmit voice over data networks.

One of the key parameters we will discuss at appropriate points in this chapter is the *latency,* or *delay* inherent in different voice coding methods. If this book was titled *Wall Street Voice* and this author was the mythical character Gordon Gecko, we would not be able to say, "Greed is good." This is because there are certain trade-offs between the operating rates of different voice-digitization techniques and the delay in encoding voice using those techniques. Unfortunately, as we get greedier and attempt to use a lower bit rate voice coder, the delay or latency associated with the voice coding technique normally increases. Thus, the key trade-off we must consider is between the operating rate of different voice coding methods and their voice coding delay. If we remember the so-called Rosetta stone introduced in Chapter 1, we have a maximum delay of approximately 250 milliseconds before voice reconstruction sounds awkward—perhaps intolerably so. Thus, to Gordon Gecko of *Wall Street Voice,* we can say, "Greed may be bad for the health of your voice-over-data network application."

BASIC PROPERTIES OF SPEECH

Human speech is produced when air is forced from our lungs through our vocal cords and along our vocal tract. The vocal cords are formed from two pairs of folds of mucous membrane that project into the cavity of the larynx, while the vocal tract extends from the opening in the vocal cords, referred to as the *glottis,* to the mouth. In an adult male the length of the vocal tract is approximately 17 cm, with its cross-sectional area varying from zero to about 20 cm^2. As we speak our lungs function as an air supply. As air is forced from our lungs through our vocal cords and along the vocal tract, the vocal folds open and close, resulting in sound produced by the reverberation of the folds. This reverberation, which is more formally referred to as *sound resonances, or formants*, concentrates audio energy that represents the frequency spectrum of a speech sound. As the shape of the vocal tract is varied—by moving your tongue, for example—the frequencies of the formants are controlled. The resulting sound heard by another person as you talk is also controlled by factors such as the size of the opening of your mouth, your lips, and even the size and spacing of your teeth. Figure 5-1 illustrates the human vocal tract and several parts of the body that affect the production of speech.

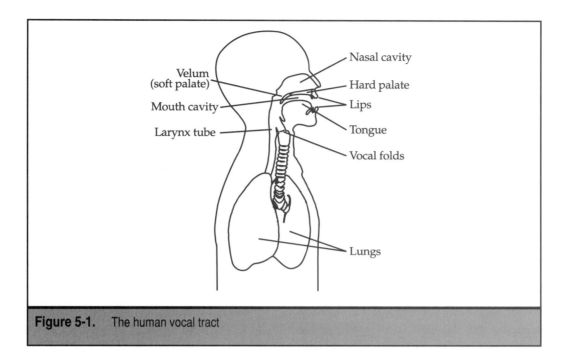

Figure 5-1.   The human vocal tract

In examining Figure 5-1, note that the nasal cavity is an auxiliary path for the creation of sound. The nasal cavity, which begins at the velum or soft palate at the root of the mouth, separates the oral cavity from the nasal cavity. When the velum is lowered, the nasal tract becomes acoustically coupled with the rest of the vocal tract and changes the nature of the sound we produce.

Classes of Speech

A majority of normal speech sounds can be categorized into one of three classes based upon their mode of excitation. Those three classes of speech include voiced sounds, unvoiced sounds, and plosive sounds.

Voiced Sounds

Voiced sounds are produced when our vocal cords vibrate as a result of our lungs generating sufficient pressure to open our vocal folds. As air flows from our lungs, our vocal folds vibrate, with the frequency of vibration based on the length of our folds and their tension. For most persons, the vibration frequency of their folds is within the 50- to 400-Hz range and is referred to as the *pitch frequency* component of voice. Voiced sounds have a high degree of recurrence at regular intervals throughout the pitch period, which is commonly between 2 and 20 ms, while the amplitude of the sound attenuates during that time period.

As you might surmise based upon conversations with many persons over the years, women and children have a higher average pitch frequency than men. The reason for this can be traced to the length of their vocal folds, which are shorter than that of a typical adult male.

Unvoiced Sounds

Unvoiced sounds refer to the period of time when our vocal folds are normally open, allowing air to pass from our lungs freely into the rest of our vocal tract. During this time, the frequency spectrum of the unvoiced sounds is relatively flat, although the resulting speech signal will include a spectral structure due to the pitch frequency and its harmonics. The unvoiced flat spectrum is passed through the remainder of the vocal tract, which can be considered to represent a spectral shaping filter. As a result of the flow of unvoiced speech through the vocal tract the filtering effect generates a frequency response on the incoming signal. That response varies based on the size and shape of one's vocal tract. Examples of unvoiced sounds include *s*, *f*, and *sh*, generated by constricting the vocal tract by slightly closing our lips.

Plosive Sounds

A third classification of sound is represented by plosive sounds. Plosive sounds result from the complete closure of our vocal tract, resulting in air pressure becoming extremely high behind the closure. Once the vocal tract opens, the result is a sound that contains a high degree of low-frequency energy. Examples of plosive sounds are *p* and *b*, which are formed by the closure and sudden release of the vocal tract.

In addition to voiced, unvoiced, and plosive sounds, there are some sounds that do not fall into a distinct class and are better categorized as a mixture of classes. One example results from the occurrence of a vocal cord vibration and a constriction in the vocal tract, causing speech produced with friction, referred to as a *fricative consonant*.

Figure 5-2 illustrates the repetitive or periodic nature of a voiced sound over a small increment of time, which typically ranges between 2 and 20 ms. Figure 5-3 shows an example of the power of a voiced sound, which, after resonating between nearly similar limits, alternates at its higher-frequency components. If we were to chart an unvoiced sound, we would note a different set of characteristics, with a flatter amplitude-versus-time plot and a higher power level across the frequency spectrum up to approximately 2,500 Hz, after which we usually encounter a more significant drop in power than is associated with voiced sounds.

Properties of Speech

As we probably note from our daily encounters with other persons, very rarely do two persons sound exactly alike. The reason for this results from differences in our lung capacity, our vocal folds, vocal cord length, lips, and even gaps between our teeth. As we

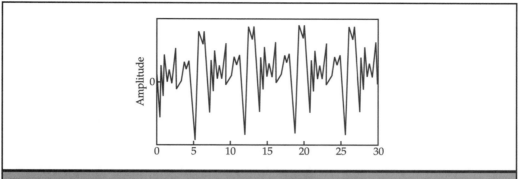

Figure 5-2. The repetitive or periodic nature of a voiced sound is observed by plotting its amplitude over time.

speak, our voiced, unvoiced, and plosive sounds merge together, resulting in a bandwidth of approximately 4 KHz.

In actuality, the frequency range of humans varies from approximately 200 Hz to about 20,000 Hz, with most conversations limited to a range between 300 Hz and 3,000 to 4,000 Hz. Only when we speak very softly or attempt to hit the high notes from the "Les Miserables" song "I Dream a Dream" do we go beyond those boundaries. Due to this, telephone companies filter the components of speech below 300 Hz and above 3300 Hz, resulting in an analog voice line having a 3,000 Hz passband.

The voiced sounds that are generated when our vocal tract acts as a resonant cavity commonly resonate at frequencies centered on 500 Hz and its odd harmonics. As a result of the resonance large peaks are produced in the resulting speech spectrum. Those peaks as we previously noted are referred to as formants and can be visually identified by the

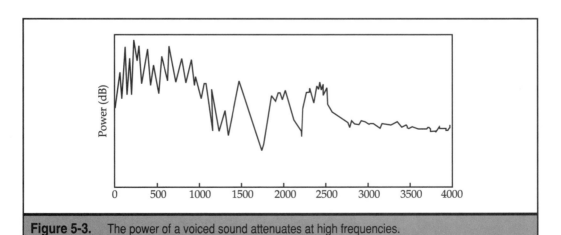

Figure 5-3. The power of a voiced sound attenuates at high frequencies.

four high spikes in Figure 5-2. The formants of the voiced sound contain a large majority of the information in a signal, which enables, as we will shortly note, the vocal tract to be modeled. While unvoiced sound has no formants the energy in the signal is much lower than that of the voiced components and can also be modeled. Thus, let's turn our attention to this topic.

Speech Modeling

When engineers examined the composition of human speech, they made two observations that resulted in different methods being used to digitize and transmit voice. First, they noted that if the composition of human speech could be sampled at appropriate intervals, it could be digitized and reconstructed into a sound that would be difficult to distinguish from the original. All that was required was the ability to take an appropriate number of samples and encode the value of each sample so its reconstructed amplitude did not significantly differ from its original amplitude. This type of encoding is referred to as *waveform encoding.*

A second aspect of speech properties noted by engineers was that in terms of milliseconds, sound did not significantly vary. This means that, to a degree, sound is sort of predictable and can be synthesized. Thus, a second method of digital encoding could be based upon modeling speech, a technique referred to as *vocoding.* In vocoding, the synthesis that reconstructs voice results from analyzing speech and determining its key parameters, such as its energy, tone, pitch, and other characteristics. Thus, the synthesis of speech first requires an analysis of a waveform to determine its parameters. Because the parameters of speech are transmitted instead of an encoded waveform, the operating rate of an analysis-synthesis vocoding technique is generally lower than that of a waveform encoding technique.

As you might expect, a third method was developed based on waveform *and* vocoding—a hybrid technique. In a hybrid voice coding technique, a small portion of a waveform is analyzed and key speech parameters are extracted. However, instead of simply passing the parameters for transmission, the hybrid coder uses the parameters to synthesize the speech sample and compare it to the original waveform. Then the coder uses the difference between the actual sample and the synthesized sample to adjust the parameters.

Prior to the development of low-cost high-processing-capability digital signal processors (DSPs), the ability to perform effective hybrid coding was more a figment of one's imagination than a reality. However, the development of low-cost high-processing-capability DSPs resulted in the potential use of a family of Code Excited Linear Predictor voice coding techniques commonly referred to as the CELP (pronounced *kelp*) family of voice coders. Now that we have an appreciation for the basic properties of speech, let's look at speech-coding methods. However, prior to doing so a slight digression concerning how different speech coding methods are scored is in order as it will assist us in evaluating different coding methods.

Mean Opinion Score

One of the more important characteristics of a voice digitization technique is the clarity of the reconstructed signal. The Mean Opinion Score (MOS) represents a voice quality assessment made by gathering opinions from a group of persons located in a sterile test environment.

MOS scoring is based upon a five-point scale whose values and value assignments are listed in Table 5-1. It is important to note that MOS scores are subjective and the background and cultural heritage of the persons placed in the sterile test environment as well as the messages they listen to play an important role in determining the average MOS score.

As indicated in Table 5-1, a higher MOS score is better than a lower MOS score. Based upon the work of Dr. Grace Rudkin, which was published in the *BT Journal* in April 1997, 64 Kbps PCM received a 4.3 MOS.

APPLICATION NOTE: The Mean Opinion Score represents a subjective opinion of the quality of reconstructed speech based upon a five-point scale. The higher the MOS score associated with a voice coding scheme, the higher the quality of reconstructed speech. Thus, you should consider the MOS of speech coders in your selection process.

There are two ITU standards that relate to MOS and deserve mention—P.800 and P.861. The P.800 standard defines a method to derive a MOS via recording several preselected voice samples over different transmission mediums and playing them back to a mixed group of persons under controlled conditions. In comparison, the P.861 standard, which is titled Perceptual Speech Quality Measurement, automates the process by defining an algorithm through which a computer can derive scores that have a close correlation to MOS scores.

Score	Meaning
1	Bad
2	Poor
3	Fair
4	Good
5	Excellent

Table 5-1.   Mean Opinion Score Scale

WAVEFORM CODING

Waveform coding is a process whereby an analog signal is digitized without requiring any knowledge of how the signal was produced. In this section, we will examine several waveform coding techniques, beginning with *pulse code modulation* (PCM) for three reasons: First, PCM is nearly universally the method of voice digitization used by communications carriers for transmission on the worldwide public switched telephone network. Second, an explanation of the manner in which PCM operates will provide a foundation for comparing its efficiency and reconstructed voice clarity with those of other waveform coding methods. Third, and for business applications of key importance, PCM is referred to as *toll-quality voice*. Due to this, its clarity, delay, and bandwidth utilization are important characteristics against which other voice coding methods are commonly compared.

Pulse Code Modulation

Pulse code modulation (PCM) is a waveform coding technique based on a three-step process: sampling, quantization, and coding. Thus, an understanding of the operation of PCM requires an understanding of each step in the PCM process.

Sampling

Under PCM, an analog signal is sampled 8,000 times per second, or once every 125 ms. The selection of the sampling rate is based on the Nyquist theorem, which requires the number of sample points to be at least equal to twice the maximum frequency of the signal for the signal to be faithfully reconstructed.

Although a standard voice channel is filtered to produce a passband of frequency from 300 Hz to approximately 3,300 Hz or a bandwidth of 3,000 Hz, in actuality the filters do not work instantaneously and allow some lower-power speech to pass below 300 Hz and beyond 3,300 Hz, as illustrated in Figure 5-4. As a result, the passband can extend to near 4,000 Hz, resulting in the selection of a sampling rate of 8,000 samples per second.

The sampling process results in a series of amplitude segments that form a pulse amplitude modulation (PAM) wave. Figure 5-5 illustrates the creation of a PAM wave resulting from the sampling of a single sine wave. In actuality, communications carriers use equipment to sample a group of voice channels at one time. In North America, the most common system samples 24 voice channels in sequence, while in Europe an equivalent system samples 32 channels.

In examining Figure 5-5, note that the PAM signal represents a series of samples that can have an infinite number of voltages. The reason those voltage samples can be infinite is because the samples represent analog heights and not discrete digitally encoded signal values. Thus, the second step in the PCM process, called *quantization*, reduces the PAM signal to a limited number of discrete amplitude values.

Quantization

The second step in the PCM process requires the coding of each PAM sample. The coding process requires a conversion of each discrete-time continuous-valued PAM sample into a discrete-time, discrete-valued sample. This conversion is referred to as *quantization*. The

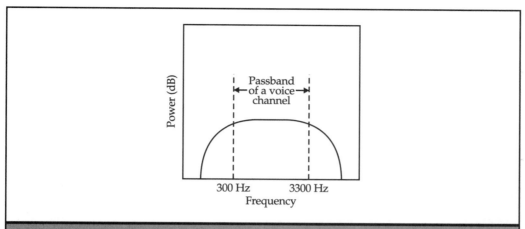

Figure 5-4. Construction of a voice channel. Through the use of filters, a passband between 300 and 3,300 Hz is established for use as a voice channel.

quantization process requires the value of each signal sample to be represented by a value selected from a finite set of possible coding values. The difference in height between the analog PAM sample and its discrete-value coded sample is referred to as the quantization noise. Thus, one objective of a good coding scheme is to minimize quantization noise.

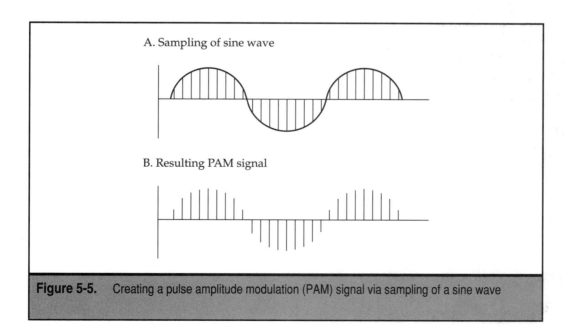

Figure 5-5. Creating a pulse amplitude modulation (PAM) signal via sampling of a sine wave

When PCM was being developed, it was recognized that it was important to limit the number of coding values for each sample while maintaining the ability to faithfully reproduce the sample. This need to limit the number of coding values for a sample was necessary to limit the transmission rate of an encoded signal. For example, a four-level binary code would provide 2^4, or 16, possible values, requiring a transmission rate of 8,000 samples per second × 4 bits/sample or 32,000 bps. Similarly, an eight-level binary code would provide 2^8, or 256, possible values and would require a transmission rate of 8,000 samples per second × 8 bits/sample, or 64,000 bps.

When examining the range of intensities of voice over an analog telephone channel, its approximately 60-dB power range would require 12 bits per sample if linear quantization was used. *Linear quantization*, which is also referred to as *uniform quantization*, results in a technique in which there is an equal distance between coding elements. For example, encoding an analog signal using a uniform or linear quantization method based upon 3 bits would result in eight levels (0 to 7) with each level having an equal distance from an adjacent coding level or element. While the use of linear quantization is both simple and economical to implement, to provide a sufficient level of clarity for reconstructed voice would require 4,096 evenly spaced levels provided by the use of 12 bits. This would result in a bit rate of 96,000 bps, which, while feasible, represents an excess amount of bandwidth that could be reduced by using nonlinear quantization. A second method that can be used to reduce the number of bits required per sample is to compress or compand the signal prior to quantization, followed by uniform quantization.

Nonuniform Quantization

Nonuniform quantization is based on the fact that there is a higher probability of occurrence of lower-intensity signals than of higher-power signals. Larger quantum steps are used for encoding larger amplitude portions of a signal. Conversely, finer steps are used to encode signals that have a lower amplitude. Figure 5-6 illustrates the assignment of larger steps for higher-intensity signals and finer steps for lower-intensity signals.

Companding

Most PCM systems today use companding followed by uniform quantization to reduce the number of bits necessary to encode each PCM sample to 8. To do so, the *compandor* (a term derived from *com*pressor-ex*pandor*) raises the power of weak signals so they can be transmitted above the noise and crosstalk level associated with a typical communications channel while attenuating very high signals to minimize the possibility of crosstalk affecting other communications channels. If you think about how humans converse it becomes obvious that companding has a minimal effect upon reconstructed speech. This is because a conversation is difficult to hear when the speaker is in effect operating at "low human power" and most persons never shout at "the top of their lungs" during a telephone conversation.

Figure 5-7 illustrates the typical operation of a compandor. The compression portion of the compandor accepts a 60-dB-input power range, typically between 20 and 80 dB.

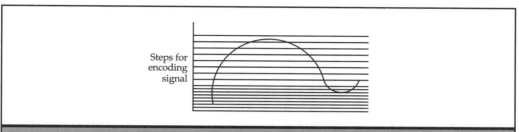

Figure 5-6. Nonuniform quantization results in the assignment of larger steps to higher-intensity signals and finer steps for lower-intensity signals.

The compressor raises the weakest sounds in power from 20 to 40 dB, while the strongest sounds are decreased in power from 80 to 70 dB, minimizing the power range to 30 dB.

The expandor portion of the compandor works in an opposite manner. That is, it expands the reduced power range back to its original form. To do so, the weakest sounds are decreased in power from 40 to 20 dB, while the strongest sounds are increased in power from 70 dB up to 80 dB, resulting in a 60-dB-output power range that is the same as the input power range.

The compression and expansion functions of the receiving compandor are logarithmic and follow one of two laws: the A-law, which is employed primarily in Europe, and the μ-law, used in North America. The curve for the A-law can be plotted from the following formula, where A is set to a value of 87.56.

$$Y = \frac{Ax}{(1 + \log A)} \quad 0 \le \frac{V}{A}$$

$$Y = \frac{1 + \log(Ax)}{(1 + \log A)} \frac{V}{A} \le v \le V$$

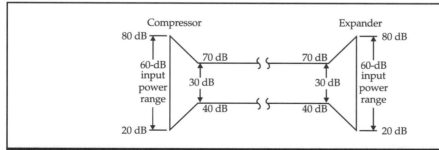

Figure 5-7. A compandor compresses a transmitted voice signal to reduce its power range and expands a received signal to reconstruct its original power range.

where v represents the instantaneous input voltage and V represents the maximum input voltage. The curve for the μ-law can be plotted from the following formula:

$$\mu = \frac{\log(1+\mu x)}{\log(1+\mu)}$$

where μ has a value of 255. In the preceding formulas x has the value v/V and varies between –1 and 1, Y has the value i/B, with i representing the number of quantization steps commencing from the center of the range and B representing the number of quantization steps on each side of the center of the range.

Figure 5-8 illustrates a general plot of a μ-law compandor, showing the compressed input resulting from a range of input signals. This curve represents a value of 100 being used for μ, which was the value used for the original North American μ-law encoding system. Later, that value was changed to 255.

Both the A- and μ-laws define the number of quantizing levels used to describe a sample and how those levels are arranged. Under the μ-law, the quantization scale is divided into 255 discrete units of two different sizes, called *chords* and *steps*. Chords are spaced logarithmically, with each succeeding chord larger than the preceding one. Within each chord are 16 steps spaced linearly. Thus, steps are larger in larger chords. Figure 5-9 illustrates an example of the spacing of two chords and the steps within each chord. In actuality, the μ-law uses 16 chords, 8 for the positive portion of a signal and 8 for the negative portion of a signal. Since there are 16 steps in each chord and the zero level is shared, the number of levels used becomes $16 \times 16 - 1$, or 255.

Coding

Instead of directly encoding a step representing the power level of a sample, each PCM word is segmented into three parts: a polarity bit, 3 bits for the chord value, and 4 bits to

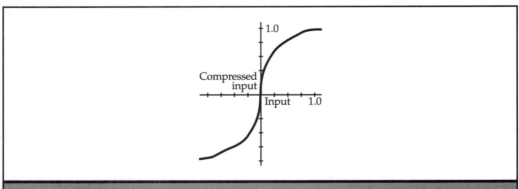

Figure 5-8. The logarithmic curve for the μ-law

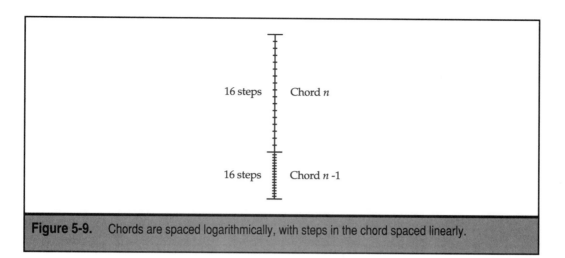

Figure 5-9.   Chords are spaced logarithmically, with steps in the chord spaced linearly.

represent one of the 16 possible steps within a chord. The format of the PCM word is as follows:

| P | C | C | C | S | S | S | S |

where P is the polarity bit and indicates whether the sample is above or below the origin, C represents the chord value bits, and S represents the value of a step within the indicated chord.

In comparison to the use of 16 chords under the μ-law, the A-law uses 13 segments. Six are used to represent the positive portion of a signal, while a zero chord results in the 13th segment. For both the A-law and the μ-law, coding results in each sample being encoded in 8 bits. Since there are 8,000 samples per second, the encoding process results in a digital data stream of 64 Kbps transporting one PCM analog voice conversation.

APPLICATION NOTE: Although PCM has the highest bandwidth of all voice coding techniques, its encoding latency is practically negligible, with an encoding delay under 1 microsecond, which for comparison purposes is several thousand times lower than low-delay hybrid coders.

PCM Multiplexing

As mentioned earlier in this section, PCM is commonly performed by communications carriers through the use of equipment at their central offices that encodes groups of either 24 or 30 voice conversations. The encoding of groups of voice conversations dates to the 1960s, when PCM systems were installed to relieve cable congestion in urban areas. At that time, communications carriers installed equipment referred to as *channel banks* in

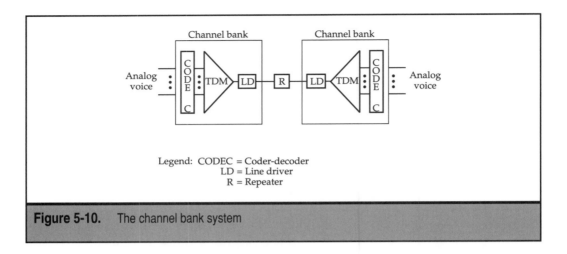

Figure 5-10.   The channel bank system

their central offices. Each channel bank consisted of a codec, a time division multiplexer (TDM), and a line drive as illustrated in Figure 5-10.

The codec (a term derived from *co*der-*dec*oder) accepts a group of 24 analog voice signals in North America and samples each of those signals 8,000 times per second, producing a series of PAM signals that are quantized and coded into 8-bit bytes. The TDM combines the digital bit stream from each of the 24 data sources into one high-speed serial bit stream, adding a framing bit to every sequence of 24 8-bit groups. This framing bit, which occurs 8,000 times per second, provides for the synchronization of transmission and results in each multiplexing frame consisting of 24 × 8 + 1, or 193 bits. As frames are transmitted 8,000 times per second, the multiplexer produces a serial bit rate of 193 bits/frame × 8,000 frames/second, or 1.544 Mbps, which results in the operating rate of a T1 circuit. In fact, the channel bank system illustrated in Figure 5-10 can be considered a forerunner of the T-carrier transmission system.

The line driver in the channel bank converts the electrical characteristics of the serial bit stream for transmission on the digital transmission facility. This device converts unipolar signals into bipolar signals, and its modern equivalent is the *channel service unit/data service unit* (CSU/DSU). The repeater shown in Figure 5-10 actually represents a series of devices installed approximately 6,000 feet from one another on span lines consisting of copper cables. Each repeater examines the digital pulses flowing on the span line linking channel banks and regenerates a new pulse, removing any prior distortion to the pulse. In comparison, amplifiers used with analog transmission systems boost the strength of the analog signal to include increasing any prior distortion of the signal. Thus, in addition to relieving cable congestion, digital channel bank systems also provide a higher level of signal quality than analog transmission. When T1 circuits became available for transmitting data, users were able to obtain a lower error rate than with analog transmissions.

PCM System Operation

PCM was standardized by the International Telecommunications Union (ITU) as Recommendation G.711 and represents by far the most commonly used method of waveform encoding on a worldwide basis. Both North American μ-law companding and European A-law companding result in a high quality of reproduced speech, and the differences between the two methods are so slight that a call between two locations results in a reconstructed signal that, to the human ear, is essentially indistinguishable from the original. Although used worldwide, the technology was developed during the 1960s, and improvements in voice encoding methods resulted in other techniques that provide what is referred to as *near-toll quality* reproduced voice at far lower data rates per digitized conversation.

Since communications carriers invested hundreds of billions of dollars over the years in PCM-based technology, it is reasonable to expect the PSTN to continue to use that technology for the foreseeable future. In fact, the popular .au audio files that are often used to transport sound over the World Wide Web are PCM files. However, there are expensive international communications facilities as well as private networks that can more practically employ other coding techniques, which has resulted in the development of both standardized and non-standardized methods of lower-bit-rate voice encoding methods. One of those standardized methods is ADPCM, which we will now examine.

Adaptive Differential PCM

As discussed in the first section of this chapter, speech has a degree of repeating waveforms due to the vibration of the vocal cords. This degree of correlation between speech samples makes it possible to design a waveform coding scheme based on the prediction of samples if the error between the predicted samples and actual speech samples have a lower variance than the original speech samples. If so, the difference between the actual sample and the predicted sample could be quantized using fewer bits than the original speech sample. This technique forms the basis for a series of differential PCM methods, including *adaptive differential PCM* (ADPCM), in which the predictor and quantizer adaptively adjust to the characteristics of speech being coded. ADPCM was standardized by the ITU in the mid-1980s as Recommendation G.721.

Operation

ADPCM uses a sampling rate of 8,000 samples per second, which is the same as that used in PCM. Instead of quantizing the actual signal sample, ADPCM uses a transcoder that includes an adaptive predictor and compressor, with the compressor subtracting the predicted value from the actual value of the sample and encoding the difference as a 4-bit word. At the receiver, another transcoder provides a reverse operation, using summation circuitry and the value generated by a predictor to subtract the predicted value from the received 4-bit difference, regenerating the actual value of the sample. In the event that successive samples vary widely, the predictive algorithm adapts by increasing the range represented by the 4 bits. However, the adaptation process reduces somewhat the accuracy of voice-frequency reproduction. Figure 5-11 illustrates the basic operation of an ADPCM system.

Through the use of ADPCM, the data rate of a digitized voice conversation becomes:

8,000 samples/second × 4 bits/sample = 32,000 bps

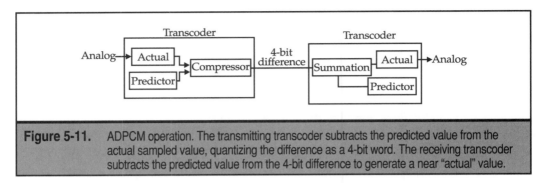

Figure 5-11. ADPCM operation. The transmitting transcoder subtracts the predicted value from the actual sampled value, quantizing the difference as a 4-bit word. The receiving transcoder subtracts the predicted value from the 4-bit difference to generate a near "actual" value.

Thus, the use of ADPCM results in a transmission rate that's one-half the PCM rate of 64 Kbps.

The first standardized operating rate of ADPCM was the 32-Kbps rate just described. Later, the development of additional predictive techniques resulted in operating rates of 16, 24, and 40 Kbps being standardized.

Implementation

The most popular implementation of ADPCM is by communications carriers on long-distance international circuits at a data rate of 32 Kbps. Since ADPCM is based on the transmission of the difference between the actual sample and the predicted sample, it is not suitable for use by high-speed modems. This explains why an international modem connection often results in transmission occurring at a modem's lower fallback operating rate.

Using ADPCM is also popular with organizations that have established their own internal voice communications networks. Through the use of ADPCM adapter cards in T1 multiplexers, they can double the voice-carrying capability of T1 circuits between organizational locations when using ADPCM at 32 Kbps.

ADPCM is a popular option supported by many voice-over-IP and voice-over-frame-relay products. This is because ADPCM can operate at as low as one-quarter the rate of PCM while providing toll-quality reconstructed voice. Similar to PCM, ADPCM is considered to represent toll-quality speech. According to the previously referenced article by Dr. Grace Rudkin that appeared in the April 1997 issue of *BT Journal*, 32 Kbps ADPCM received a MOS of 4.0. It should also be noted that there are several versions of ADPCM. Although the most popular version operates at 32 Kbps there are also versions of ADPCM that operate at 16 Kbps, 24 Kbps, and 48 Kbps.

APPLICATION NOTE: The predictor and summation operations performed by ADPCM increase its latency over PCM by approximately 25 percent. However, this results in only a 125 ms delay, which is relatively insignificant when compared to the delay associated with many hybrid coding techniques.

Continuously Variable Slope Delta Modulation

In the continuously variable slope delta (CVSD) modulation technique, the analog input voltage is compared to a reference voltage. If the input is greater than the reference, a binary 1 is encoded, while a binary 0 is encoded if the input voltage is less than the reference level. This method of voice encoding permits a 1-bit data word to be used to represent the digitized voice signal.

Operation

The use of CVSD is based on a succession of 0 or 1 bit being used to indicate the change in the slope of the analog curve representing a voice conversation. Thus, the key to the fidelity of the reproduced signal is a sampling rate fast enough so that the sequence of 1-bit words can faithfully reproduce the analog signal.

Most CVSD systems sample the input at 32,000 or 16,000 times per second, resulting in a bit rate of 32 or 16 Kbps representing a digitized voice signal. Another popular CVSD rate is 24 Kbps, which results from a sampling rate of 24,000 times per second.

Utilization

The original use of CVSD dates to military systems, as the generation of a digital data stream was much easier to encrypt than using filters to move segments of frequencies of an analog signal. Today, CVSD is primarily offered as an option on some T1 multiplexer voice-compression modules that can be set for sampling rates of 8, 16, 24, or 32 thousand times per second. Although a sampling rate of 8,000 times per second enables 192 voice channels to be carried on a T1 circuit, the quality of voice is so poor that this level of compression is rarely used. Instead, organizations using CVSD on their private internal networks commonly use operating rates of 24,000 or 32,000 bps for their voice channels. Similar to ADPCM, the use of CVSD presents a barrier to the use of high-speed modems. In addition, ADPCM commonly delivers a higher voice quality of speech than CVSD, which will probably result in the gradual elimination of the use of this waveform encoding method. Because of its lack of popularity, MOS scores for CVSD are conspicuous by their absence.

Digital Speech Interpolation

Digital Speech Interpolation (DSI) is a technique that enhances the ability of a trunk circuit to transport additional voice conversations instead of using a voice-compression method. Although technically it does not belong in a section covering waveform coding methods, it is frequently used by equipment supporting different waveform coding methods and thus deserves coverage.

DSI recognizes the fact that human speech is half-duplex—unless we are rude and talk at the same time as the other party. In addition, because humans periodically pause when speaking, it becomes possible to take advantage of periods of silence in one conversation by placing a portion of another conversation in a trunk slot routed between two multiplexers or similar equipment. For example, a DSI system might be configured to support 36 PCM voice conversations over a T1 circuit that has 24 time slots and normally is limited to supporting 24 voice conversations. While DSI configured for a 36:24 ratio normally works very well due to the half-duplex nature of voice and the periods of silence in conversations, suppose 25 persons begin to talk. Because there are only 24 time slots, one sample must be dropped. This dropping occurs randomly in small segments over all 24 slots and is referred to as *clipping*. Although DSI properly configured at a ratio of 3:2 or less is normally imperceptible to the human ear, it will cause havoc to modem transmission. Thus, DSI and modem transmission are mutually exclusive.

VOCODING

As discussed in the first section of this chapter, a second method used to digitally encode speech is based on the characteristics of the human voice. This method is called *vocoding*, which is an acronym for *voice coding*, and is based on the modeling of speech production.

Operation

Vocoding is based on the assumption that speech is produced by exciting a linear system through a series of periodic pulses for voiced sounds or random noise for unvoiced sounds. Figure 5-12 illustrates the general speech production model employed by vocoders.

The general vocoder speech production model illustrated in Figure 5-12 assumes that when voiced sound occurs, the distance between the periodic series of impulses represents the pitch period. The model also assumes that unvoiced speech resulting from the air pressure of our lungs blowing through a constriction in our vocal tract can be modeled via a sequence of random noise.

The vocoding process results in the transmitter analyzing the various properties of speech to select appropriate parameters and the excitation or power level that best corresponds to the source voice input. Instead of digitizing the analog signal, the vocoder digitizes the voice model parameters and excitation level, which are then transmitted to the receiving vocoder. At the receiving vocoder, the model parameters and excitation level are used to synthesize speech.

The primary advantage associated with the use of vocoders is their ability to produce intelligent speech at very low bit rates. Depending on the model used, the resulting synthesized sounds can seem unnatural and awkward. The reason that vocoders generate less than toll-quality voice results from the fact that the human ear is very sensitive to pitch and human vocal tract characteristics vary widely, which makes modeling accuracy difficult. Another problem associated with the use of vocoders results from the fact that the speech

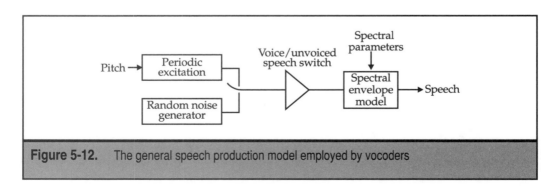

Figure 5-12. The general speech production model employed by vocoders

modeling used assumes either voiced or unvoiced sound and does not include any intermediate states. In spite of these problems, there are several popular types of vocoders, and their development has resulted in the knowledge used to build hybrid encoding devices that produce a more natural sound while maintaining a relatively low bit rate.

Types of Vocoders

Over the past 50 years, a number of different types of vocoders have been developed. Two of the more popular types are homomorphic vocoders and linear predictive vocoders.

Homomorphic Vocoders

Homomorphic vocoders are based on the use of nonlinear signal processing techniques that provide a better model of human speech. A homomorphic vocoder assumes that speech represents a convolution in time of the vocal tract's impulse response and excitation functions. Since the impulse response of the human vocal tract varies slowly in comparison to the variation of the excitation level, it becomes possible to separate the two components via the use of a low-pass filter. This enables a more accurate modeling of pitch information, which is transmitted along with certain coefficients of the signal generated by applying an inverse Fourier transformation of the frequency components of the signal. The resulting coefficients, which describe the vocal tract along with a more accurate representation of pitch information, can be transmitted at a data rate of approximately 4 Kbps.

Linear Predictive Vocoder

The linear predictive vocoder is the most popular type of vocoder and one of the most useful methods for encoding good-quality speech at a very low bit rate. Although it uses the same speech production model as other vocoders, a linear predictive vocoder employs a different model of the vocal tract. That model assumes that each voice sample represents a linear combination of previous samples, enabling an infinite impulse response filter to be used to model speech. The linear predictive vocoder computes the coefficients of the filter to minimize the error between the prediction and the sample.

Linear predictive vocoders operate on blocks of speech of approximately 20-ms duration. Each block is stored and analyzed so the vocoder can determine the appropriate set of predictor coefficients that minimize the error between the sample and the predictor. Once the coefficients are determined, they are quantized and transmitted to the receiving vocoder. That vocoder uses the received coefficients to generate 20 ms of predicted speech, which is then passed through an inverse of the vocal tract filter, allowing the prediction error to be generated. This technique enables the predictor to remove the correlation between adjacent voice samples and allows the pitch period of speech to be more accurately reconstructed. The resulting effect results in a more natural-sounding synthesized voice.

The first generation of linear predictive vocoders was actually limited to generating synthesized human speech. During the 1970s, Texas Instruments marketed an "educational" calculator. Pressing a button on the calculator generated a word such as *dog* or *cat* through a miniature speaker. A child using the calculator would then enter the characters necessary to spell the pronounced word. A green light would indicate that the spelling was correct. Today, linear predictive vocoders can be used to provide a transmission mechanism for synthesized speech at bit rates as low as 2.4 to 4.8 Kbps, which results in a reasonable quality of reconstructed speech. Unfortunately, certain speech sounds are not very well reproduced, so certain portions of speech appear better than others.

Second-generation linear predictive coders are designed to assume that speech signals are generated by a buzzer at the end of a tube. The space between the vocal cords, known as the glottis, generates the buzz, which is characterized by its intensity or loudness and frequency or pitch. Our throat and mouth form the tube, which is characterized by the previously described voice formants or resonances. LPC analyzes speech by first taking 20-ms segments of speech and estimating the formants in the sample. Once this is accomplished the formants are removed and the algorithm estimates the intensity and frequency of the remaining buzz. The process of removing the formants from the voice sample is referred to as *inverse filtering*, while the remaining signal after the formants are removed is known as the *residue*.

One version of LPC, denoted as LPC-10, partitions speech into 54-bit frames. Each frame contains 41 bits for the reflection coefficients (formants), 7 bits for pitch and voiced/unvoiced parameters, and 5 bits for gain, while a single bit is used for synchronization. At a frame rate of 44.44 frames per second, this results in a data transmission rate of 44.44 fps × 54 bits/frame, or 2,400 bps. Another version of LPC, referred to as LPC-10e, is described in Federal Standard (FS) 1015.

One of the key problems associated with the use of LPC is to determine the formants from the speech signal. This is accomplished by using a difference equation, which expresses each sample of the signal as a linear combination of previous samples, with the equation referred to as a linear predictor. This also explains why this method of voice digitization is referred to as Linear Predictive Coding.

Because the coefficients of the difference equation define the formants, LPC needs to estimate those coefficients. To do so, an algorithm minimizes the mean-square error between the actual signal and the coefficients used to generate a predicted signal. While it is possible to obtain a reasonable estimate of the formants, the general assumption that speech represents a buzz at the end of a tube introduces other problems. For example, while our vocal tract can be reasonably well represented by a single tube when we produce ordinary vowels, for nasal sounds the single tube representation deteriorates. Similarly, when a human generates certain consonants via turbulent airflow from his or her lungs, the result is a combination of fricatives and consonants that produces a hissing sound rather than a buzz. This requires an LPC encoder to decide if the sound source is buzz or hiss for each speech sample. While LPC encoders are capable of estimating the frequency if the sound source is buzz and encoding information correctly for a decoder to operate upon, when a sample has a combination of hiss and buzz the encoding will not result in an accurate reproduction by an LPC decoder.

HYBRID CODING

In examining the operation of waveform coders, we noted that they provide a mechanism to reconstruct the signal based on samples taken from the original waveform. This technique, which is the key to the operation of PCM, ADPCM, and CVSDM, can produce very high quality reproduced speech at data rates ranging from PCM's 64 Kbps to ADPCM's 32 Kbps, and even at a CVSDM data rate of 24 Kbps. Unfortunately, none of these techniques are capable of being used to code speech at a very low data rate. In comparison, vocoders attempt to extract parameters of speech that can be used to synthesize its reproductions. Although vocoders can be used to transmit speech at very low bit rates, down to 2.4 Kbps, the resulting speech, while being highly intelligible, tends to have periods that sound very synthetic. In an attempt to fill the gap between waveform and vocoders, a series of hybrid coding methods were developed. Such methods combine portions of waveform and vocoding with the most successful hybrid coders using an analysis-by-synthesis method. Such coders use the same or a very similar linear prediction filter model of the vocal tract that is used in linear predictive coding-based vocoders.

The key difference between LPC vocoders and hybrid coders is in the method of modeling speech. LPC-based vocoders use a model that concentrates on voiced and unvoiced portions of speech. In doing so, the isolation of the predictable portion of speech requires the removal of other components, referred to as *speech residue,* containing information that can be important for the effective reconstruction of sound. Since the goal of LPC is to minimize the data rate of encoded speech, the transmission of the residue would defeat the purpose of the encoding technique. In an analysis-by-synthesis hybrid coder, the selection of an excitation signal compensates for the residue problem by attempting to match the analysis parameters via synthesis as closely as possible to the original speech waveform. As you might expect, the ability to perform this analysis-by-synthesis process is highly computation-intensive, and although some hybrid coders were introduced during the early 1980s, it wasn't until the availability of powerful digital signal processors (DSPs) at reasonable prices during the late 1980s and early 1990s that it became practical to implement different hybrid coding methods. Two of the more popular types of hybrid encoders are the Regular Pulse Excited coder and the Code Excited Linear Prediction coder, with the latter using a code book in its operation.

Regular Pulse Excited Coding

The Regular Pulse Excited (RPE) hybrid coding method forms the basis for speech encoding used by the Global System for Mobile Communications (GSM), a digital mobile radio system used extensively throughout Europe. Under the RPE process, input speech is subdivided into 20-ms frame segments. For each frame, a set of eight short-term predictor coefficients are computed, after which the frame is further subdivided into four 5-ms subframes. For each subframe, the encoder computes a delay and a gain for its long-term predictor. Next, the coder uses 40 residual signal samples (eight per subframe) and converts those samples into three groups of excitation samples, each 13 samples in length. The computation of excitation samples can be considered an interactive process, with the selected coefficients used by a synthesis filter to reconstruct a waveform that is matched against the original waveform. After the difference between waveforms is noted, the process is repeated using a new set of coefficients until the error between synthesized speech and the original waveform reaches a minimal level. Three samples that, when synthesized, produce a minimal difference between the original and synthesized waveform are then selected. The sequence that has the highest energy level is then selected as the best representation of the excitation sequence, and the amplitude of each pulse in the sequence is quantized using 3 bits. This technique results in the generation of a 13-Kbps data rate. At the decoder, the received signal is fed through long- and short-term synthesis filters to reconstruct speech. Finally, a postfilter is used to improve the perceived quality of the reconstructed sound.

The RPE coder provides a good quality of reconstructed speech and does not require a very high level of processing to implement. Based upon research by this author, various studies indicate MOS value ranging from 3.7 to 4.0 for RPE. While RPE represents a viable hybrid coding method, a higher quality of reproduced speech can be obtained by using several versions of a series of hybrid coders known as CELP hybrid coders. Thus, let's turn our attention to CELP.

Code Excited Linear Predictor Coders

The Code Excited Linear Predictor (CELP) speech coders were first proposed in 1985, and they represent a relatively recent addition to the various methods used to encode speech. CELP also represents a hybrid coding technique that employs both waveform and vocoding techniques, resulting in an analysis-by-synthesis process to code speech.

CELP differs from RPE in that the excitation signal is vector-quantized after speech is passed through a vocal tract and pitch predictor, and an index from a code book is used in place of the actual quantization of the excitation signal. The code book contains typical residue signals, with a different code assigned to each source frequency (pitch of the voice sample). Under this technique an analyzer compares the residue in a speech sample to the entries in the codebook and selects the closest match, sending the code for the entry instead of the actual residue. For CELP to work well requires a sufficient number of codebook entries to store all kinds of residues. However, if the codebook gets too large the time required to search the entries for a match will become excessive. In addition, as the number of entries in the codebook increases the resulting number of bits required to define a code value will increase, resulting in another limitation. One technique allows a code book of 1,024 entries to use a 10-bit index, which can substantially reduce the data rate required to transmit the excitation signal that best represents the original waveform. That is, an analysis-by-synthesis approach minimizes the difference between the original waveform sample and a synthesized waveform by using an index selected from the code book.

Another technique uses two small codebooks instead of a very large one. One codebook is fixed by the system designers and contains a sufficient number of codes to represent one pitch period of residue. The second codebook is adaptive. This codebook is initially empty and is filled in during a conversation with copies of the previous residue delayed by time.

As a result of using a code book index, CELP coders operate at data rates between 4.8 and 16 Kbps. The key difference between various CELP coders is in the adaptability of code book entries and the delay associated with selecting and transmitting a code book index that represents the excitation signal.

FS 1016

The U.S. Department of Defense standardized a version of CELP, FS 1016, that operates at 4.8 Kbps. This version of CELP uses a code book in which one portion is adaptively changed, while the other portion is fixed. The fixed portion of the code book contains entries that represent one pitch of speech residue, while the adaptive portion is filled in during operation with copies of the previous residue delayed by variable amounts of time.

Although the FS 1016 CELP method is widely used by the U.S. Department of Defense for secure telephone communications via the encryption of its digitized data stream, it introduces an end-to-end transmission delay that at times during a normal voice conversation may appear slightly awkward. This is caused by delays associated with processing a block of sampled voice and searching the code book for a suitable index, during which time the analyzed sample is synthesized until an entry that produces a minimum amount of error is selected. The delay associated with the FS 1016 CELP system can range up to approximately 100 ms.

The G.728 Recommendation

Recognizing the delay problem associated with previously developed versions of CELP, the ITU released a set of requirements for a 16-Kbps voice digitization method in 1988 to provide toll-quality speech comparable to the 32-Kbps ADPCM standardized under its G.721 Recommendation. The ITU requirements were satisfied by the development of a backward-adaptive CELP coder that was developed at AT&T Bell Laboratories and standardized in 1992 as Recommendation G.728.

The G.728 CELP Recommendation uses a backward-adaptive process to calculate short-term filter coefficients. This enables the computation of filter coefficients from past reconstructed speech instead of from 20 to 30 ms of buffered input and allows a much shorter frame length than that used by other CELP coding techniques. Under the G.728 Recommendation, a frame length of five samples is used, which results in a delay of approximately 20 ms. The algorithm delay of the G.728 coder is approximately 0.625 ms, resulting in an achievable end-to-end delay of approximately 2.5 ms. This represents the lowest latency of all CELP coding techniques.

APPLICATION NOTE: G.728-compatible coders are referred to as *Low Delay* CELP (LD-CELP) because their 2.5-ms latency is the lowest of all members of the CELP family of coders.

Another interesting aspect of the G.728 Recommendation is its use of a high-order, short-term predictor, thus eliminating the need for a long-term predictor. This technique allows 10 bits to be used per sample, with 7 representing a fixed code book index and 3 representing an excitation gain, resulting in a data rate of 16 Kbps.

The G.728 Recommendation results in a relatively good quality of reconstructed voice and low delay. According to press reports the MOS associated with G.728 is 4.0, which is the same as the score for the 32 Kbps version of ADPCM. Two additional CELP-related standards warrant attention: the G.729 Recommendation and the G.723.1 Recommendation. Let's turn to these now.

The G.729 Recommendation

The G.729 Recommendation and a reduced-complexity alternative specified in Annex A to that standard represent two voice coding methods that produce both high-quality voice and a high compression ratio that result in a low data rate for the transmission of digitized speech. The G.729 Recommendation is based on the CELP algorithm, but it uses a speech frame length of 10 ms. The more formal name of the G.729 Recommendation is Conjugate-Structure Algebraic Code Excited Linear Prediction (CS-ACELP), and the original standard was approved by the ITU in November 1995.

CS-ACELP uses a 10-ms frame size plus a 5-ms lookahead, which results in a total of a 15-ms algorithmic delay, slightly exceeding the G.728 low-delay coding method. However, instead of requiring a 16-Kbps data rate, CS-ACELP uses a data rate of 8 Kbps to encode a voice conversation.

For each 10-ms frame the speech signal is analyzed to extract linear prediction filter coefficients, adaptive and fixed-codebook indices, and gains that are encoded and transmitted. In actuality, the 10-ms speech frame is subdivided into two 5-ms subframes to prevent the spectral transition from one frame to another being abrupt and results in an improvement in reconstructed voice. Table 5-2 indicates the allocation of G.729 coder parameters. Note that all the parameters except line spectrum pairs are encoded for each subframe. Also note that since the G.729 coder operates on a speech frame of 20 ms, this corresponds to the transmission of 100 samples of 80 bits per sample per second, or a bit rate of 8 Kbps.

One interesting item concerning the use of G.729 is the fact that its MOS in several reports is equivalent to the 4.0 MOS obtained by the G.728 LD-CELP standard. Because LD-CELP operates at 16 Kbps while G.729 operates at half that rate, if latency is not an issue the use of G.729 will conserve bandwidth as well as enable more simultaneous calls on a common access line into a packet network.

The G.729 Annex A CS-ACELP speech-compression algorithm represents a reduced-complexity version of the G.729 coder. The G.729 Annex A speech coder was developed for use in simultaneous voice and data multimedia applications and is primarily employed in modems that provide simultaneous voice and data transmission capability.

Due to the fact that voice digitization based on the G.729 Recommendation provides a near-toll quality roughly compatible to the ADPCM G.721 standard but at one-quarter of its bandwidth, another popular use of G.729 is in adapter cards for multiplexers, frame relay access devices (FRADs), and similar products. Its use enables voice to be transported at

Parameter	Subframe 1	Subframe 2	Total/Frame
Line Spectrum Pairs	-	-	18
Adaptive Codebook Delay	8	5	13
Pitch-Delay Parity	1	-	1
Fixed-Codebook Index	13	13	26
Fixed-Codebook Sign	4	4	8
Codebook Gains (Stage 1)	3	3	6
Codebook Gains (Stage 2)	4	4	8
Total			80

Table 5-2.   G.729 Bit Allocation for a 10-ms Frame

a very low data rate. In fact, some vendors have added a proprietary silence-suppression capability to the G.729 coding mechanism that reduces the average amount of bandwidth required to transport a voice conversation to approximately 4 Kbps.

APPLICATION NOTE: Although the use of a CS-ACELP coder results in half the bandwidth of LD-CELP, the coding delay increases from approximately 2.5 ms to 15.0 ms.

The G.723.1 Recommendation

When transporting voice-encoded packets over public packet networks, there exists a probability that packets will be lost. This is especially true for transmission occurring on frame relay networks since the design of this network was based on certain frames being discarded during periods of high utilization. When transmission occurs on other types of packet networks, such as an IP network, the occurrence of an error condition could result in the retransmission of a packet. If the packet contains digitized voice, the delay associated with retransmission would alter the ability to receive the packet at the time it was needed for the correct reconstruction of the portion of a voice conversation carried in the packet, and it might just as well be dropped. Thus, the transmission of packets containing digitized voice requires a coding method with a degree of robustness to handle lost packets.

This robustness, also referred to as *frame erasure ability*, became an important consideration when the ITU was examining several competing methods to standardize the audio portion of videoconferencing and public telephony. As a result, the ITU selected a dual-rate coding technique now known as Recommendation G.723.1.

A G.723.1 coder supports both 5.3- and 6.3-Kbps coding rates, with the higher bit rate providing a higher quality of reproduced voice. Support for both rates is a mandatory part of the specification, and an option exists to enable implementers to suppress periods of silence using a voice activity detection (VAD) technique, which when used provides a variable operating rate that can result in an average data rate of between 2.65 and 3.15 Kbps.

The G.723.1 Recommendation was approved by the ITU in March 1996 and, during 1997, was recommended by the International Multimedia Teleconferencing Consortium's Voice over IP Forum as the default low bit rate audio coder for the ITU H.323 standard. The ITU H.323 standard defines the method for voice and video communications over packet-based networks, which makes the G.723.1 standard suitable for Internet videoconferencing, Internet telephony, voice-over-frame relay, and other applications. The fact that two bit rates are associated with the G.723.1 speech coder provides additional flexibility in comparison to the use of a fixed rate coder. Similar to other hybrid coders, G.723.1 encodes frames of speech using linear predictive analysis-by-synthesis coding. The 6.3 Kbps coder uses Multiple Maximum Likelihood Quantization (MP-MLQ) for the excitation signal. In comparison, the 5.3 Kbps coder uses Algebraic-Code-Excited Linear Prediction (ACELP). For both rates the frame represents a 30-ms period of speech, which is longer than most other speech coder standards.

For both operating rates the encoder starts with a 30-ms frame of 240 samples. The frame is high-pass filtered and subdivided into subframes, with a 10th-order Linear Prediction Coder filter computed on each subframe. This is followed by the computation of a perceptually weighted speech signal from the LPC components. The weighted speech signal is then used to compute an open-loop pitch period over two 120 sample subframes. Once this is accomplished all remaining processing occurs on 60 sample frames to compute a closed-loop patch predictor. Finally, either ACELP or MP-MLQ is used to compute the non-periodic component of the excitation in the encoder, with the method used based upon the operating rate of the encoder.

The quality of reproduced voice encoded and decoded based on the G.723.1 Recommendation is best judged by the results of its MOS. On the MOS scale, a score of 4.0 is considered to represent toll quality, the quality of speech you would hear on a call routed through a typical PSTN call. Based on several MOS tests, G.723.1 coders were rated at 3.8, which is a variation of only 0.2 point from full toll quality while requiring approximately 1/17th the bandwidth of PSTN toll-quality communications. Due to this capability, it is very reasonable to expect that the use of G.723.1 coding will be adopted by a large number of hardware and software developers.

However, like Gordon Gecko's greed, the low bandwidth of the G.723.1 Recommendation can get you into trouble. This is because the multirate CELP coding technique has the longest delay of all CELP coding techniques. Because G.723.1 operates upon 30-ms segments of speech, its delay is 30 ms without any processing. When the coding or algorithmic delay is added to the 30-ms period of speech the total delay is approximately 67.5 ms, which is a considerable portion of the 250 ms end-to-end delay a human ear can tolerate. Due to this extended delay, the anticipated migration to the G.723.1 Recommendation has not reached its apparent level of original interest.

ALGORITHM SELECTION

Although your first impression might be to select a voice coding algorithm that uses the least amount of bandwidth, the use of that metric might not result in the selection of an appropriate algorithm. While bandwidth is an important selection criteria, there are additional areas you should investigate. Table 5-3 lists seven algorithm selection criteria questions you may wish to answer. To facilitate answers to the fifth question, Table 5-4 provides a comparison of five popular G-series voice coding recommendations. By carefully comparing the answer to each question contained in Table 5-3 with the information contained in Table 5-4 against the requirements of your application, you can select an appropriate voice coding algorithm to satisfy your requirements. However, in doing so a few words of advice are in order. First, the MOS values listed in Table 5-4 are opinions and not a scientific fact. Secondly, the coding delay represents the delay resulting from the gathering of a sample of speech and its processing to include the extraction of parameters and any applicable codebook searching. While a relatively long delay may be

- What coding bandwidth does the algorithm require?
- Does the algorithm generate high-quality voice or just intelligent speech?
- Do other vendor products support interoperability based on the algorithm being considered?
- Is the algorithm standardized?
- What is the end-to-end delay associated with the algorithm?
- Is the algorithm suitable for use on packet networks?
- Does the algorithm pass fax and/or modem modulation and call progress or similar signaling tones?

Table 5-3.   Voice Coding Algorithm Selection Criteria to Consider

Standard	Description	Bandwidth (Kbps)	MOS	Coding Delay
G.711	PCM	64	4.3	1.0 µs
G.721	ADPCM	32, 16, 24, 40	4.0	1.25 µs
G.728	LD-CELP	16	4.0	2.5 ms
G.729	CS-ACELP	8	4.0	15.0 ms
G.723.1	Multi-rate CELP	6.3	3.8	67.5 ms
		5.3	3.6	67.5 ms

Table 5-4.   G Series Voice Coding Comparison

unsuitable for some application environments it is important to note that the voice coding delay is only one component of the end-to-end delay. If the other components are nominal it may be possible to use a longer delay but lower bandwidth coder, which might be a better solution if bandwidth is a constraint. However, if latency is a constraint, then the selection of a higher bandwidth coder could represent a better solution.

APPLICATION NOTE: The selection of a voice coder represents a tradeoff between bandwidth, MOS, and latency. The bandwidth required by a voice coder is generally proportional to its MOS and inversely proportional to its coding delay.

CHAPTER 6

Telephone Operations

In Chapter 5, we focused on various methods used to digitize human speech. In this chapter, we turn our attention to the manner in which digitized voice is conveyed between source and destination over the telephone company infrastructure or a private network formed by the use of leased lines to interconnect corporate PBXs. To do so, we will examine the operation of the telephone and PBX and learn how signal information is passed between telephone instruments. This will provide the foundation necessary to understand how to configure different types of voice-over-data network products so they can correctly interoperate with existing and planned voice and data communications equipment.

SIGNALING

The purpose of signaling in a voice network is twofold. First, signaling is necessary to convey information about a potential connection. Second, signaling enables a connection to be established through a voice network.

Types

There are two primary types or categories of signaling that are necessary to establish a telephone call. When you pick up a telephone handset, your action must be recognized by a central office or PBX serving your instrument. This type of signaling is commonly referred to as *subscriber loop* or *station loop signaling,* as it occurs on the loop connecting the telephone subscriber to the central office or PBX.

A second type of signaling occurs at the central office or PBX when the destination telephone number is not serviced by the central office switch or a corporate PBX. In such situations the call must be forwarded to another switch or PBX via a trunk line linking central office switches or company PBXs. The ability to access the trunk involves trunk signaling and represents a second type of signaling required to establish a voice call.

In this chapter, we will examine both types of signaling. Because all manual calls commence with a telephone instrument, let's turn our attention to the manner in which it is used to initiate a telephone call.

INITIATING A TELEPHONE CALL

When you pick up a telephone handset in your home or office, a sequence of predefined operations occurs that provides you with the ability to receive a dial tone and initiate a telephone call. The ability to make a telephone call begins when you remove the handset from the telephone. Thus, let's examine the manner by which a telephone is connected to a communications carrier's central office switch or a corporate PBX, as both devices function in a similar manner by directly interoperating with the connected telephone instrument. In doing so we will focus our attention on the operation of analog telephones. If digital handsets are used the connection between a subscriber and a central office switch will be based upon the use of the ISDN D channel. If a digital handset is connected to a digital PBX, ISDN signaling may also be used; however, it is also possible that the connection between the handset and the PBX is proprietary to the PBX manufacturer.

The Telephone Connection

A telephone is connected via a pair of wires, called the *ring* and the *tip*, to a connector on a communications carrier's central office switch or a similar connection on a corporate PBX. An example of this connection is illustrated in Figure 6-1.

The terms "tip" and "ring" date to the era when switchboards were used at a central office and an operator used patch cords to provide an interconnection between a caller and the called party. The tip and ring at the ends of the patch cord provided a circuit that lives on in terminology long after the manual switchboards were replaced by automation. Today the tip and ring lives on in the two leads used in RJ-11 modular jacks and plugs. Such plugs have metal rails as illustrated in the left portion of Figure 6-2. In comparison, RJ-11 jacks have metal pins as illustrated in the right portion of Figure 6-2. The metal rails in the plug connect to the wire leads in twisted-pair cabling and the plug is inserted into a jack to establish a telephone connection. The jack in turn is wired to a PBX or telephone company central office with the tip lead at 0 volt direct current (VDC) while the ring provides –48 VDC in the United States. This voltage is used to power the phone instrument as well as for signaling, such as activating the ringer.

Each local telephone number consists of a prefix and extension. The prefix, such as 477, identifies an exchange served by the central office switch or PBX. The four-digit extension identifies a unique device in the exchange. Originally, most central office switches and PBXs were limited to supporting a single exchange; however, the rapid growth in the installation of telephones resulted in the development of switches and PBXs capable of handling multiple exchanges. In the example illustrated in Figure 6-1, we will assume for simplicity that the switch or PBX serves one exchange whose prefix is 477, while the extension assigned to the telephone is 0293.

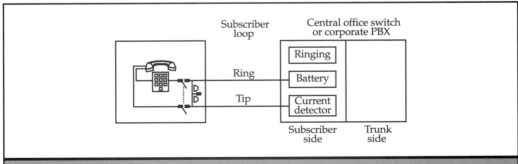

Figure 6-1. Directly connecting a telephone to a central office switch or corporate PBX

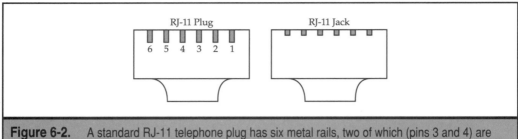

Figure 6-2. A standard RJ-11 telephone plug has six metal rails, two of which (pins 3 and 4) are used to mate with metal pins in an RJ-11 plug.

The Subscriber Loop

The wire pair routed from the switch or PBX to the telephone is referred to as a *subscriber loop*. A 48-VDC potential is placed across the loop by the switch or PBX to power the telephone instrument as well as to provide a mechanism for monitoring telephone activity. Concerning the latter, when you lift the telephone handset to initiate a call, this action results in the closure of a switch hook in the telephone, enabling current to flow through the local loop. The flow of current signals to the switch or PBX that the instrument has gone off-hook and the user wants to initiate a call. This type of signaling is referred to as *loop start signaling* and represents the most common method used to alert a PBX or telephone company switch that the telephone is activated or off-hook.

Dial Registers and Dial Tones

On detecting the presence of loop current, a switch or PBX performs two associated actions. First, it searches for an available buffer area to store dialed digits as they are received, an area more formally referred to as a *dial register*. Once an available dial register is associated with the local loop, the switch or PBX transmits a dial tone to the subscriber, enabling the telephone user to dial the desired number.

Intraswitch Communications

If the dialed exchange is served by the switch or PBX, the device will attempt to make a cross-connection through the switch. To do so, the switch or PBX first checks the line associated with the dialed number to ascertain if it is in use. Assuming it isn't, the switch or PBX transmits a ringing voltage down a different local loop, which causes the bell in the telephone attached to that loop to ring. The telephone ringer is activated by the switch or PBX generating a ring tone. That tone consists of an AC signal generated between 20 and 47 Hz on the subscriber loop. An AC signal is used as it provides a mechanism to activate the ringer without requiring the switch hook to be closed. The ring tone crosses the leads of the ringer, activating a bell on legacy telephone instruments or an electronic circuit on more modern phones, with the latter generating a ring tone. If the called party is in, he or she lifts the handset, which causes current to flow in the loop. This action informs the switch or PBX that the ringing voltage should cease and the audio connection between calling and called parties should be established. Now that we have a basic appreciation for the method in which a call is established through a central office switch or corporate PBX, let's examine in additional detail the components of the telephone infrastructure and the manner in which signaling information is passed through the telephone network.

The Telephone Set

Although the switched telephone network is a two-wire network, the connection between switches and PBXs as well as the operation of the telephone are based on the use of four wires representing a pair of wire pairs. To understand why the telephone requires a pair of wire pairs requires a brief examination of the telephone handset. Then we can briefly discuss how the four-wire to two-wire conversion is accomplished.

The Handset

The telephone handset has a *transmitter* and a *receiver* better known as its *mouth* and *earpieces*. Two wires are connected to the mouthpiece and two to the earpiece, as each operates independently of the other and requires two wires to form a circuit. Those four wires are routed into the base of the telephone where they are connected to a conversion device known as a *hybrid*. The hybrid converts the four-wire interface into a two-wire interface, which enables the telephone to operate via a two-wire connection to the telephone company switch or corporate PBX. Other components of the telephone that warrant a degree of discussion include its switch hook, side tone, dialer, and ringer.

The Switch Hook

The switch hook is placed in an open position by pressure from the handset. Thus, when the handset is lifted, the switch hook closes, enabling current to flow through the telephone and local loop and signaling the telephone company switch or corporate PBX.

The Side Tone

The side tone represents a design of the hybrid within the telephone that enables a portion of speech to "bleed" over the earpiece or receiver. The purpose of the side tone is to enable people to hear themselves talk so they have a better ability to adjust the tone of their conversation.

The Dialer

There are two types of telephone dialers: rotary and push-button or tone. In rotary dialing you rotate a circular dial to the digit position to generate the dialed digit. In comparison, push-button dialing results in the generation of a unique combination of frequencies or tones when a button is pushed, hence the name *tone dialing*. When we discuss signaling later in this chapter, we will include both pulse and tone dialing in that discussion.

The Ringer

The ringer is activated by a switch or PBX generating an AC signal between 20 and 47 Hz. As previously discussed, this signal activates a bell in legacy phone instruments or a circuit in more modern phones, with the latter generating a ringing tone.

The Local Loop

As previously discussed, the connection between the telephone instrument and the switch or PBX is the *subscriber loop*, more commonly known as the *local loop*. This connection is a two-wire circuit, with one wire known as the *tip* and the other referred to as the *ring*. As previously noted in this chapter, this nomenclature dates to the use of switchboards, in which two wires terminated in a plug that was inserted into a circular tube in the switchboard to make a connection. One wire in the plug is attached to the top of the connector, while the other wire is electrically connected to the ring on the connector, resulting in the terminology "tip" and "ring." In the more modern communications environment, the tip wire is connected to the groundside of the battery at the central office switch or corporate PBX, while the ringside is connected to the negative terminal.

The Switch Hybrid

The use of a two-wire circuit to connect a telephone instrument to a serving telephone company switch or corporate PBX was primarily based on economics. A two-wire circuit is less expensive than a four-wire circuit and provides a full-duplex transmission path suitable for a variable distance based on the wire gauge of the conductor. Usually, two-wire circuits can be used at distances of up to approximately one to two miles.

When the telephone network was established, the routing of calls between exchanges required the transmission of voice over much longer distances than those used in local loops. Since a signal attenuates as it traverses a circuit, the telephone company inserted amplifiers within long-distance circuits to rebuild attenuated signals. However, since amplifiers are unidirectional, voice had to be separated into two different paths for transmission between central office switches, with one path used for transmission and the other used for reception. This signal splitting resulted in the use of a four-wire circuit for transmitting voice between central offices, with the circuit commonly referred to as a *trunk*. This four-wire circuit requirement is also applicable when leased lines are used to create a private voice network by interconnecting corporate PBXs.

The conversion of the two-wire local loop to a four-wire trunk is accomplished by a hybrid in a manner similar to the hybrid in a telephone. However, unlike the telephone-set hybrid that is lightly imbalanced to generate side tones, the switch hybrid must be balanced. Otherwise, it would reflect energy that would result in an echo. On a four-wire circuit connecting PBXs or central office switches, the transmit path conductors are referred to as tip and ring, while the receive path conductors are referred to as tip 1 and ring 1 to differentiate them from the transmit path. Figure 6-3 illustrates the operation of a central office hybrid. To alleviate the echo that results from a portion of speech energy being reflected back toward the talker, communications carriers use echo suppressors or echo cancelers. Both devices are designed to minimize the effect of reflected speech energy.

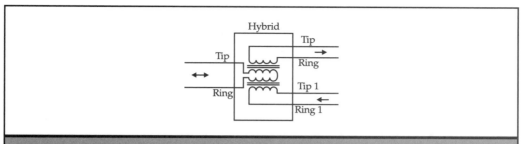

Figure 6-3.   The central office hybrid converts the two-wire local loop to the four-wire trunk and vice versa.

When transmitting voice over IP or frame relay, it is important to note that the telephone system normally does not perform echo cancellation on local calls, even when a call is routed between two switches and results in a two-wire to four-wire conversion and a reconversion from a four-wire to a two-wire circuit. For relatively short transmission distance, typically less than 50 miles, latency is not long enough for the echo to return as a separate transmission. Because local connections are used for voice-over-IP and voice-over-frame relay, the telephone system will not perform echo cancellation. However, because long-distance calls are being made, the echo from two-wire to four-wire hybrids will propagate through the network back to the speaker and would become very disruptive unless canceled. Thus, voice-compliant routers, FRADs, and gateways should provide an echo-cancellation capability when used for transmission beyond a 50-mile range.

APPLICATION NOTE: To prevent disruptive echo from being returned to the speaker, voice-compliant routers, FRADs, and gateways should perform echo cancellation when used to establish a voice-over-IP or voice-over-frame-relay transmission path that exceeds 50 miles in length.

SIGNALING METHODS

In this section, we turn our attention to the types of signaling used to establish a telephone call and the manner in which those signals are routed via subscriber or local loops onto trunks. This information builds on the prior section, and the goal of both sections is to provide you with the information necessary to acquire and configure equipment to transmit voice over data networks.

Types of Signaling

There are three types of signaling that must be conveyed to establish a telephone call: supervisory, address, and call progress. In this section, we will examine each type of signaling; however, due to key differences between local loop and trunk supervisory signaling, we will discuss each separately.

Supervisory Signaling

Supervisory signaling is used to inform the telephone instrument and ports on the telephone company switch or corporate PBX of the status of the local loop and any connected trunks between switches and PBXs. There are several methods used for supervisory signaling, with the loop-start signaling method representing the most common method used on the local loop.

Loop-Start Signaling

Under the loop-start signaling method, the lifting of the telephone handset, a condition referred to as *off-hook*, results in the switch hook closing. This action results in a current flowing through the local loop, which the switch or PBX detects and responds to with a dial tone. When the switch or PBX generates the dial tone, it also places its port connected to the telephone via the local loop in an off-hook indication mode. This action results in the local loop becoming active. When a call is terminated by a subscriber placing the headset back into its cradle, the switch hook opens, stopping the flow of current, and the local loop returns to its *on-hook* state. Regardless of the type of signaling used, the terms "on-hook" and "off-hook" are universally used to describe the state of the local loop with respect to the presence or absence of a current caused by the state of the switch hook. When loop-start signaling is used, the terminating end of the loop in the form of a central office switch or corporate PBX supplies the battery and current-detection circuitry. In comparison, the originating end provides only a DC path via the positioning of the switch hook.

Address Signaling

A second type of signaling used in a telephone system is address signaling. Address signaling provides the telephone number that enables a call to be routed to its appropriate destination. There are two commonly used methods of address signaling: dial pulse and dual-tone multifrequency (DTMF) signaling.

Dial Pulse

Under dial-pulse signaling, digits are transmitted from the originating telephone instrument by the opening and closing of the local loop. In actuality, as you turn the rotary dial, your action winds up a spring. When you release the dial to select a digit, the spring rotates back to its originating position and opens and closes the loop a number of times, which corresponds to the value of the dialed digit. Figure 6-4 illustrates the generation of three digits via a rotary dial.

In examining Figure 6-4, note that each dialed digit must be generated at a specific rate and must be within a predefined tolerance for it to be recognized. To accomplish this, each pulse consists of two parts referred to as *make* and *break*. The make segment represents the period when the circuit is closed, while the break segment represents the period of time that the circuit is placed in the open condition during a dialed-digit period.

In North America the make/break ratio is 39 percent make to 61 percent break, while in the United Kingdom the ratio used is 33 percent make to 67 percent break. Figure 6-5 illustrates the pulse dialing of the digit 2 in North America. Note that the 39 ms time period

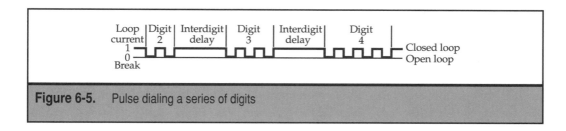

Figure 6-5.   Pulse dialing a series of digits

when the circuit is closed or off-hook is the make period, while the 61 ms period when the circuit is on-hook is the break period. Because each dialed digit opens and closes the loop a number of times that corresponds to the number dialed, dialing 2 results in two make/break cycles. A second metric that governs the operation of pulse dialing is pulse rate. In North America, digits are pulsed at a constant rate of 10 pulses per second (PPS). The rate may vary in other locations.

APPLICATION NOTE: When planning to interconnect a voice gateway to the switched telephone network in a foreign country, you should determine the make/break ratio and pulses-per-second rate required for pulse dialing.

Tone Dialing   A second type of address signaling results from the transmission of tones instead of pulses. Tone dialing, more formally known as DTMF signaling, results in the use of a 12-key keypad that has two frequencies associated with each key. Each row and column of keys on a tone-generation instrument is associated with a predefined frequency, as illustrated in Figure 6-6. Pressing a key on the keypad results in the generation of a low- and high-frequency sine wave tone pair. For example, pressing the 8 key shown in Figure 6-6 would result in the generation of the frequency pair 852 and 1336 Hz, which informs the connected switch or PBX port that the digit 8 was dialed. This also explains the *dual* in DTMF.

Informational Signaling

A third type of signaling can be categorized as informational, as it informs the originator of the status of a dialed call and rings the called party to alert him or her to the presence of

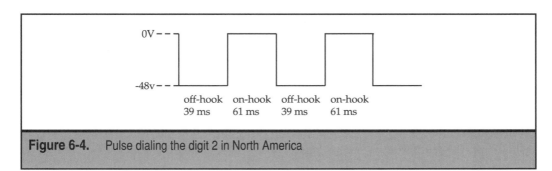

Figure 6-4.   Pulse dialing the digit 2 in North America

697 Hz	1	2	3
770 Hz	4	5	6
852 Hz	7	8	9
941 Hz	*	0	#
	1209 Hz	1336 Hz	1477 Hz

Figure 6-6.   Tone assignments for the DTMF keypad

the call. This type of signaling is also commonly referred to as *call-progress signaling*, since most of the signaling is used to inform the originator of the call's progress. Common types of informational signaling include busy and fast busy signals, a dial tone signal, a ring signal, a ringback signal, and an off-hook notifier signal.

Dial Tone   In North America, certain predefined frequency pairs are used to convey informational signaling. For example, a dial tone results from the generation of a continuous 350- and 440-Hz frequency pair by the telephone company switch or the corporate PBX.

Ringing Signal   When a call is routed to its destination, the serving switch or PBX commonly sends a 20-Hz, 86-VAC ringing signal that cycles for two seconds on followed by four seconds off.

Ringback Signal   To inform the originator that the call is in a ringing state, a ringback tone is generated by the local switch or PBX. This ringback tone consists of a 400- and 480-Hz frequency pair, also cycled at two seconds on followed by four seconds off.

Busy Signal   If the called party is using the telephone, the local switch or PBX will generate a busy signal to the call originator. This signal consists of the frequency pair 480 and 620 Hz cycled for 0.5 second on followed by 0.5 second off. If the call cannot be completed because of trunks between switches being busy (for example, on Mother's Day), the local switch or PBX will generate a fast busy signal.

Fast Busy Signal   The fast busy signal consists of the frequency pair 480 to 620 Hz cycled for 0.2 seconds on followed by 0.3 seconds off. As previously noted, this signal is generated immediately by the local switch when it cannot find a trunk through which to route a call whose destination is beyond the local switch. All of the preceding informational signaling tones represent frequencies and cycling periods used in North America. These will vary in other locations throughout the world.

Off-Hook Notifier Signal   If you are like this author, upon occasion you may forget to hang up your phone or alternatively, you might take your phone off-hook and get distracted from dialing. After approximately 45 seconds, in an off-hook condition without dialing the PBX or switch will generate an off-hook notifier signal. This signal consists of the frequencies 1400, 2060, 2450, and 2600 cycled for 0.1 second on followed by 0.1 second off, which is irritating enough to usually get your attention.

> **APPLICATION NOTE:**   Prior to connecting equipment to a PBX or the switched network in a foreign location, it is important to verify the compatibility of the equipment with call-progress signaling frequencies. Doing so ahead of time can alleviate a considerable amount of potential frustration.

Interface Terminology   In addition to noting the effect of different types of signaling upon a telephone call, there are several types of interfaces that govern the connection of a telephone to a PBX or switch port. Because the voice interface on routers and gateways also use these interfaces, let's turn our attention to the terminology associated with two popular voice interfaces and their meaning.

FXS   The term "FXS" references Foreign eXchange Station and represents the standard analog telephone interface in the form of an RJ-11 two-wire jack. An FXS interface is used to connect telephones, fax machines, analog PBXs, and PC modems.

FXO   The term "FXO" is a mnemonic for Foreign eXchange Office and can be considered to represent the opposite point to an FXS. That is, an FXO is the port on a standard telephone or fax machine that communicates with a PBX or central office switch. Later in this book when we examine the configuration of a voice-over-IP gateway we will note that one of the configuration parameters of the gateway is to select the voice interface in the form of either an FXS or FXO type interface. It is also important to note that FXS and FXO represent two-wire or single pair interfaces. Later in this chapter when we discuss E&M signaling used by a PBX or switch as a trunk signaling method, we will note that this signaling method supports both two-wire and four-wire operations.

Trunk Signaling

The composition of a telephone number is used by a switch as a decision metric to determine if the destination is local or requires the establishment of a path to a distant switch. For example, if you enter the numeric 1 followed by other digits when making a call over the PSTN, the 1 indicates that the called party is not local to the switch serving the subscriber. If you are connected to a corporate PBX, the entry of a 9 is commonly used to inform the PBX you need an outside line, and the digits that follow are then routed to a telephone company switch. Similarly, entering a 7 might be used by the PBX to identify local PBX calls and allow you to enter a four-digit extension, resulting in the PBX making a cross-connection directly between the caller and called party without routing the call beyond the PBX.

The ability to program a PBX by assigning prefix numbers to predefined trunk groups or PBX ports that function as a rotary group enables organizations to connect their PBXs to different communications devices and even different communications carriers. For example, you could connect a group of PBX ports to voice card ports on a frame relay access device and configure the group as a hunt or rotary group. This means that multiple users connected to the PBX could each dial a prefix, such as 5, and have the PBX establish a cross-connection to the first available port in the group of ports designed to function as an entity.

As previously discussed, the path between two central office switches is referred to as a *trunk*. When corporate PBXs are interconnected to form a voice network, those connections are also referred to as trunks. Unlike the local loop, both central office and PBX trunks are shared by transmissions from different subscribers, although only one subscriber can use a trunk at any point in time.

Originally, trunks represented individual physical lines, with each trunk carrying one conversation at a time. The development of the T1 circuit as a mechanism to reduce cable congestion in urban areas resulted in multiple logical trunks occurring on one physical trunk. That is, the T1 circuit was developed to convey 24 voice-digitized conversations between switches via the assignment of one voice conversation to each of 24 channels derived by time. Thus, the T1 circuit is a modern-day grouping of 24 logical trunks onto one physical trunk.

Avoiding Glare

Similar to the local loop, trunks require supervisory signaling. Although loop-start signaling, which is primarily used on local loops, can be used on trunks, its use on certain types of trunks can result in several problems. First, when loop-start signaling is used, only the originating instrument can release the connection. A second problem associated with the use of loop-start signaling is the fact that a trunk could be simultaneously seized from both sides, a condition referred to as *glare*.

Although glare can be tolerated on dedicated local loops, the occurrence of this condition on trunks that users are contending for cannot be tolerated. Therefore, it was necessary to develop a two-way handshaking mechanism to coordinate access to a trunk, with one end requesting access and the remote end acknowledging the request prior to the originating end grabbing the trunk. Trunk supervisory signaling methods have been developed to alleviate the problems associated with the loop-start method when applied to trunks. Some of the more popular ones are the start-dial, the ground-start, and the E&M signaling methods. As we will note later in this book when we examine the configuration of a voice-over-IP gateway, in addition to configuring a voice interface on a router or gateway as FXS or FXO we must also configure the signaling method supported by the port. Thus, let's turn our attention to each signaling method.

Start-Dial Supervision

The start-dial supervision signaling method actually refers to a series of signaling methods, including wink-start, immediate-start, and tone-start signaling. Each of these trunk-signaling methods occurs in response to a subscriber going off-hook and results in the seizure of a trunk so digits can be transmitted to a remote switch.

Wink-Start Signaling Wink-start represents a commonly used trunk-signaling method. Under this signaling method, the originating trunk is placed in an off-hook condition,

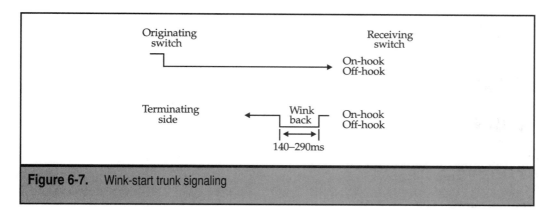

Figure 6-7.   Wink-start trunk signaling

which results in the remote switch responding by transmitting an off-hook pulse of between 140 and 290 ms in duration, after which the switch returns to an idle or on-hook
state. This off-hook pulse is referred to as a *wink-back,* which when detected by the originating switch results in the switch waiting for at least 210 ms and then transmitting the
address digits to the remote switch. The remote switch then returns to an off-hook condition to answer the call. Figure 6-7 illustrates the sequence of operations at the originating
and receiving switch during a wink-start trunk-signaling operation.

Immediate-Start Signaling   Immediate-start signaling represents one of the most basic
trunk-signaling methods. When an immediate-start signaling method is used, the originating switch places the trunk off-hook and maintains that condition for a minimum of
150 ms, after which the switch outputs the address digits. Since there is no real handshake
between switches, this type of trunk signaling is appropriate only when a dedicated logical or physical trunk is used between switches. Figure 6-8 illustrates an example of immediate-start signaling.

Tone-Start Signaling   In tone-start signaling, the originating trunk circuit is placed in an
off-hook condition. At the receiving switch, the recognition of this condition generates a
dial tone. This dial tone can be used by the switch to directly output previously stored

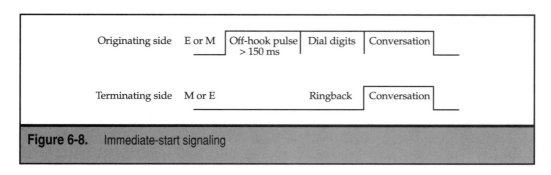

Figure 6-8.   Immediate-start signaling

digits, or, if the switch is in a cut-through mode, the distant dial tone is passed to the user for dialing. Tone-start signaling is normally used in a private voice network formed by the interconnection of PBXs and not on the switched telephone network. Via the use of the cut-through mode, a user in New York could dial a predefined code, such as 66, and receive a dial tone from the company's PBX in Chicago, allowing the employee to dial a number as if he or she were in the Chicago office.

Ground-Start Signaling

The ground-start signaling method is a modification of local loop-start signaling. It eliminates the potential for two switches to simultaneously seize both ends of a trunk. To avoid dual trunk seizure, the ground-start method provides a current-detection mechanism at both ends of the trunk, enabling each end to agree prior to the trunk being seized at one end.

To illustrate the operation of ground-start signaling, consider a PBX that needs to route a call to a telephone company switch, requiring the PBX to contend for a trunk. The PBX will ground the ring lead, causing current to flow from the PBX to the central office switch module across the ring circuit. At the telephone company's central office, the switch module senses the current and interprets it as a trunk-seizure request. Assuming the trunk is available, the module acknowledges the request by closing its tip switch, which in turn generates a ground on the tip lead. The flow of current across its tip lead serves as an acknowledgment to the PBX request. Then PBX will close the loop by holding a coil across the tip and ring leads and removing the ring ground. Once this is accomplished, the circuit acts like a loop-start circuit. Figure 6-9 illustrates the sequence of steps associated with ground-start signaling when a PBX initiates the seizure of a trunk. The seizure is bidirectional, allowing the central office switch to initiate the process.

When the switch initiates ground-start signaling, it requests the trunk by closing the tip switch and generating a ringing voltage over the ring lead. The PBX must recognize the incoming seizure within 100 ms to prevent glare. In response to the voltage on the ring lead, the PBX will place a holding coil across tip and ring, which completes the loop and is followed by the central office switch removing the ringing voltage. Once this is accomplished, the trunk acts in a manner similar to loop-start signaling.

E&M Signaling

E&M signaling is the most commonly used method of analog trunk signaling. Under E&M signaling, separate paths are used for voice and signaling. For the voice path, the trunk can consist of either two or four wires. In comparison, for the signaling path there are five methods that can be used, referred to as Types I through V, while a sixth popular method is a British Telecom standard that, as you might expect, is popular in the United Kingdom.

Under E&M signaling, a PBX requests a trunk by raising its M lead. The distant end will honor the request by initiating the flow of the current on the E lead to the requestor. This method of signaling is used by each type of E&M signaling, with the manner in which the flow of signaling occurs used to differentiate one type of signaling from another.

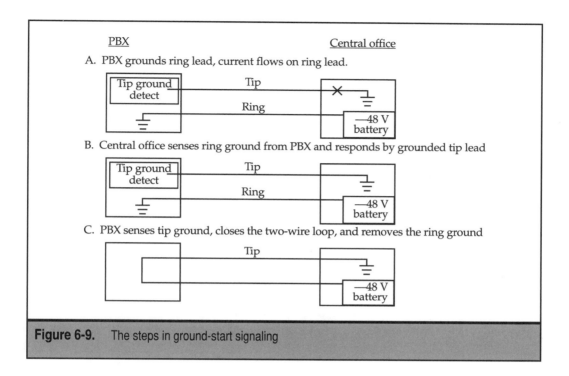

Figure 6-9.   The steps in ground-start signaling

Terminology   The letters "E&M" are derived from the words *ear* and *mouth*, where the M lead is used to transmit signaling and the E lead is used to receive signaling information. When defining trunk signaling, it is common for vendor specification sheets to indicate both the type of trunk, based on the wiring (two- or four-wire circuit), as well as the type of E&M signaling. For example, a specification for 4W E&M TI would refer to a four-wire voice interface to a trunk that requires E&M Type I signaling.

In some respects, E&M signaling is similar to a serial port used for data transmission. That is, like serial ports, E&M signaling has a DTE/CDE type of reference. The trunking side can be viewed as being similar to the DCE and is normally associated with central office functionality. Thus, the connection of a router, FRAD, or voice gateway port commonly re-sults in the port representing the trunking side of the interface. In comparison, the other side of the E&M interface is the signaling side. That side can be viewed as functioning similarly to a DTE. The PBX usually represents the signaling side of an E&M interface.

In addition to five types of E&M signaling, there are two types of audio interface you must consider: two-wire and four-wire. Adding a bit of confusion to an already large number of combinations is the fact that a four-wire E&M interface cable can have six to eight physical wires. This results because the difference between two-wire and four-wire circuits depends on whether the audio path is full-duplex on one pair or two pairs of wires. Table 6-1 summarizes the possible E&M interface signals.

Signal	Description
Ear (E)	Signal wire from trunking (CO) side to signaling (user) side.
Mouth (M)	Signal wire from signaling (user) side to trunking (CO) side.
Signal ground (SG)	Applicable to certain types of E&M signaling, provides –48 V or ground.
Signal battery (SB)	Applicable to certain types of E&M signaling, provides –48 V or ground.
Tip/ring (T/R)	Used on four-wire circuit to carry audio from signaling (user) side to the trunking (CO) side. Not used on a two-wire circuit.
Tip-1/ring-1 (T-1/R-1)	Used on four-wire circuit to carry audio from the trunking (CO) side to the signaling (user) side. On a two-wire circuit, used to carry full-duplex audio.

Table 6-1. E&M Interface Signals

Now that we have some familiarity with the terminology, let's turn our attention to the types of E&M signaling.

Type I Signaling　Under E&M Type I signaling, the battery for both E and M leads is supplied by the PBX. At the PBX, an on-hook condition results in the M lead being grounded and the E lead open. In comparison, an off-hook condition results in the M lead providing the battery and the E lead being grounded. Type I signaling is the most commonly used four-wire trunk interface in North America.

Table 6-2 summarizes E&M Type I signaling. Note that if you connect a PBX directly to a voice port on a router or voice gateway, the router or gateway should normally be

Condition	M Lead	E Lead
On-hook	Ground	Open
Off-hook	Battery	Ground

Table 6-2. E&M Type I Signaling

Condition	M Lead/SB	E Lead/SG
On-hook	Open	Open
Off-hook	Battery	Ground

Table 6-3. E&M Type II Signaling

configured to ground its E lead to signal a trunk seizure. The PBX should apply battery to its M lead to signal a seizure.

Type II Signaling   One of the problems associated with Type I signaling is the fact that the interface can cause a high return current through the grounding system. If two PBXs were improperly grounded, this could result in current flowing down the M signaling lead, which results in a remote PBX detecting the current on the E lead. This in turn results in the occurrence of a false seizure of a trunk. To address this problem, E&M Type II signaling added two additional signaling leads: battery (SB) and signal ground (SG). Under Type II signaling, the E lead works in conjunction with the SG lead, while the M lead is strapped to the SB lead. This results in the grounding of the trunk at each end and eliminates potential grounding problems from occurring.

Table 6-3 summarizes E&M Type II signaling. Note that the connection of a voice port on a router or voice gateway is similar to the connection to E&M Type I signaling. That is, the router or voice gateway port grounds its E lead to signal a trunk seizure, while the PBX would apply battery to its M lead to signal a seizure. E&M Type II signaling is also used in North America. This type of signaling is symmetrical and enables signaling endpoints to be directly connected to one another via a crossover cable.

Type III Signaling   Type III signaling is similar to Type I, the key difference being in the use of transmission equipment to supply the battery and ground source, which results in loop current flowing on the M lead when an off-hook condition occurs. Type III signaling was primarily used with older central office equipment and is now in very limited use because most older central office switches have been replaced. Under E&M Type III signaling a PBX indicates an off-hook condition by disconnecting the M lead from the SG lead and connecting it to the SB lead from a central office switch.

Table 6-4 summarizes E&M Type III signaling. Note that the connection of a voice port on a router or voice gateway requires the router or voice gateway port to sense the flow of current on the M lead for an inbound seizure and ground its E lead for an outbound seizure.

Type IV Signaling   Type IV signaling is similar to Type II, but the operation of the M lead differs. In Type II signaling, the M lead states are "open" and "battery." Under Type IV signaling, the states are "ground" and "open." Table 6-5 summarizes E&M Type IV signaling. Under Type IV signaling, both devices—for example, a PBX and a voice port on a

Condition	M Lead/SB	E Lead/SG
On-hook	Ground	Open
Off-hook	Loop current	Ground

Table 6-4.　E&M Type III Signaling

router or voice gateway—ground their leads to indicate a trunk seizure. The key advantage of Type IV signaling is the fact that an accidental shorting of the SB lead will not result in an excessive current flow.

Type V Signaling　Under Type V signaling, both the switch and the transmission equipment supply a battery. Here the battery for the M lead is located in the signaling equipment, while the battery for the E lead is located in the PBX.

Table 6-6 summarizes E&M Type V signaling. Under Type V signaling, the PBX side grounds its M lead to seize the trunk. In comparison, a voice port on a router, FRAD, or voice gateway supporting Type V signaling would ground its E lead to seize the trunk. Type V signaling is the ITU E&M signaling standard and represents the most common method of E&M signaling outside of North America.

Interface Considerations　The ability to correctly connect a voice port on a router, FRAD, or voice gateway to a PBX port depends on the correct configuration of each device to be interconnected. Both the PBX and the port of the router, FRAD, or voice gateway must be configured to support the same signaling method. In addition, many products can be configured for either two- or four-wire operation and need to be compatible at both ends of the link.

Cabling

E&M signaling uses six to eight of the pins on an RJ-48 modular jack, with the number of pins dependent upon the type of signaling employed. Table 6-7 provides a summary of

Condition	M Lead/SB	E Lead/SG
On-hook	Open	Open
Off-hook	Ground	Ground

Table 6-5.　E&M Type IV Signaling

Condition	M Lead/SB	E Lead/SG
On-hook	Open	Open
Off-hook	Ground	Ground

Table 6-6.   E&M Type V Signaling

the use of E&M signals on the RJ-48 jack and their applicability to different E&M types of signaling. In examining the entries in Table 6-7 note that the signal flow on pin 7 is from the central office switch to a PBX. Thus, if a router is used as the termination for the PBX the signal would flow from the router port to the PBX. Then, pin 2 represents the signal flow in the opposite direction. That is, the signal on pin 2 flows from a PBX to the central office switch or, when a router is used, to the router port.

Name	Pin	Description	Signaling Type Supported
E (ear)	7	Signal from trunking side to signaling side	I, II, III, IV, V
M (mouth)	2	Signal from signaling side to trunking side	I, II, III, IV, V
SB (signal battery)	1	Connects to 48 VDC battery	II, III, IV
SG (signal ground)	8	Connects to electrical ground	II, III, IV
T (tip)	6	Audio path from signaling side to the trunking side on four-wire circuit	I, II, III, IV, V
R (ring)	3	Only used on four-wire E&M	I, II, III, IV, V
T1 (tip 1)	5	Audio path from signaling side to the trunking side on four-wire circuits	I, II, III, IV, V
R1 (ring 1)	4	Full-duplex audio path on two-wire circuit	I, II, III, IV, V

Table 6-7.   E&M RJ-48 Pin Utilization

Dialing

While the different types of E&M signaling identify on-hook and off-hook conditions they do not identify when a dialing string can be transmitted. The latter is normally accomplished via the use of immediate-start, wink-start, or delay-start signaling. Both immediate-start and wink-start signaling were previously described when we examined trunk signaling. The third signaling method used, delay-start, results in a switch delaying the originating PBX's transmission of dial digits until the switch is ready to process those digits. To do so, the switch will sense an off-hook condition generated by the PBX and will respond with a constant off-hook signal until the switch is ready to process dialed digits. A PBX configured for delay start will wait 200 ms after initiating its off-hook signal. After that period of time the PBX will check the E-lead. If its still off-hook the PBX continues to wait; however, if it is on-hook the PBX will then transmit dialing digits. Thus, the switch simply delays the sending of digits from the PBX by holding the line off-hook until it's ready to receive the dialing digits.

Configuration Considerations

When configuring a PBX, it is important to note that many such products use dip switches rather than consoles to configure port interfaces. Even when a PBX appears to be correctly configured, it is important to check the dial plan of the PBX with the numbering plan supported by the router, FRAD, or voice gateway. For example, assume your PBX uses a four-digit-extension dial plan, while an attached voice-capable FRAD configured for four digits requires users to dial a 6 to select the FRAD trunk group. In this situation, the PBX trunk group needs to be configured for a five-digit dial plan to forward the number properly.

T1 Signaling

As previously mentioned in this chapter, T1 circuits can be viewed as representing a logical group of trunks transported on a common physical transmission facility. When used for voice transmission, the T1 circuit consists of 24 channels or time slots that repeat 8000 times per second. Each time slot represents an 8-bit encoded PCM sample. A framing bit is added to each group of 24 samples, resulting in a frame length of $24 \times 8 + 1$, or 193 bits. Since the frame repeats 8000 times per second, this results in the operating rate of the T1 line becoming 193 bits/frame $\times$ 8000 frames/second, or 1.544 Mbps.

To convey 24 analog E&M signals we would normally require 24 wire pairs. Since a T1 circuit consists of two wires in each direction, E&M signaling cannot be directly transferred over a T1 circuit. Instead, an in-band signaling technique is used in which busy and idle information is periodically intermixed with digitized voice samples.

CCS   There are two methods that can be used to transmit supervisory signaling over a T1 circuit. Those methods are referred to as *common channel signaling* (CCS) and *common associated signaling* (CAS). Under the CCS method, signaling information is transmitted along the same path as the voice signal; however, it flows on a separate channel that is multiplexed with the digitized voice signals. This type of signaling is primarily used on the 30-PCM channel circuit known as an E1 circuit line, which uses one channel for signaling and another for frame alignment, resulting in the E1 circuit having 32 channels, each operating at 64 Kbps, for a composite transmission rate of 2.048 Mbps. Although CCS signaling is primarily used with European E1 circuits, it is sometimes used on T1 circuits. When used on a T1 circuit, CCS requires one voice slot to be dedicated to transmitting signaling information.

CAS　To eliminate the necessity of using a voice slot, CAS signaling is primarily used with T1 circuits. Under CAS signaling, the seventh bit position in frames 6 and 12 in a 12-frame framing sequence are "robbed" to convey signaling information. Hence, CAS signaling is informally referred to as *bit robbing*.

Bit position 7 in frames 6 and 12 is used to convey supervisory signaling between PBXs on a private voice network between a PBX and a communications carrier's central office switch or between two switches or two PBXs. The actual pattern represented by bit 7 in frames 6 and 12 is based on the type of signaling, such as loop-start, ground-start, or E&M, and the signal being conveyed. For example, under loop-start signaling, such signaling information as loop open, loop closed, ring present, and ring removed must be conveyed. In comparison, E&M signaling only has two states: idle and busy. To differentiate between the signal bits, the seventh bit in the sixth frame is referred to as the "A" bit, while the signal bits in the twelfth frame are referred to as the "B" bits.

A second T1 framing method, called *extended superframe format* (ESF), extends the T1 frame sequence to 24 frames. Under ESF, signal bits are conveyed in frames 6, 12, 18, and 24, with the bits labeled A, B, C, and D, respectively. Due to the fact that T1 lines can be provisioned as either a 12-frame sequence known as D4 framing or a 24-frame sequence known as ESF, you must consider the compatibility between the signaling method used and the framing format when acquiring equipment to construct a voice network.

Digital Voice Signaling

When a PBX is connected to a central office switch via a digital circuit, such as a T1 line, the endpoints or nodes on the circuit must inform each other when they need to use the circuit. Similar to analog circuits, digital circuits thus need a signaling method. In the wonderful world of digital signaling you can use different types of E&M, loop-start, and ground-start signaling for in-band operations or ISDN signaling when the T1 circuit is provisional for that method of operation.

E&M Signaling

There are three different types of E&M signaling used with digital circuits for in-band signaling operations. Those signaling methods are referred to as E&M Immediate Start, E&M Feature Group B, and E&M Feature Group D. Each signaling method supports direct inward-dialing (DID) as well as two-way dialing, setting the A and B bits to 0 for on-hook and 1 for off-hook.

E&M Immediate Start

In a digital environment E&M Immediate Start is similar to an analog immediate start. That is, under immediate-start signaling a PBX indicates an off-hook condition and then transmits its dial digits. The central office switch then goes off-hook after receiving the dialing digits.

E&M Feature Group B

E&M Feature Group B is similar to the wink-start signaling method used in analog communications. However, instead of pulses the A and B bits are used. Under E&M Feature Group B the transmitting side goes off-hook to indicate that a call is to be placed. The terminating side responds with an off-hook wink lasting approximately 200 seconds. The terminating side then goes back on-hook after the wink and listens for the dial digits. Once the digits are received and the call is established, the terminating side then goes off-hook for the duration of the call.

E&M Feature Group D

E&M Feature Group D functions in the same manner as E&M Feature Group B until the dial digits are received. At this point in time the terminating side generates a second 200 ms duration wink to the source to denote it received the digits. After the second wink the terminating side will remain on-hook until the call is answered. When this condition occurs the terminating side then goes off-hook for the duration of the call.

Loop-Start Signaling

In a digital circuit environment the circuit is always active. Therefore, the ability to close the local loop is not possible and must be simulated. Under FXS loop-start signaling the customer premise equipment (CPE) sets the A bit to 0 for on-hook and to 1 for off-hook. At the network side the B bit is set to 0 for no ring and to 1 for ring.

When the CPE needs to originate a call it sets the A bit to 1 and then transmits its dial digits. In comparison, when the network needs to deliver a call it toggles the bit between 1 and 0 to generate a two-second on, four-second off ringing pattern. After the CPE answers the call it will set the A bit to 1 and the network will reset the B bit to 0 for the duration of the call.

Ground-Start Signaling

A third digital signaling method involves using the A and B bits to simulate analog ground start signaling. In an FXS ground-start environment the customer premise equipment sets the A bit to 0 to denote an on-hook condition and a 1 for off-hook. The network will respond with the A bit set to 0 to indicate an open circuit or set the A bit to 1 to indicate a closed circuit. The network will also use the B bit for ring indication; however, it uses values opposite that of loop start. That is, the B bit is set to 0 for ring and to a 1 for a no-ring condition.

Out-of-Band Signaling

As previously noted, the use of A and B bits is referred to as in-band signaling obtained via bit-robbing. A second type of digital signaling method occurs through the use of a separate channel for signaling. In an ISDN environment the D-channel is used to support out-of-band signaling. When ISDN is provisioned on a T1 circuit one 64-Kbps D-channel provides signaling for the 23 64-Kbps B channels on the circuit. The actual signaling to include call establishment, call information, call clearing, and conveying of call status information is governed by the ISDN Q.930 and Q.931 standards. Thus, when considering the transmission of voice over a data network, you must also consider the type of signaling supported by your organization's PBX and the manner in which such signaling can be conveyed over a data network, including communications equipment you may use to obtain a voice-over-data network transmission capability.

PBX INTERFACE CONSIDERATIONS

In concluding this chapter, we will briefly discuss several items you should consider when connecting voice-compliant router, FRAD, or gateway ports to a PBX. Because most products currently on the market interface with a PBX via an analog signaling method, the key to the successful configuration of equipment is to trick the PBX into viewing the attached device as a bank of ordinary analog telephone lines. Those lines would normally be routed to a central office, and thus the PBX configuration should be set as if the PBX were connected to ordinary analog CO lines.

The connections from ports on the attached device to the PBX should behave in the same manner as analog trunk lines. That is, they should present a ringing voltage when a call is received and DTMF or pulses with appropriate make/break cycle and digits when calls arrive. In addition, when calls are forwarded through the PBX, the attached device should provide appropriate call-progress tones, such as fast busy, busy, and ringback.

If you are connecting a PBX via digital signaling you must configure your router or voice gateway T1 line to be compatible with your PBX. Thus, you must consider whether or not your PBX supports in-band or out-of-band signaling. If the former, your voice-compliant router, FRAD, or gateway port must be configured to support the same signaling method used by the PBX.

As we noted earlier in this chapter, it is also important to consider the dial plan used by both the PBX and the attached device. Doing so will result in the correct number of digits being passed between devices. Last but not least, it is extremely important to configure the PBX and attached ports to support the same signaling method. While these points may appear self-explanatory, they contribute to a majority of the interoperability problems that users encounter.

CHAPTER 7

Voice-over-IP Networking

Until this chapter, we have purposely avoided any detailed discussion of a technology that for many readers is perhaps the most common method of transmitting voice over a data network: Internet telephony. The reason for this avoidance is twofold. First, Internet telephony is a term used primarily to refer to the use of software in conjunction with a sound card and microphone to provide individual PC users with the ability to initiate and receive calls over the Internet. This is a very important application and one that can be expected to continue to grow and is therefore covered in this chapter. However, it normally relies on the use of a separate "instrument" instead of a PC user's existing telephone, which in most organizations is connected to a PBX and provides the focal point for incoming and outgoing calls via an organization's private interconnected PBX network or the public switched telephone network (PSTN). Thus, a second reason for having avoided a detailed discussion of Internet telephony until now is that it represents more of an individual solution than an organizational solution for the transmission of voice over the Internet. However, as we will note in this chapter, there are several methods by which voice can be transported via an IP network, and for many organizations a mixture of methods may be required to satisfy both internal and external requirements, with external requirements primarily driven by potential and actual customers using the Internet to access a help desk, an order desk, or another facility operated by the organization.

In this chapter, we will first discuss the differences between Internet telephony and what this author prefers to refer to as "telephony on the Internet," by which I mean the indirect connection of standard telephones to the Internet. As we examine the basic hardware and software required to support each method of communication, including the advantages and disadvantages, we will consider the reliability issues associated with transmitting audio on the Internet and on private IP networks. From this foundation, we will review the operation and utilization of several vendor products as well as the construction of economic models to illustrate why individuals and businesses are viewing the Internet as the new audio frontier.

In concluding this chapter, we will turn our attention to a technique that can be used to obtain the reliability and predictability required to expedite the transport of voice into a wide area network. This technique is obtained through the use of router queuing. Because Cisco Systems has over 70 percent of the market for IP routers, examples in this chapter will be oriented towards the configuration of different queuing methods on Cisco routers. However, because routers manufactured by other vendors have many features in common with Cisco routers you will more than likely be able to apply the router queuing techniques described in this chapter to other vendor products.

INTERNET TELEPHONY VS. TELEPHONY OVER THE INTERNET

Many years ago, the playwright Oscar Wilde, referring to the United States and the United Kingdom, noted that they were two great countries separated by a common language.

If we fast-forward to our present era and discuss Internet telephony and telephony over the Internet, we could paraphrase his words as "two great techniques separated by technical incompatibilities." While the use of both techniques can be expected to significantly increase in usage, they are designed to satisfy different user and organizational requirements, and they incorporate different technologies to satisfy those requirements. To better understand the major differences between the two voice-transmission techniques, let's examine each technique in detail.

Internet Telephony

In this book, we will use the term *Internet telephony* to refer to the transmission of digitized voice conversations over the Internet by individual PC users. The technology associated with Internet telephony is primarily based on the use of sound cards installed in a PC, a microphone connected to the sound card, and appropriate software. However, later in this chapter we will note that there are many flavors of Internet telephony to include phone-to-phone transported over an IP network as well as PC-to-PC and PC-to-phone. Because of the difficulty in classifying these techniques as an individual method used by a consumer or as individual techniques cumulatively used by businesses, we will refer to all three techniques later in this chapter as Net phone communications.

Overview

The basic operation of an Internet telephony system commences when a person talks into a microphone. The microphone is in turn connected to a sound card installed in the computer, which accepts an analog waveform and converts it into a digital data stream. Internet telephony software operating on the computer takes the digitized voice data stream, which normally represents a 64-Kbps PCM or a 32-Kbps ADPCM-encoded voice, and compresses the standard encoding data stream into a lower data rate based on the use of a proprietary or standardized voice-compression technique. Once this is accomplished, the software packages the digitized and compressed data stream into packets using a protocol for transmission over the Internet. Most Internet telephony products were originally developed primarily to support modem connections; however, modern products also support LAN-based operations when the LAN is connected to the Internet.

There are two primary transport protocols used for an Internet telephony session. TCP is used to transport addressing or directory information, while UDP is used for the actual transfer of voice-digitized packets. Although the actual ability to digitize voice entered through a microphone is a relatively simple process, differences in the manner in which connections are established over the Internet, voice-digitization methods, and the framing of digitized voice samples result in a high degree of incompatibility between vendor products. Before turning our attention to the operation of specific products, let's digress a bit and discuss the economic issues associated with Internet telephony and its basic operation.

Economic Issues

Any discussion of economic issues associated with Internet access must consider the method of access used. Thus, prior to discussing the economic issues associated with Internet access, let's examine the two basic methods used to obtain such access.

Basic Access Methods   There are two basic methods associated with Internet access: dial-up and direct connection. Dial-up access is based on transmission using the Serial Line Interface Protocol (SLIP) or the Point-to-Point Protocol (PPP) to an Internet Service Provider's (ISP's) network access device.

SLIP vs. PPP   A key difference between SLIP and PPP is the fact that the Serial Line Interface Protocol requires the user to know the IP address assigned to their computer by their Internet Service Provider as well as the IP address of the remote system the computer will dial into. If the ISP assigns IP addresses dynamically the SLIP software operating on the local computer must be able to adjust to automatic IP address assignments. In addition, the local computer operator may have to configure such TCP/IP parameters as the MTU (maximum transmission unit); MRU (maximum receiver unit); the use of Van Jacobson (VJ) compression, which results in SLIP functioning as Compressed SLIP (CLSIP); and other features. Recognizing that the configuration of such parameters was beyond the area of knowledge of the growing base of non-technical dial-up users of the Internet resulted in the rationale for PPP, which represents a newer protocol.

PPP has several benefits over SLIP. Those benefits include negotiating configuration parameters at the start of a connection and the support of two security methods for login to a remote system. Because PPP negotiates configuration parameters at the beginning of a session its use considerably simplifies the configuration of a PPP connection. Concerning security, PPP supports the Password Authentication Protocol (PAP) and the Challenge-Handshake Authentication Protocol (CHAP). The selection of either method permits the local computer to automatically transmit a previously entered user-ID and associated password to the remote system. Thus, a majority of serial point-to-point Internet communications occurs via the use of PPP today. We can obtain an appreciation for its capability by briefly examining the structure of the PPP frame.

Pricing Structure   The PPP frame represents a frame structure first standardized by the International Standards Organization (ISO) for High Level Data Link Control (HDLC). In fact, PPP uses the HDLC protocol as a basis for encapsulating datagrams over point-to-point circuits. Figure 7-1 illustrates the composition of the PPP frame.

In examining Figure 7-1 any familiarity with HDLC will allow you to shortly note that a PPP frame is a streamlined version of HDLC. The Flag field represents the bit sequence 01111110 and indicates the beginning or end of a frame. Similar to HDLC, PPP will insert a binary 0 when a sequence of five 1s occurs naturally in the data to prevent its misinterpretation as a Flag field, a process referred to as *zero insertion*. PPP will also remove the inserted 0 at the receiver, performing insertion and removal transparent to the user.

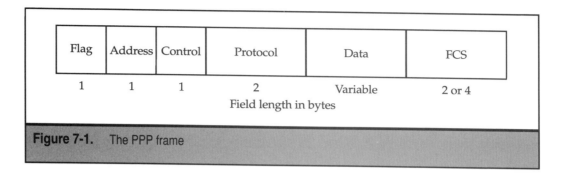

Figure 7-1.   The PPP frame

Unlike an HDLC frame that will transport different addresses in its Address field, PPP always uses the fixed binary sequence 11111111, which represents a standard broadcast address. Because PPP is used on point-to-point circuits, it does not assign individual station addresses. In comparison, HDLC can be used to address multiple devices and normally conveys a specific device address.

The PPP Control field contains the fixed binary sequence 00000011. This value denotes the transmission of user data in an unsequenced frame. Because PPP is restricted to point-to-point circuits, it does not need to transmit sequenced frames or supervisory frames, with the latter two types of frames supported by HDLC.

The two-byte protocol field results in PPP having the capability to encapsulate other protocols besides IP. This field is followed by a variable length Data field. The default maximum length of this field is 1500 bytes, which is also the maximum length of the Information field in an Ethernet frame. Finally, the Frame Check Sequence (FCS) field terminates the frame. This field is normally 2 bytes; however, PPP supports a 32-bit (4-byte) FCS for improved error detection.

Regardless of the protocol used, both SLIP and PPP operating devices will dial an ISP access device. The ISP's network access device physically consists of a series of rack-mounted modems connected to a communications server. The server represents one of several devices connected to a local area network, with a router connected to the LAN, while the router's serial port is used to provide a high-speed communications connection from the ISP to an Internet network service provider (NSP). The NSP typically operates a high-speed backbone connection that provides interconnectivity between ISPs.

Dedicated access is normally associated with the connection of a group of subscribers located within a building or university campus. Under this access method, subscribers are connected to a corporate LAN, and the local area network is in turn connected via the use of a router and leased line to an ISP. Instead of the line terminating at an ISP's communications server, the leased line terminates at a multi-port router connected to a LAN. Each subscriber PC commonly operates a browser on top of a TCP/IP protocol stack, either purchased from a third-party provider or obtained from operating systems that include built-in stacks, such as Windows 95, Windows 98, or Windows NT/Windows 2000. Figure 7-2 illustrates the two primary methods used for accessing the Internet. Now we will turn our attention to the pricing structure of each method.

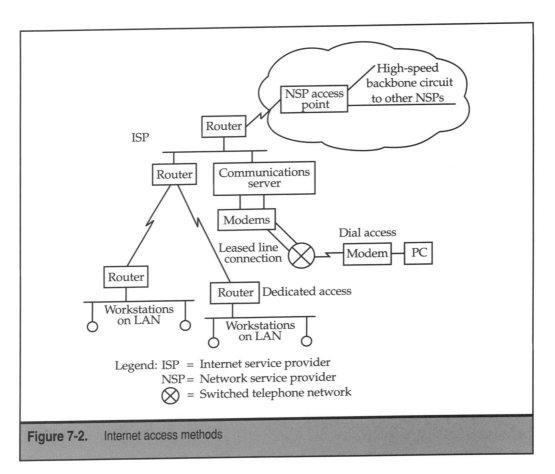

Figure 7-2.   Internet access methods

Pricing Structure   On an individual dial-up connection, many ISPs offer a flat-fee pricing structure, typically $19.95 to $21.95 per month for unlimited use. When this type of pricing structure is used, there is essentially no additional cost associated with transmitting voice over the Internet, other than the one-time cost for hardware and software and fees charged by certain vendors that now offer a voice gateway service. The voice gateway service enables calls routed via the Internet to be dialed to their ultimate destination via the public switched telephone network, as illustrated in Figure 7-3. In examining Figure 7-3, it should be noted that the voice gateway is programmed to accept calls from predefined accounts or a pay-as-you-use account based on the use of "digital cash" or through the use of a credit or debit card. Once access is authorized, the voice gateway will out-dial the desired telephone number, billing the user for a local call and surcharge or for a long-distance call that's more economical because the gateway operator can purchase a block of minutes at a lower rate than individuals can obtain.

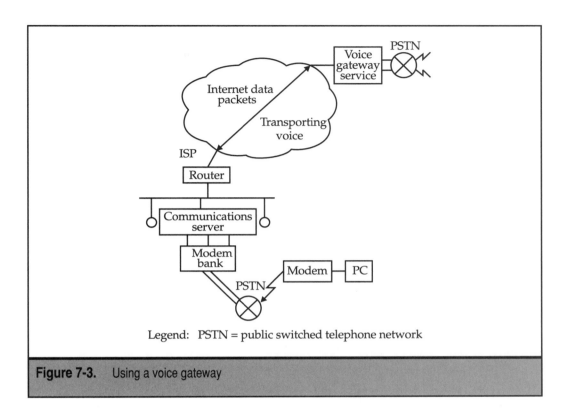

Figure 7-3. Using a voice gateway

The use of a voice gateway is most effective for conducting international long-distance calls. For example, a voice gateway provider might bill calls received via the Internet for a gateway service in London at $.10 per minute. If you were calling via an ISP connection in New York, you could avoid a long-distance international call between New York and London that could cost between $.25 and $1.10 or more per minute, depending on the time of day the call is made. Thus, you might be able to save between $.15 and $1.00 per minute for this type of call.

One new "wrinkle" concerning ISPs that deserves mention is the growth in advertiser-supported free Internet access. During 2000 several popular totally free Internet access providers had accumulated approximately 10 million subscribers. In fact, both this author and his wife were Bluelight fanatics, using the joint Kmart-Yahoo! advertiser-sponsored free Internet access service on a daily basis.

While the ability to reach the Internet at 56 Kbps was equivalent to paid ISP services, there is no free lunch. To pay for free Internet access the subscriber must view advertisements that are continuously displayed on a small portion of their screen. An example of this is shown in the lower portion of Figure 7-4. If you examine the lower-left portion of the screen you will note an advertisement for MasterCard. While you can learn to live

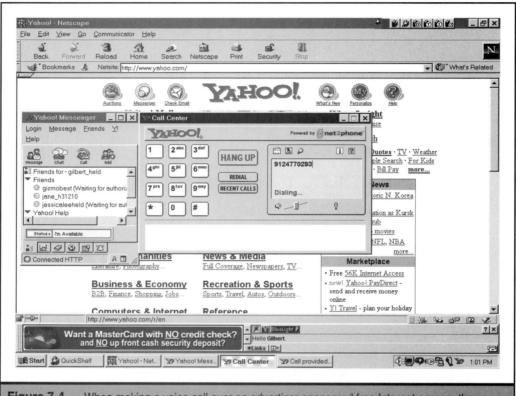

Figure 7-4. When making a voice call over an advertiser-sponsored free Internet access, the periodic updating of advertisements both delays and distorts the voice conversation.

with this minor visual inconvenience, it can play havoc on any attempt to use voice calling over this type of Internet connection.

If you attempt to use Yahoo!Messenger's voice calling facility, which went from the beta version described in Chapter 1 to an operational state when this chapter was written, you would encounter periodic voice distortion. The reason for this distortion results from the fact that the advertising on your screen is periodically updated. If you are listening to a distant party as advertising is updated, the advertising will delay the arrival of packets transporting digitized voice, which may make the conversation difficult to understand. Hopefully, Bluelight and other free services will modify their code to prioritize packets transporting digitized voice over advertisements. However, until this is done the ability to use the Yahoo! Call Center and similar products on advertiser-sponsored Internet access may remain unsuitable for prime time.

Flat-Fee Billing   There are two primary methods ISPs use for billing a dedicated connection. The first is on a monthly flat-fee basis, which is based on the operating rate of the line connection. As you might expect, a 256-Kbps fractional T1 (FT1) line connection costs more than a 56-Kbps line connection, while a T1 line connection operating at 1.544 Mbps costs more than a 256-Kbps FT1 line connection. When based on the line operating rate, the cost associated with transmitting voice over the Internet depends on the capacity and cost of the current line connection. If there is sufficient capacity on the current line connection for the transmission of voice, then the cost of voice transmission is limited to the one-time expense associated with required hardware and software. If sufficient capacity is not available on the current line connection, then the cost associated with voice transmission requires you to determine the additional costs of replacing the existing connection with a higher-speed connection.

Line-Utilization Billing   The second popular billing method used by ISPs is based on line utilization. Some ISPs will install a T1 connection that is billed at a monthly base cost for the line plus a utilization charge, with the latter based on either the average percentage, a fixed percentage, or a similar metric computed over a 24-hour period. This type of billing is more difficult to use for estimating the cost of transmitting voice, as you now have to determine the potential effect of transmitting and receiving digitized voice on the utilization level of your organization's Internet connections.

For example, if you obtain equipment that results in the digitization of speech at an 8-Kbps data rate, then every minute of speech activity results in the additional transfer of 8 Kbps × 60 seconds/8 bits/byte, or 60 Kbytes per minute. On an hourly basis, this results in the additional transfer of 3.6 Mbytes of data, without considering the effect of the overhead of frames used to transport digitized speech or the potential effect of silence suppression. The former results in additional data transfer, while the use of silence suppression results only in periods of speech actually being conveyed and reduces the amount of data transfer. Since many Internet telephony applications lack the sophistication necessary to suppress periods of silence, each hour of voice transmitted will probably result in approximately 4 Mbytes of additional activity over an Internet connection if you include frame overhead. Whether or not this level of additional activity adversely affects your organization's existing Internet connection will depend on the current operating rate of your organization's Internet connection, its current level of utilization, and the anticipated level of digitized voice expected to be transported over the connection.

APPLICATION NOTE: For cost estimation purposes a voice digitized conversation at 8 Kbps results in approximately 4 Mbytes per hour of traffic.

Upgrade Considerations   When a dedicated connection to the Internet is used, the economics associated with transmitting voice over this IP network make its use harder to justify when a line upgrade is required. This is because a line upgrade might easily increase your organization's monthly Internet usage charge by $250 to $500 per month or more. This means that when calls are national, your cost avoidance might be limited to between $.10 and $.15 per minute. Thus, if your organization's monthly Internet access billing increases by $250 per month, you would have to transmit 2500 minutes of voice per month at \$.10 per minute for PSTN usage simply to break even—without even considering the one-time cost of equipment. This means you must carefully estimate your anticipated volume of voice transmission to determine if using the Internet to transmit voice via certain types of dedicated connections is economically sound. Fortunately, corporate PBX reports and communications carrier bills represent two viable sources for obtaining the information required to make a sound economic decision. Now that we have an appreciation for the general economic issues associated with the transmission of voice on the Internet, let's turn our attention to the operation of Internet telephony, including several performance issues you might wish to consider.

Internet Telephony Operations

The actual PC configuration required for Internet telephony operations depends on the software program you are using. The software program operates under a specific operating system (such as a version of Microsoft Windows or the Apple Computer Macintosh System 7.x), interfaces with a sound card to compress voice input entered through a microphone connected to the card, and coordinates a TCP/IP SLIP or PPP communications connection via the use of a modem. Originally, most sound cards were half-duplex, but today most manufactured sound cards can operate in a full-duplex mode. Thus, obtaining a full-duplex telephone connection requires the use of a full-duplex sound card as well as software that supports full-duplex operations.

Hardware and Software Requirements

Table 7-1 lists the general categories of hardware and software required for Internet telephony.

Processor and RAM   Concerning specific hardware, at a minimum, most programs require a high-performance 486 processor or equivalent since the use of the processor to

Software	
Telephony program	Operating system
TCP/IP stack	Macintosh, Windows
Hardware	
Sound card (half- or full-duplex)	Microphone
Speaker	Computer platform
Microprocessor	RAM

Table 7-1.   General Categories of Hardware and Software Required for Internet Telephony

perform voice compression is processor-intensive. The actual process and RAM memory required is commonly noted in the program specification sheet for a product.

Modem Since the transmission of digitized voice involves some overhead resulting from the framing of voice packets, including header and trailer fields, most Internet telephony programs require the use of a modem that operates at a minimum data transfer rate of 14.4 Kbps. To put this operating rate in perspective, it is equivalent to 1800 bytes per second, while voice encoding using PCM requires 8000 bytes per second of bandwidth, which is reduced to 4000 bytes per second of bandwidth when ADPCM coding is used. Clearly, then, Internet telephony depends on the use of a vocoding or hybrid coding technique to enable 14.4-Kbps modems to support a digitized voice transmission capability. In fact, the method of voice digitization can differ between vendor products, and this is one of several issues that currently result in a high degree of non-interoperability between different vendor products.

While most Internet telephony products can operate at a data transfer rate of 14.4 Kbps, the quality of reconstructed voice considerably improves as the modem operating rate increases. If you use a V.90 or the newer V.92 modem you will obtain the ability to download data at approximately 42 Kbps for each modem while uploading can be accomplished at 33.6 Kbps for a V.90 and 40.0 Kbps when a V.92 modem is used. Because both the V.90 and V.92 modems provide more than double the operating rate of a symmetrical 14.4 Kbps modem, their use will significantly reduce latency on the access line. This may be sufficient to convert a marginal voice-over-IP call into one that provides a high quality of audio reconstruction.

Sound Card Through the use of a full-duplex sound card and software support, you can enable both parties to talk at the same time. In actuality, a full-duplex communications capability is slightly better than a half-duplex communications capability because you are able to gracefully back out of a conversation instead of having to awkwardly wait to ascertain if a person is done speaking as in half-duplex communications. Some programs enable the use of two sound cards, one for playback and one for recording the conversation. If you have an extra sound card and your program supports the use of two cards, you can avoid the purchase of a full-duplex card.

Although you might be tempted to avoid the use of a full-duplex sound card because of possible bandwidth problems, you should note that the actual bandwidth required may increase by only a few percent over the use of a half-duplex sound card. This is because many Internet telephony programs that support full-duplex operations also use silence suppression, transmitting data only when a person actually speaks.

Directory Services

Originally, many Internet telephony software products were based on the use of a "directory service" server. To initiate a call, you use the program operating on your computer to access the software vendor's directory and double-click on an entry. The other party must be logged on to an Internet Service Provider and running the same software on his or her computer for the call to be received. Other Internet telephony software products provide a directory with a list of online users and chat rooms as a mechanism to allow users to meet new friends around the globe. Other programs either include a direct access capability or allow direct access after you access their central server. If you know the IP address of a user you want to call directly, you can simply enter his or her IP address and click on a call button on the program interface to initiate a call. If the called party is logged on to the Internet and running the same software program, the call will be received.

A relatively new addition to Internet telephony is the incorporation of telephone software into different Internet messenger products, such as Yahoo!Messenger, which was illustrated in Figure 7-4 earlier in this chapter and which was discussed in some detail in Chapter 1. Yahoo!Messenger and similar products that added a "Call" capability were using the facilities of Net2Phone or another native communications carrier which operates a series of gateways in larger cities across the globe. In the case of Yahoo!Messenger, users of this service could make free telephone calls to any telephone in the United States without charge. Calls to overseas locations could be accomplished at a substantial discount to the cost associated with the use of one of the big three long-distance companies. Now that we have an appreciation for the basic components required to implement Internet telephony, let's look at some of the constraints and compatibility issues associated with this relatively new technology.

Constraints and Compatibility Issues

There are a number of constraints associated with the use of Internet telephony, and they are solved in a variety of ways by different vendors. Unfortunately, the lack of a uniform approach to solving these constraints results in a number of compatibility issues that make interoperability between different vendor products difficult, if not impossible. Table 7-2 lists five key areas of constraints. We will examine each of these groups in this section.

Bandwidth Conservation Method   The ability to reproduce a natural-sounding conversation requires a trade-off between the speech-coding scheme and processing power of the PC. In general, the ability to highly compress speech while enabling the reproduction of it to produce natural-sounding conversation requires more processing power than encoding schemes that reproduce either synthetic sound or produce digitized speech at a higher data rate.

There are three voice-coding techniques used in different popular Internet telephony applications:

▼ Low Delay Code Excited Linear Prediction (LD-CELP), standardized by the ITU as Recommendation G.728, resulting in a 16-Kbps data stream

■ Conjugate-Structure Algebraic Excited Linear Prediction (CS-ACELP), standardized by the ITU as Recommendation G.729, resulting in an 8-Kbps data stream

▲ The dual-rate 6.3/5.3-Kbps voice-coding method, recently standardized by the ITU as Recommendation G.723.1

Bandwidth conservation method

G.729   G.728

G.723.1

Packet delay and loss handling

Repair lost packets with silence   Repair lost packets with synthetic speech

LAN connectivity operation

Requires modification to firewall   Products use different ports
and router access lists

Connection method

Directory- or IP address-based   Gateway-based

Protocols used

Table 7-2.   Internet Telephony Constraints

Until recently the G.729 coding method was very popular. However, the standardization of the G.723.1 dual-speed coding method is gaining in popularity and is also recommended as the low-bit-rate speech coder for the ITU H.323 standard for video and voice communications over packet-based networks. At one time, it was expected that most Internet telephony products would adopt the G.723.1 Recommendation. However, from a real-world operational perspective, the delay or latency associated with the G.723.1 dual-speed coding method is approximately 60 ms, which is 12 times the delay associated with the use of the LD-CELP coding method. In many situations, the additional delay is of sufficient duration to cause reconstructed voice to sound awkward. Thus, the trade-off between bandwidth and delay can represent a key item you may wish to consider when configuring an Internet telephony product that supports multiple voice-coding methods. Many times, a voice-coding method that results in awkward-sounding reconstructed voice at one operating rate may provide a higher quality of reconstructed voice when you select a higher operating rate.

APPLICATION NOTE: Consider varying the use of different audio codecs as a mechanism to determine if a more suitable codec is available for use.

Windows Codec Support   Although several Windows-compliant products use proprietary codecs, other products permit you to select a codec supported directly by Windows or add a codec. By using the Control Panel, you can both determine the audio codecs supported by the version of Windows used on your computer as well as observe and obtain the ability to alter the manner by which a codec is selected for many applications.

To check the codecs supported on your computer, go to Start | Control Panel and double-click the icon labeled Multimedia Properties. Next, select the tab labeled Advanced, choose the Audio Compression codecs entry, and click the plus mark (+) to its left. This action will result in the display of the audio compression codecs supported on your computer. Figure 7-5 illustrates the display of the Multimedia Properties Advanced tab on this author's computer after the previously described sequence of operations was performed.

You can display information about a codec by clicking on it to display a dialog box with a tab labeled General. The left side of Figure 7-6 shows an example of the dialog box displayed as a result of selecting the Microsoft CCITT G.711 A-Law and u-Law CODEC. If you click the About button, a description of the codec is displayed. The right portion of Figure 7-6 illustrates the display of information about the Microsoft G.711 codec. In the left portion of Figure 7-6, note the box with the down arrow that has a current priority value of 2. This value is used by Windows as a criterion for using a particular audio codec. When two or more audio codecs are equally capable of compression, Windows will use them in the order of the priority assigned to them.

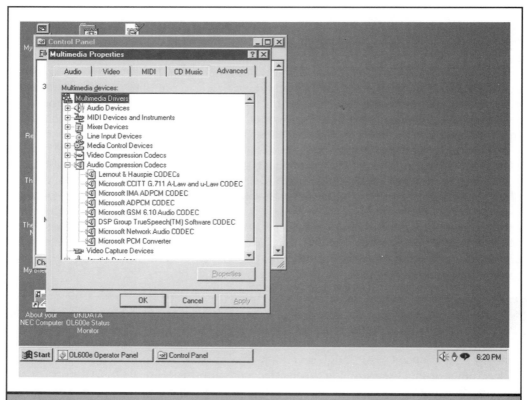

Figure 7-5. Viewing currently installed audio codecs on a Windows-based computer

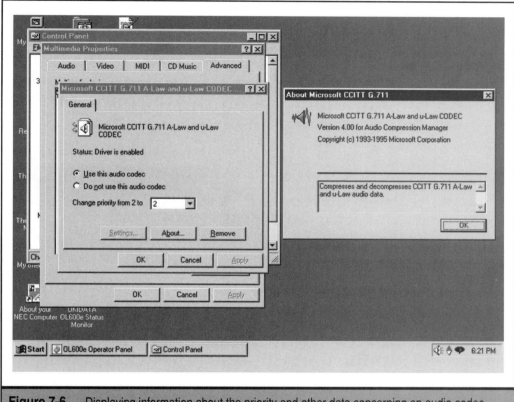

Figure 7-6.   Displaying information about the priority and other data concerning an audio codec

Under Windows 2000 access and utilization of audio codecs slightly changed. You can continue to go to the Control Panel; however, the applicable icon to access codecs is now labeled Sounds and Multimedia. Double-clicking that icon displays a dialog box with three tabs labeled Sounds, Audio, and Hardware. This dialog box with its Hardware tab selected is shown in the top-left portion of Figure 7-7.

Selecting Audio Codecs from the Hardware tab displays the dialog box labeled Audio Codecs Properties, which lists the audio codecs built into Windows 2000. This dialog box is shown in the upper-right portion of Figure 7-7. If you compare the built-in audio codecs shown in Figure 7-7 to those shown in Figure 7-5 you will note Windows 2000 added two new codecs (msg723.acm and Indeo audio software) and dropped the Lermout & Hauspie codec used by Windows 95 and 98. However, we will shortly note that Lermout & Hauspie is still included in Windows 2000 in its Sound Recorder program.

Returning to Figure 7-7, double-clicking on any audio compression codec in the upper-right corner of that illustration will display information about the codec as well as the ability to enable or disable the use of the codec. In the example shown in Figure 7-7 this

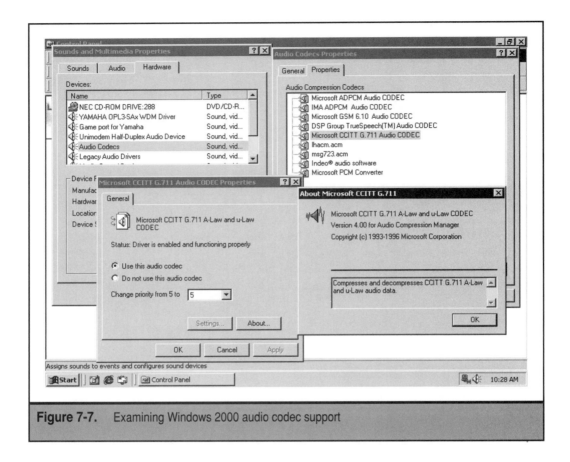

Figure 7-7.   Examining Windows 2000 audio codec support

author selected the CCITT G.711 codec and moved the resulting display to the lower-left portion of the screen. Once this was accomplished the About button was selected and the information about the selected codec was positioned in the lower-right corner of Figure 7-7. Thus, while similar to the use of codecs under earlier versions of Windows, the latest version has a few differences concerning the support of different codecs.

As previously mentioned, Windows 2000 includes a Sound Recorder program. What was interesting about this recorder was the fact that some audio codecs no longer supported by Sounds and Multimedia are still supported by the Sound Recorder.

The top-left portion of Figure 7-8 illustrates the Windows Sound Recorder, while the top-right portion of the display shows the Save As dialog box after this author sang a song (which explains the filename to be saved). By clicking the Change button you can change the format of the audio recording from its default as a .WAV file to another format. The resulting dialog box labeled Sound Selection is shown in the lower-right portion of Figure 7-8. Note in the pull-down menu that the Lermout & Hauspie codecs are still

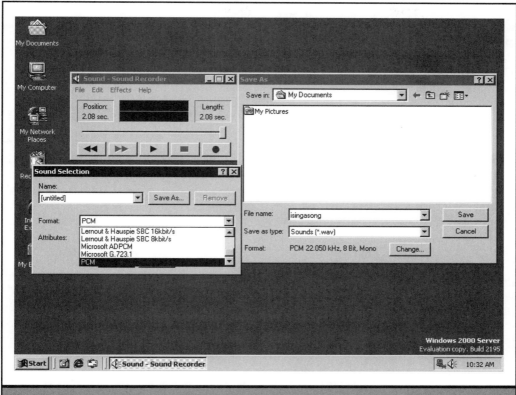

Figure 7-8. The Windows 2000 Sound Recorder supports some audio codecs that are not supported in Sounds and Multimedia.

supported. Now that we have an appreciation for the audio codecs supported by different versions of Microsoft Windows, let's continue our examination of constraints and compatibility issues associated with Internet telephony.

Packet Delay and Loss Handling   Two key considerations associated with the use of packet networks affect voice transmission: packet delay and packet loss. The congestion of routers and gateways—resulting from either processing packets or the inability to transfer packets onto communications facilities due to heavy line utilization—causes delay or loss of packets.

When data is being transferred, a slight delay in the arrival of one or more packets is usually not noticeable. At worst, it might result in a person waiting an extended period of time for a file transfer to complete, but the content of the transferred file would not be affected. Similarly, the loss of packets as they flow through an IP network resulting from congestion and routers, workstations, or gateway discarding packets is compensated for by the retransmission of discarded packets. However, when packets transport digitized voice, normal data transmission methods cannot be used. This is because the loss or delay of packets results in the disruption of speech intelligibility.

There are two methods that can be used to compensate for the loss or delay of packets. Those methods involve repairing lost or delayed packets with periods of silence or with synthetic speech.

Silence Generation   Currently, most Internet telephony applications simply generate periods of silence to compensate for lost packets and reproduce delayed packets. This results in the clipping of speech and a loss of its intelligibility when packets are lost in the network, and a distortion of speech when delayed packets are used to reproduce speech. At the time this book was being prepared, several Internet telephony vendors were considering the possibility of using previously received packets as a mechanism to generate synthetic speech to fill in periods of silence. In doing so, there are several methods that can be used for voice reconstruction, based on where the reconstruction process occurs.

Voice Reconstruction   Voice reconstruction can occur by the receiver attempting to reconstruct the missing segments of speech from correctly received packets preceding the packet or from packets that are lost or delayed. This can be accomplished by the repetition of a portion of the last correctly received speech waveform or via the interpolation process. When a combined transmitter and receiver method is used, extra information is included within each transmitted packet to facilitate the reconstruction and interpolation process at the receiver. Another combined transmitter and receiver technique involves the adjustment of packet sizes dynamically, based on packet delay and packet loss metrics. Making packets smaller enhances their ability to flow through a packet network, since many routers and gateways use queues that favor small-size packets. Currently, no standards exist concerning the handling of lost or delayed packets, and it may be several years until an approach is standardized. This means that the selection of a product that provides this capability should have an option to turn off lost and delayed packet handling if it is to interoperate with other products that do not offer this feature or that implement it in a different manner.

LAN Connectivity Operation

Although most Internet telephony products were originally developed to use SLIP and PPP dial connections to ISPs, many products now support the use of LAN connections, enabling the product to recognize gateway and Domain Name Server (DNS) addresses configured with the TCP/IP protocol stack operating on a computer. Since Internet telephony products can use both TCP and UDP ports for establishing connections to a directory server or directly to a distant party, the use of those ports may cause conflicts with existing firewalls or router access lists designed to provide a level of security to organizational computational equipment located behind those devices.

APPLICATION NOTE: Coordinate the use of Internet telephony products on a LAN with your organization's firewall and router administrators. Many times, Internet telephony products will fail to operate due to the ports they use being blocked by a router access list and/or a firewall.

Security Considerations   To illustrate the effect of security implemented in the form of access lists on voice traffic, consider Figure 7-9, which shows the use of a firewall to protect a corporate LAN. DMZ is an acronym for *demilitarized,* and a DMZ LAN represents a local area network that has no workstations connected to the network. This means that inbound packets from the Internet must first flow onto the DMZ LAN, from which they are received on one port of the firewall for processing prior to being placed onto the corporate LAN. By using a DMZ LAN, the firewall is able to process every inbound packet prior to the packet being able to be received by another corporate network device.

Both routers and firewalls process packets based on access lists as well as other metrics; however, the access list can be considered the initial qualifier, determining whether the packet reaches the next processing step or is discarded. Thus, it is extremely important for router and firewall access lists to be configured to enable the use of an Internet telephony product. For example, the Vocal Tec Internet Phone product uses two channels. It uses TCP port 6670 to connect to the vendor's Internet Phone Server's directory service, while audio is passed through UDP port 22555 on both local and remote computers. Internet Phone also uses TCP port 25793 to connect to the Vocal Tec addressing server and TCP port 1490 for whiteboard, chat, and file transfer when its conferencing option is

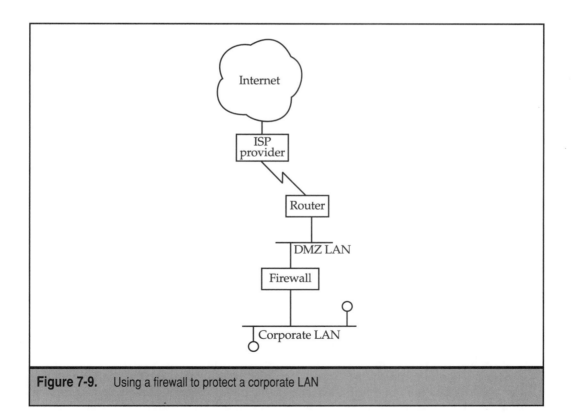

Figure 7-9.   Using a firewall to protect a corporate LAN

used. This means that organizational firewalls and routers must be configured to enable data transfer on those ports if the Vocal Tec Internet Phone is used. Unfortunately, there is no standard port assignment, and not only do certain products differ in their use of TCP and UDP ports between vendor products, but there are also differences between versions of certain vendor products with respect to the use of TCP and UDP ports.

As mentioned in our discussion of TCP/IP earlier in this book, the placement of statements in an access list will affect the delay associated with allowed packets flowing through a router. In general, the router or firewall administrator should place voice-related access control list statements as high as possible in the list, usually directly after antispoofing statements.

Connection Method   As mentioned earlier in this section, some Internet telephony products are directory-based, requiring a user to first access a directory prior to establishing a call. Other products permit users to enter an IP address or, increasingly more common, a telephone number to establish a call. For the first two methods, the destination or called party must be online and operating the same software program.

When a telephone number is used the destination represents a PSTN subscriber. Thus, normal telephone operations apply at the distant end, only requiring a person to be in close enough proximity to the dialed telephone to hear its ringer being activated.

One version of a directory service that is receiving a considerable degree of interest is the Lightweight Directory Access Protocol (LDAP), which allows a collection of directories—such as e-mail, voicemail, security, and even a PBX function—to function as a single integrated directory service.

Figure 7-10 illustrates an example of the use of an LDAP directory service. In this example, information about employees (including their e-mail address, single logon system password, telephone number, and any other directory-related information) is placed on a centralized LDAP directory. When a new employee is added to a LAN, employee information is furnished only to the LDAP directory. The LDAP protocol then disseminates appropriate information to each communications system operated by the organization that supports the LDAP. In the example shown in Figure 7-10, one entry to the LDAP directory results in the automatic update of five directories, including an Internet telephony directory.

Gateway Service

A third connection method enhances the use of the Internet, as it provides a gateway service that converts an Internet call to the PSTN. One popular example of a gateway service product is the IDT Corporation's Net2Phone program. Net2Phone functions similarly to other Internet telephone products in that it operates in conjunction with a sound card, microphone, and speakers to initiate and receive calls via the Internet. Where Net2Phone differs is in its use of a central telephone switch, which functions as a gateway for out-dialing via the PSTN. This use of a centralized switch located in the United States results in all gateway calls originating in the United States. In addition, through the use of a gateway, users of Net2Phone are not limited to calling a party who is online and using a multimedia-equipped PC. Instead, users can call any person who has a standard telephone and the outbound call will ring the distant phone.

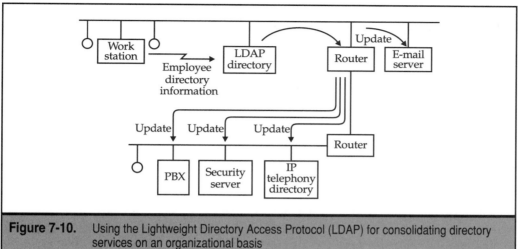

Figure 7-10.   Using the Lightweight Directory Access Protocol (LDAP) for consolidating directory services on an organizational basis

Utilization Economics   Net2Phone, like many relatively young Internet telephone companies, offers subscribers several methods for making Net phone calls. Net2Phone supports phone-to-phone calling based upon a prepaid account as well as free PC-to-PC and PC-to-phone calling in the United States and Canada, with significant discounts for international calling. For our discussion of economics we will focus our attention upon Net2Phone's prepaid phone-to-phone calling. To use this calling option you would either access the Net2Phone secure Web server and purchase via credit card $25, $50, $100, or $250 in prepaid calling or send them a check or wire transfer to purchase a prepaid level of service. Once your prepaid account is established you can dial an access number located in major cities around the globe. You would be prompted to enter a ten-digit account number and four-digit pin. For calls to the U.S. and Canada you would then dial 1 + area code + number when prompted. For other countries you would use the international prefix of 011 instead of the long distance prefix of 1 and follow the international prefix with the country code, city code, and local number. Thus, international calls would have the format 011 + country code + city code + local number.

While Net2Phone was probably the largest Internet telephony vendor marketing to consumers during 2000, its global reach was limited to major cities. For example, if you live in Atlanta, Georgia, it was a local telephone call to connect to a node on the Net2Phone network in the form of connecting to a local access number. However, if you live in Macon or Savannah, Georgia, which are not quite rural locations, you would have to make a long-distance call to Atlanta or Athens, Georgia, where the only local access points into the Net2Phone network were in place in this author's home state when this book was written. Thus, it is important to consider rates as well as access fees to obtain discount rates. One other item worth noting concerning Net2Phone is the fact that your account will be debited $0.99 each month as a processing fee. Even with calls costing $0.039 per minute in the continental United States appearing to be a bargain, readers should note that if you can structure your calls to the evening or weekend, other dialing plans may be as attractive or even more so. For example, Sprint was marketing 1,000 long-distance U.S. continental minutes for $20 per month during the Fall of 2000.

The current voice-digitization method used by Net2Phone requires a voice bandwidth of approximately 13 Kbps, which, with protocol overhead, results in a minimum operating rate of 14.4 Kbps. Net2Phone operates under Windows 95, Windows 98, and Windows NT version 4.0.

Net2Phone's key economic value is for communications between foreign locations. In the past, a number of callback companies took advantage of the high tariffs associated with the cost of calling between different non-U.S. locations by enabling users to first dial the United States. After receiving the caller ID number, but prior to answering the call, the callback service would disconnect the incoming call and dial the distant party. The distant party would enter an access code for billing purposes and receive a dial tone that allowed them to initiate a call from the United States to the distant party. Typically, the cost of the two U.S.-originated calls, one to the calling party via a callback and one to the called party, was half the cost of directly calling from one country to another. In fact, you can judge the popularity of international callback by reading a copy of the *International Herald Tribune*. In each issue you'll find a large number of advertisements for callback services.

In addition to providing a callback service for non-U.S. destinations, callback services also support calling throughout the United States. This enables users in foreign countries to call a person or business in the United States, with the cost based on prices for the callback and the long-distance call in the United States. In addition, through the use of Net2Phone, it becomes possible for U.S. businesses with 800, 888, and 887 numbers to expand on a global basis. With Net2Phone, most toll-free 800, 888, and 887 numbers become global toll-free numbers.

By applying the callback principle to Internet-originated calls, Net2Phone subscribers can use their PCs to call any city in the United States from anywhere in the world for as little as $.10 per minute during 1999, with certain rates declining to as low as $.039 per minute during 2000. Significant savings can also be obtained when calling persons in other countries, because IDT purchases blocks of communications capacity at wholesale prices and resells it at significant discounts. For example, Net2Phone users could dial Australia for $.10 per minute, Austria for $.15 per minute, Finland for $.17 per minute, Germany for $.10 per minute, Israel for $.17 per minute, and Tel Aviv for $.10 per minute, and the United Kingdom for $.10 per minute during 1999. However, for certain countries listed in the Net2Phone rate chart, it could be more economical to use an international calling plan offered by one of the major communications carriers or the facilities of a "1010" bypass carrier. Concerning the latter, since 1995 a number of small communications resellers have been established that purchase long-distance transmission in bulk and resell international calls at rates significantly below the rates of the major communications carriers. Each of these services are accessed by dialing 1010 followed by a three-digit suffix. Since the 1010 bypasses the normally selected long-distance communications carrier, the term "1010" is commonly referred to as a *bypass carrier*. Until 1998, the digit pair 10 was used to bypass the subscriber's default long-distance carrier. However, because the number of carriers offering bypass has expanded considerably, the prefix was changed to 1010. This allows other prefixes, such as 1010, to be used once all 999 bypass carriers are assigned to the 1010 prefix. In late 1999, this author could dial Argentina and Egypt using a bypass carrier for $.35 and $.50 per minute, respectively. In comparison, Net2Phone's cost was $.42 per minute for calling Argentina and $.60 per minute for calling Egypt. Thus, for some persons, it may pay to comparison shop, especially if you anticipate frequent calling to a particular country.

Another important consideration is the fact that both domestic and international long-distance rates are anything but static. For example, by October 2000 Net2Phone rates for long-distance international calls to Australia or Germany were $0.079 per minute, while calling Austria or Finland cost $0.099 per minute. Calls to all phones in Israel cost $0.09 per minute, while calls to phones in the United Kingdom cost $0.079 per minute. If you compare 1999 and 2000 calling costs you will note a significant decline in calling costs on a percentage basis. For example, calling Australia was reduced from $0.10 per minute to $0.079 per minute, a reduction of 21 percent. Similarly, calling Austria went from $0.15 per minute to $0.099 per minute, a reduction of 51.5 percent. With reductions of this magnitude it is relatively easy to understand why AT&T and MCI Communications were discussing spinning off their consumer long-distance operations at the time this book was written.

Net Phone Vendors   In the previous edition of this book I made a distinction between Internet telephony and telephony over the Internet. However, due to the evolution of Internet telephony another distinction is required. That distinction results in the added classifications of phone-to-phone, PC-to-PC and PC-to-phone calling to Internet telephony. While this distinction is to a degree trivial, it is important as some vendors of Internet telephony only support the use of conventional telephones to voice gateways where a conversation is digitized and routed through an IP network to the city where the called party resides. At that city another gateway connected to the PSTN converts the digitized voice conversation compressed using the codec supported by the vendor to the 64 Kbps PCM format used by the PSTN. Other vendors permit PC-to-PC calling via the use of a directory service while a third classification involves PC-to-phone transmission. The latter usually involves the use of a vendor's private IP network where the call is routed through a gateway connected to the PSTN in a manner similar to the previously described phone-to-phone classification. Because many vendors now offer two or more of the previously described services, I will simply classify all such services oriented towards individual users as Net phones, which explains the heading of this section.

Table 7-3 lists eight Net phone vendors that were known to be operating when this author was preparing the revised edition of this book. As I review the major characteristics of each vendor it is worth noting that as this is a rapidly evolving field, you may wish to explore each vendor's Web site to obtain the latest available information about one or more specific vendors.

Deltathree   At the time this book was revised, Deltathree had changed its name to iConnectHere; however, its Web site URL remained **http://www. deltathree.com.** Deltathree or iConnectHere provides phone-to-phone calls via a toll-free number in the U.S., Canada,

Vendor/URL	Calling Card Phone-to-Phone	Free PC-to-Phone Calls in U.S.	Free PC-to-PC Calls
Deltathree (www.deltathree.com)	Yes	Yes	Yes
Dialpad (www.dialpad.com)	Yes	Yes	Yes
Firetalk (www.firetalk.com)	No	No	Yes
HotTelephone.com (www.hottelephone.com)	No	Yes	No
MediaRing (www.mediaring.com)	No	Yes	Yes
Net2Phone (www.net2phone.com)	Yes	Yes	Yes
PhoneFree.com (www.phonefree.com)	No	Yes	Yes
TalkFree (www.i-link.com)	Yes*	Yes (1 limited availability)	No

*After PC connection places calling and called numbers, the connection may be terminated to allow a phone-to-phone call.

Table 7-3. Representative Net Phone Vendors

Western Europe, and Hong Kong, with calls in the U.S. and Canada at $.069 per minute. This vendor also supports free PC-to-PC and PC-to-phone calling in the U.S. and Canada.

Figure 7-11 illustrates an example of the iConnectHere dialing pad. Note that you can access a phone book of predefined numbers as well as call and hang up a call by clicking on a button. If you want to make calls outside the U.S. and Canada you must first fund your account.

Dialpad Similar to iConnectHere, Dialpad supports phone-to-phone as well as free PC-to-phone and PC-to-PC calling. In fact, Dialpad's phone-to-phone rate was $.069 per minute, which was the same as the iConnectHere rate. One of the key features of Dialpad is its use of a Java plug-in, which after you complete a lengthy questionnaire permits you to directly dial telephone numbers similar to the Net2Phone examples described in Chapter 1 and earlier in this chapter.

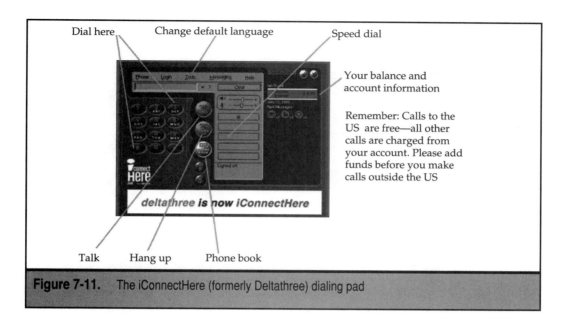

Dial here   Change default language   Speed dial

Your balance and
account information

Remember: Calls to the
US are free—all other
calls are charged from
your account. Please add
funds before you make
calls outside the US

deltathree is now iConnectHere

Talk   Hang up   Phone book

Figure 7-11.   The iConnectHere (formerly Deltathree) dialing pad

Figure 7-12 illustrates an example of the Dialpad phone book containing an entry for this author in the background of the figure while the foreground shows Dialpad's dialpad. As you will note from a careful examination of Figure 7-12, there are advertisements in each dialog box. This is to be expected since there is no free lunch and someone must pay for the calls we make.

In examining the dialpad shown in Figure 7-12 note the two slidebars in the upper-right corner under the Amazon.com advertisement. The top slidebar allows you to control the volume of your speakers, while the lower slidebar controls your microphone.

Firetalk   Unlike iConnectHere and Dialpad, which offer all three types of Net phone services, Firetalk specializes in PC-to-PC communications. Firetalk provides free PC-to-PC phone service and includes banner advertisements. A second service, referred to as Firetalk VQ, represents a premium service that supports both two-party and conference calling with better voice quality and without banner adds. The cost of Firetalk VQ was $19.95 for a six-month subscription when this book was prepared.

HotTelephone.com   Similar to Firetalk, HotTelephone.com only provides one of the three categories or types of Internet telephony services. In the case of HotTelephone.com its service is free PC-to-phone calls in the U.S. and to 29 other countries. You can use HotTelephone.com to call friends in places ranging from Australia to Wales; however, you will have to view flashing adds as the price of free calling.

MediaRing   MediaRing offers a series of voice calling for both businesses and individual consumers. For businesses MediaRing provides PC-to-PC services, instant voice messaging,

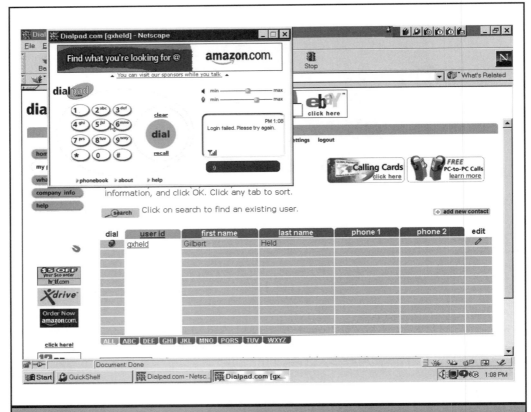

Figure 7-12. The Dialpad phonebook in the background with one entry while the foreground dialog box illustrates the vendor's dialpad

and PC-to-phone calls. For consumers MediaRing provides free PC-to-phone and PC-to-PC calling in the United States as well as discount calling to over 200 cities and countries around the globe.

Net2Phone As discussed earlier in this chapter, Net2Phone offers all types of Net phone calling options. Since we previously discussed the use of Net2Phone we will not repeat what was already presented.

PhoneFree.com PhoneFree.com provides free PC-to-phone calls throughout the United States and free PC-to-PC calling on a worldwide basis. In addition to the previously mentioned services PhoneFree offers numerous additional services. Those services include voicemail, video calling, video mail, teleconferencing, instant messaging, and bubble chat. Concerning the latter, bubble chat represents a chat area with pictures and bubbles that allows users to locate persons that they may wish to contact via a free PC-to-PC call.

TalkFree   Although TalkFree was providing free PC-to-phone calls in the United States, there was a limit on call duration. At the time this edition was revised calls were limited to 4 minutes and could be extended to up to 20 minutes, 2 minutes at a time. TalkFree is operated by I-Link, which explains the URL shown in Table 7-3. Currently, TalkFree represents a hybrid commercial free PC-to-phone and phone-to-phone service. That is, you use your PC to access the TalkFree Web site and enter your phone number and the number to be called. If your modem and telephone share the same line, which is common for many Internet surfers, you must disconnect your Internet connection. Then the TalkFree server will call your number and the called party.

Features   Several common features recently added to Internet telephony products include call-progress displays and an Internet call-waiting feature. Although Internet call waiting can be implemented as a directory service by an Internet telephony company, the feature would then be limited to informing you that another caller was attempting to call you using the same directory service. Because most homes that have Internet subscribers are limited to one telephone line, the previous method of call waiting does not provide the ability to service conventional PSTN-based calls. Although many persons subscribe to call waiting, you usually turn off the service to avoid disconnecting an existing Internet connection. Recognizing this problem, Nortel Networks introduced an Internet call-waiting (ICW) service that enables persons on an existing Internet connection to receive a pop-up window on their computer that alerts them to the incoming call. The user can then accept the call, forward or redirect the call, or play an outgoing message. If the subscriber decides to take the incoming call, software on the PC will accept the call via a voice-over-IP connection while allowing the subscriber to maintain his or her Internet connection.

Deployment of Nortel Networks' ICW is based on installing a call-waiting server in a central office and configuring the central office switch to forward calls to the server when the subscriber line is busy. The server will then communicate with the subscriber through the central office via a TCP/IP connection that pops up the previously mentioned window on the subscriber's PC. As of mid-1999, ICW service was deployed in approximately 16 cities in North America. During 2000 Sprint began marketing a combination of long-distance dialing and Internet access via Earthlink. As part of its marketing effort Spring was providing a six-month period of free call-waiting service.

Protocols Used   As mentioned earlier in this section, most Internet telephony applications use TCP for addressing information and UDP for the actual transmission of packets containing digitized and compressed voice. UDP is a best-effort connectionless protocol that does not include a negotiated flow control capability. This creates two related problems. First, there is no guarantee that a sent packet will reach its destination or will reach it in a timely manner. This is because a packet can be dropped at any point in the network due to error or network congestion. Second, there is no way to control the flow of information when you are using UDP. This lack of flow control could cause UDP packets carrying Internet telephony to flood the Internet—to the detriment of other Internet applications. Based on the preceding, a mechanism is required that would reserve bandwidth from source to destination through the Internet. If this could be accomplished, voice-encoded packets could flow in an orderly manner from end to end, eliminating the possibility of packets being dropped due to network congestion or a non-steady-state arrival, with the accompanying variances in time between packets that cause distortion to reconstructed voice.

The mechanism required to accomplish this is the ReSerVation Protocol (RSVP), described earlier in this book. Unfortunately, the actual full-scale implementation of RSVP is probably many years distant, as two key issues remain to be resolved. First, RSVP requires the upgrade of all gateways and routers between source and destination. This means that one Internet user calling another may have the connection routed between one or more NSPs and two ISPs, all of which would have to have equipment upgraded to support RSVP. This upgrade could conceivably require a decade, because many routers and gateways are not upgradable and would require replacement. Second, bandwidth is not free. This means ISPs and NSPs can be expected to bill for the use of reserved bandwidth. How this will occur, how multiple providers along a reserved path will coordinate billing, and the ultimate manner of billing to ISP subscribers remains to be determined.

While RSVP may be many years away from implementation there are other Quality of Service (QoS) techniques used by various vendors to expedite the flow of traffic through a network which minimizes latency. Such techniques include the IEEE 802.1p standard, MultiProtocol Label Switching (MPLS), router queuing, and the compression of packet headers. Although most of those techniques were previously described in this book, we purposely deferred router queuing to this chapter. Although router queuing is applicable to vendors operating Internet telephony service and businesses constructing and operating telephony over the Internet solutions, we will defer a discussion of this topic to the next section in this chapter. This is because more readers will probably use router queuing in constructing business telephony over Internet solutions than the handful of vendors using router queuing to develop egress and ingress traffic-expediting methods.

TELEPHONY OVER THE INTERNET

Telephony over the Internet is a voice-over-data-network transmission technique that allows a person's existing business telephone system to be used for calling persons via the Internet. The most common method used to initiate telephony over the Internet system is to connect PBXs at each corporate location to equipment that behaves in a manner similar to an analog trunk. That is, equipment presents a ring voltage when a call is received, responds to DTMF and/or rotary dialing, and passes caller ID data from an incoming call. In addition, the equipment presents each PBX with call-progress tones such as ringback and busy when outbound calls are made.

Figure 7-13 illustrates the proverbial black-box approach to telephony on the Internet, in which equipment required to perform signaling, voice digitization and compression, and packeting is labeled as a black box. Access to the black box from the local PBX is accomplished by an employee dialing a predefined prefix such as 7, followed by the extension of the called party. The analog call is routed to the black box, where it is digitized and compressed. In addition, the black box provides appropriate addressing for each packet

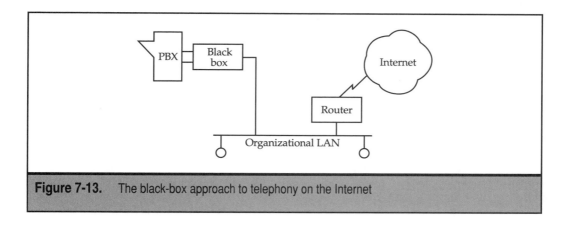

Figure 7-13. The black-box approach to telephony on the Internet

so that the packets are routed to a similar black box at the called location. At that location, voice-digitized packets are converted back into their original analog form for routing via the destination PBX to the called party.

Although there is no technical reason to preclude the delivery of digitized voice directly to a PC equipped with a sound card and appropriate software, it is far easier to ensure compatibility between calling and called parties by routing digitized voice between pairs of black boxes manufactured by the same vendor than it is to deal with the multiple conversion operations that might be required to provide compatibility among different vendors' products, which use different packet-formation and voice-compression techniques.

In Figure 7-13, the black box represents a conversion device that interfaces with the organizational PBX, converting either analog or standard PCM-encoded digital output into a compressed digital data stream that is encapsulated in IP packets addressed to a distant black box. In doing so, the black box acts as a communications server and becomes a participant on the organizational LAN. Thus, voice-encoded packets will flow over the LAN to access the router, which simply views those packets as an additional series of packets requiring routing onto the Internet or a private IP network. This means that prior to using this approach for telephony on the Internet, you should ascertain the utilization level of your organization's LAN.

LAN Traffic Constraints

In an Ethernet environment, a utilization level above 50 percent would indicate that voice-encoded packets will more than likely flow to the router for transmission on the Internet with unpredictable delays between packets. This is because of the access protocol associated with Ethernet LANs, in which a random exponential backoff algorithm is used after a collision is detected. This means collisions that occur more frequently as utilization levels increase result in a higher probability of voice packets being delayed from their required time sequence. This in turn can be expected to result in the occurrence of distortion at the destination, when voice-encoded packets arrive with delays that preclude their reconstruction into natural-sounding speech.

The unpredictable transfer of packets transporting digitized speech from local area networks into an IP network, including processing delays at nodes, causes the transmission of a uniform data stream with supposedly predefined intervals between packets to arrive at their destination with random delays between packets. These random delays result in awkward-sounding reconstructed voice. To avoid this problem, manufacturers of the proverbial black box incorporate a *jitter buffer* into their products. The jitter buffer can be viewed as a temporary holding area, enabling packets carrying digitized speech that arrives at the destination with random delays between packets to be removed from the buffer at a uniform rate. This produces a more natural sound for the reconstructed speech.

APPLICATION NOTE: Many gateways and other Internet telephony products include a selectable jitter buffer, with users able to set a delay from 0 (disabled) to 255 ms or more. When setting a selectable jitter buffer, it is important to remember that while an increase in the jitter buffer setting can improve the clarity of small blocks of reconstructed speech, the increase adds to the overall one-way delay. If there is too much delay, a person on one side of the conversation may believe the other party has stopped speaking and so may begin to talk, thus creating a "voice clash" and a requirement for both parties to listen.

Using Router Queues

As previously mentioned in this chapter, we can expedite traffic into and through a network by favoring one type of traffic over another. To do so we would use the various types of queuing capabilities supported by different routers to obtain a traffic-expediting capability. Although we can refer to the use of router queues as a mechanism to implement a QoS capability, a true QoS is currently not possible as there is no current method available in a TCP/IP environment to guarantee bandwidth on an end-to-end basis via transmission on a public network. The reason for this results from the fact that RSVP does not scale well to the size of the Internet as well as the fact that there is currently no mechanism available for billing both subscribers and ISPs for reserved bandwidth. Thus, in a public network environment we need to consider alternatives to RSVP. One alternative is to employ router queuing to expedite traffic into a public network when transit delay through the network is not unreasonably long. Another alternative is to employ router queuing as a traffic-expediting method into and through a network when we control the network. As an alternative, we can use router queuing into a network operated by another company that provides a Service Level Agreement (SLA) covering latency through the network. With careful planning all methods can be used to minimize packet delay, which in turn will allow the use of a jitter buffer at the destination to provide the capability to obtain a good level of reconstructed voice without having to guarantee bandwidth. In examining router queuing methods we will focus our attention on queuing methods supported by Cisco's Internetwork Operating System (IOS). Although our focus is upon Cisco due to its market share of routers, the topics covered in this section are normally applicable to other vendor products, however, specific commands used in the examples provided in this section may change.

Rationale

The rationale for router queues results from the fact that traditional TCP/IP networks operate on a best-effort basis. In this type of environment all user packets compete equally for network resources. While this operating environment was acceptable through the late 1990s, the rise in the use of TCP/IP networks and the development of time-sensitive applications to include voice and video services resulted in a need to expedite certain types of traffic over other types of traffic. While many ISPs simply added bandwidth to satisfy the growth in user traffic, in a business environment one of the weakest links is the access line into an ISP. Because many users share the access line it becomes necessary to differentiate traffic at the router, expediting some types of traffic over other types of traffic. The easiest way to accomplish this is through router queues, which is the focus of this section.

Cisco's IOS currently supports four different queuing algorithms—first in, first out (FIFO); priority queuing; custom queuing; and weighted fair queuing. As you might expect, and as we will shortly note, there are certain advantages and disadvantages associated with each queuing method.

FIFO

The first, simplest, and default queuing method for interfaces operating above 2 Mbps is FIFO. The term FIFO is descriptive of both the manner by which accountants can compute the cost of goods sold and the flow of data when there is no method available to differentiate traffic. FIFO queuing is illustrated in Figure 7-14. As indicated, packets are forwarded out of a router's interface in the order in which they arrive.

The key advantage of FIFO is the fact that it requires the least amount of router resources. However, the simplistic nature of FIFO queuing is also its key disadvantage. That is, because packets are output to the interface in their order of arrival it is not possible to prioritize traffic nor to prevent an application or user from unfairly over-utilizing available bandwidth.

In the example shown in Figure 7-14 assume the first packet contains a 20-ms segment of digitized voice while the packet labeled number 2 contains 1500 bytes representing a

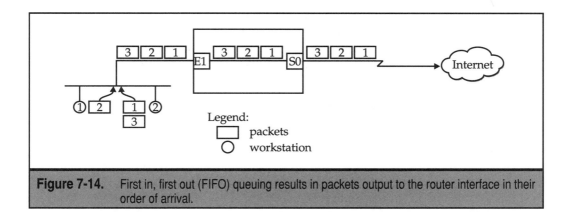

Figure 7-14. First in, first out (FIFO) queuing results in packets output to the router interface in their order of arrival.

portion of a file transfer. If the leased line connecting the router to the Internet operates at 56 Kbps and assuming no protocol overhead, the 1500 byte packet will require 1500 bytes × 8 bits/byte/56000 bps, or .21 seconds, to be placed on the line. Thus, if packet number 2 arrives at the router slightly before packet number 3, which represents another 20-ms portion of digitized voice, a 210-ms delay between the two voice digitized packets will occur. If this situation continues where long packets periodically arrive before short packets containing digitized voice the conversation will appear very awkward to the other party. Thus, FIFO is usually unacceptable when you are transmitting a mixture of time-dependent and time-independent traffic into a router.

Priority Queuing

A second type of queuing supported by Cisco routers is referred to as priority queuing. Under priority queuing traffic can be directed into up to four queues—high, medium, normal, and low. Traffic in the highest priority queue is serviced prior to traffic in lower priority queues. This means that through priority queuing you can configure a router to place traffic that is relatively intolerant to delay into an appropriate queue that favors its extraction onto a WAN.

In a Cisco router environment there are several methods available for both identifying traffic to be prioritized as well as for placing identified traffic into appropriate queues. While it is possible to use the IOS priority-list command by itself to assign traffic to predefined queues, you can also associate an access list with a priority list in order to use the considerable flexibility of access lists in filtering data as a mechanism to control the flow of traffic into priority queues. To illustrate an example of priority queuing as well as obtain a basis for describing some of the limitations of this queuing technique, let's assume traffic from a file transfer and voice gateway's digitized voice packets were flowing into a router, with both data sources destined to locations beyond the router, resulting in a requirement for each type of traffic to flow over a common WAN link. If we now assume priority queuing is enabled, the use of the queues might be configured such that file transfer packets are directed into the low-priority queue, while digitized voice packets are directed into the high-priority queue. An example of this packet direction based upon priority queuing is shown in Figure 7-15, where voice gateway generated packets are directed into the high-priority queue and file transfer packets flow into the low-priority queue.

The assignment of traffic to priority queues is accomplished through the use of the priority-list command. That command can be used by itself or in conjunction with an access-list command to direct traffic to applicable queues. When used by itself, the format of the priority-list command is as follows:

```
priority-list list-number protocol protocol-name
[high|medium|normal|low] keyword keyword-value
```

Here the list-number is in the range of 1 to 10 and identifies the priority list created by a user. The protocol-name identifies the protocol type, such as ip, ipx, and similar layer-3 protocols. Next, you would select one of the four priority queue levels. This would be followed

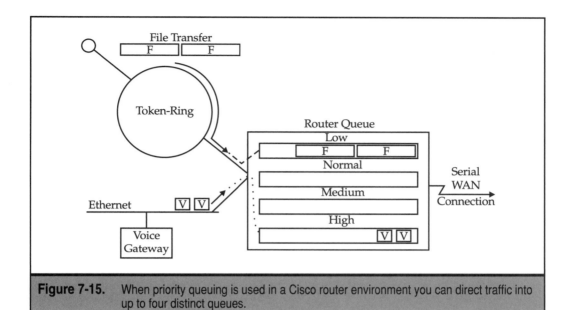

Figure 7-15.   When priority queuing is used in a Cisco router environment you can direct traffic into up to four distinct queues.

by a keyword that can further define the protocol by specifying a transport protocol carried by the network protocol, such as TCP or UDP. Then, you can specify a keyword-value that can be used to identify a specific TCP or UDP port or range of ports.

A second version of the priority-list command can be used to reference an access list. That version of the command has the following format:

```
priority-list list-number protocol protocol-name
{high|medium|normal|low} list list number
```

When using the second version of the priority-list command, the list-number variable references an extended access-list number.

Because digitized voice is transported via UDP without the use of a standardized port, let's assume that in the example shown in Figure 7-15 UDP port 3210 is used. Since FTP uses TCP port 21, the priority-list statements in a Cisco router environment required to configure priority queuing for the two data sources would be as follows:

```
priority-list 1 protocol ip low tcp 21
priority-list 1 protocol ip high udp 3210
```

To effect the filtering of traffic into applicable ports you would use a priority group command that would assign the specified priority list to the router WAN interface. Assuming the

IP address of the serial port connected to the WAN is 198.78.46.1, the applicable set of commands would be as follows:

```
interface serial 0
ip address 198.78.46.1 255.255.255.0
priority-group 1
!
priority-list 1 protocol ip low tcp 21
priority-list 1 protocol ip high udp 3210
```

Although only two types of traffic were assigned to queues it should be noted that any traffic that does not match the priority-list entries is by default placed in the normal queue. If you wish to use access lists to differentiate traffic you could associate separate access lists to a common priority list. An example of this is shown below:

```
interface serial 0
ip address 198.78.46.1 255.255.255.0
priority-group 1
!
access-list 100 permit tcp any any eq 3210
access-list 101 permit udp any any eq 21
priority-list 1 protocol ip low tcp 101
priority-list 1 protocol ip high udp 100
!
```

One of the disadvantages of priority queuing is the fact that it is possible to literally starve the ability of certain applications to obtain access to an interface, such as the WAN shown in Figure 7-15. The reason for this resides in the fact that any time there are packets in the high-priority queue they will be extracted first. Therefore, if the voice gateway is heavily utilized and generates a sustained, heavy flow of traffic to the router, it is quite possible to exclude the servicing of traffic entering other queues. In this type of situation the other queues could fill to capacity, resulting in packets flowing to those queues being dropped. This in turn could result in the retransmission of packets until a threshold is reached that would terminate the application. Perhaps recognizing the previously described limitation of priority queuing, Cisco added support for two additional queuing methods to its router platforms. One of those methods is custom queuing.

Custom Queuing

Custom queuing provides a mechanism to allocate the bandwidth of a transmission facility based upon specifying a byte count in a series of queue-list commands. By defining the number of bytes to be extracted from a queue prior to having the router process the next queue, you obtain the ability to indirectly allocate bandwidth. For example, assume you want to allocate 60 percent of the bandwidth of the WAN connection previously shown in Figure 7-15 to the voice application, 30 percent to file transfers, and the remaining 10 percent to all other traffic. To do so you would use the queue-list command to classify traffic as well as assign the byte counts to specify the maximum quantity of data to be extracted from each queue prior to the next queue being serviced.

There are two versions of the queue-list command you could use. One version established queuing priority based upon the protocol type, while the second version designates the byte size serviced per queue. The format of the queue-list command used to establish queuing priority is shown below:

queue-list list number **protocol** protocol-name queue-number keyword keyword-value

The first four entries in the queue-list command function in the same manner as the first four entries in the priority-list command. The fifth entry, queue-number, is an integer between 1 and 16 and represents the number of the queue. The keyword and keyword values function in the same manner as their counterparts in the priority-list command.

The second version of the queue-list command you would use is the queue-list-byte-count command. The format of this command is shown below:

queue-list list-number **queue** queue-number **byte-count** byte-count-number

In this version of the queue-list command the list-number and queue-number continue as identifiers for the number of the queue list and number of the queue, respectively. The key change is the addition of the keyword "byte-count," which is followed by the byte-count-number, the latter specifying the normal lower boundary concerning the number of bytes that can be extracted from a designated queue during a particular extraction cycle. It should be noted that the queue extraction process proceeds in a round-robin order and up to 16 distinct queues can be specified under custom queuing. Once you use the queue and queue-list-byte-count commands you would then use the custom-queue-list command to associate the queue list with an interface.

For the example shown in Figure 7-15, our custom queuing entries, assuming the same serial port and IP address, would be as follows:

```
interface serial 0
ip address 198.78.46.1 255.255.255.0
 custom-queue-list 1
!
queue-list 1 protocol ip 1 udp 3210
queue-list 1 protocol ip 2 tcp 21
queue-list 1 default 3
queue-list 1 queue 1 byte count 3000
queue-list 1 queue 2 byte count 1500
queue-list 1 queue 3 byte count 500
```

In the preceding example custom queuing was configured to accept 3000 bytes from the UDP queue, 1500 bytes from the FTP queue, and 500 bytes from the default queue. Note that this allocates the percentage of available bandwidth as 60, 30 and 10 percent to queues 1, 2 and 3, respectively. Also note that during the round-robin queue extraction process if there are less than the defined number of bytes in a queue the other queues can use the extra bandwidth. That use will continue until the number of bytes in the queue equals or exceeds the specified byte count for the queue.

Although the byte count plays a significant role in the allocation of bandwidth when custom queuing is used you need to carefully consider the length of each frame to obtain the desired allocation. This is due to the manner by which TCP operates as well as how queuing extraction works. Concerning the former, if you are using a TCP application where the window size for a protocol is set to 1, then that protocol will not transmit another frame into the queue until the transmitting station receives an acknowledgment. This means that if your byte count for the queue was set to 1500 and the frame size is 512 bytes, then only approximately one-third of the expected bandwidth will be obtained since 512 bytes will be extracted from the queue. Concerning the manner by which frame extraction occurs, entire frames are extracted regardless of the byte count value for a queue. This means that if you set the byte count to 512 but the frame size of the protocol assigned to the queue is 1024 bytes, then each time the queue is serviced 1024 bytes will be extracted. Thus, this would double the bandwidth used by this particular protocol each time there was a frame in an applicable queue when the round-robin process selected the queue. This means that you must carefully consider each protocol's frame size when determining the byte count to be assigned to a queue. While custom queuing can prevent the potential starvation of lower-priority queues, the actual allocation of bandwidth may not actually reach your desired metric. Another limitation of custom queuing is the fact that processing byte counts for up to 16 queues uses more processing power than the previously described queuing methods. Perhaps recognizing the first problem, Cisco provided another method that can be used to achieve a level of fairness in allocating bandwidth. That method is referred to as weighted fair queuing.

Weighted Fair Queuing

Weighted fair queuing (WFQ) represents an automated method to obtain a level of fairness in allocating bandwidth. Under WFQ all traffic is monitored and conversations are subdivided into two categories—those requiring large amounts of bandwidth and those requiring relatively small amounts of bandwidth. This subdivision results in packets queued by flow, with a flow based upon packets having the same source IP address, destination IP address, source TCP or UDP port, or destination TCP or UDP port. The goal of WFQ is to ensure that low-bandwidth conversations receive preferential treatment in gaining access to an interface while permitting the large bandwidth conversations to use the remaining bandwidth in proportion to their weights.

Under WFQ response times for interactive query-response and the egress of small digitized voice packets can be improved when they are sharing access to a WAN with such high-bandwidth applications as HTTP and FTP. For example, without WFQ a query to a corporate Web server residing on a LAN behind a router could result in a large sequence of lengthy HTTP packets flowing onto the WAN link preceding a short digitized voice packet in the interface's FIFO queue. This would cause the digitized voice packet to wait for placement onto the WAN that could induce an unacceptable amount of delay that would adversely affect the ability to reconstruct a portion of the conversation being delayed. When WFQ is enabled, the digitized voice frame would be automatically identified and scheduled for transmission between HTTP frames, which would considerably reduce its egress time onto the WAN. A second advantage of WFQ is that it requires no configuration commands. It is simply enabled by the use of the following interface command:

```
fair-queue
```

Under WFQ, traffic from high-priority queues are always forwarded when there is an absence of low-priority traffic. Because WFQ makes efficient use of available bandwidth for high-priority traffic and is enabled with a minimal configuration effort, it is the default queuing mode on most serial interfaces configured to operate at or below the E1 data rate of 2.048 Mbps.

Although WFQ does not require any configuration commands, the fair-queue command has three options that can be used to tailor its operation. To better understand those options let's first focus our attention on the basic manner by which traffic is classified under WFQ. This is illustrated in Figure 7-16.

In examining Figure 7-16 note that the WFQ classifier automatically classifies frames by protocol, source and destination address, session layer protocol and source/destination port, IP Precedence value in the Service Type byte, and RSVP flow.

Concerning configurable options in the fair-queue command, they include the congestive discard threshold, number of dynamic queues available to hold distinct conversations, and the number of queues that can be reserved by RSVP. The congestive discard threshold specifies how many packets may be queued in each flow's queue. The default is 64 and that value can be adjusted in powers of 2 from 2 through 4096 (2^{16}). The dynamic queue value controls the maximum number of conversations a router will monitor. The

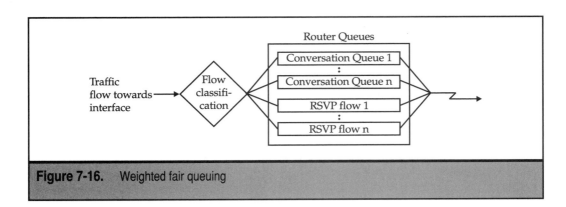

Figure 7-16. Weighted fair queuing

default value is 256. The reservable-queues option defines the number of queues RSVP can reserve. The default value is 0; however, enabling RSVP on your router automatically results in WFQ reserving 1000 queues for RSVP.

While WFQ is a good queuing method when you have a mixture of data, in its present incarnation it cannot tell the difference between a 64-byte Telnet packet and a 64-byte packet transporting digitized voice. WFQ treats both equally ahead of longer packets transporting files, which may not be your actual preference. In addition to a lack of control, WFQ consumes more router resources, which may or may not be a problem, depending upon the level of utilization of your router. On a more positive note, if your router interfaces a private network that supports RSVP, WFQ will support RSVP queues, which provides you with a mechanism to prioritize traffic into the guaranteed bandwidth on a network resulting from the network supporting RSVP.

Another type of WFQ that warrants a brief discussion is class-based WFQ (CB-WFQ). Class-based weighted fair queuing allows users to create traffic classes and assign weights to each class. CB-WFQ represents a powerful tool for backbone network connections within a layer-3 network.

Other Traffic Control Methods

In addition to establishing router queues at the entry to a public or private network, there are two additional techniques you can consider to make telephony over the Internet a reality. Those techniques include using an ISP that provides a service-level agreement that guarantees end-to-end latency and employing header compression on low-speed serial connections.

Service-Level Agreements

One example of a service-level agreement (SLA) worth noting was being provided by UUNET, a WorldCom Company when this book was revised. In actuality we will discuss two UUNET SLAs. The first SLA was offered by UUNET through October 2000. The second UUNET SLA, which represents a modification that occurred during October 2000, significantly enhanced their service-level agreement in recognition of advances in the use of optical fiber used for long-haul backbone communications. Under UUNET's first SLA, which was free of charge, customers of UUNET's North American and European operations would receive network quality in terms of latency that would not exceed 85 ms, on average, between UUNET-designated backbone measurement hubs. UUNET's Trans-Atlantic latency would not exceed 120 ms, on the average, between UUNET-designated measurement hubs located in the New York City metropolitan area and London, England. To put its wallet where the print resides, UUNET provides customers with a credit for the equivalent of one day's UUNET service fees if for two consecutive months the average latency on the North American or European network exceeds 85 ms, or if the average latency on the Trans-Atlantic network exceeds 120 ms. In the event both the North American and Trans-Atlantic latency guarantee levels are missed in the same two-month period, the credit is increased to two days of UUNET service fees.

As previously mentioned in this section, during October 2000 UUNET enhanced its service-level agreements. Under its new SLA for dedicated Internet access customers UUNET enhanced its guarantee of latency or delay over its network to 65 ms. In addition, UUNET promised a minimum packet delivery of 99 percent and a 100 percent availability level. If UUNET's network is unavailable for any fraction of an hour, customers are now credited with one day of service. If latency exceeds 65 ms or if less than 99 percent of packets are delivered UUNET will also credit customers with one day of service. For comparison purposes Table 7-4 notes the SLAs offered by UUNET, Genuity, and Cables & Wireless at the end of 2000 for latency packet delivery and availability for dedicated Internet access.

While both UUNET and other vendors also offer SLA metrics in terms of availability and time to repair outages, SLAs covering latency are the key to determining whether or not a voice-over-IP application can be expected to work at a satisfactory level. As discussed earlier in this book, the one-way delay from end-to-end should never exceed 250 ms and network delay is usually a major portion of end-to-end delay.

Header Compression

One of the problems associated with the use of RTP for time-stamping packets is the fact that when you consider the addition of UDP and IP headers to RTP the three add a significant overhead to the short portion of digitized speech conveyed in the payload. Recognizing this fact, RFC 2508 defines a method of header compression for slow-speed serial links.

One of the key problems associated with the use of RTP is the size of its header. RTP uses a 12-byte header that represents a significant overhead for a typical 20-byte payload when operating over low-speed serial lines. When you consider the fact that IPv4 has a 20-byte header and UPD has an 8-byte header, the overhead of IP, UDP, and RTP headers with respect to a 20-byte payload is 200 percent of the payload. Thus, the ability to compress the combination of IP, UDP and RTP headers provides the capability to obtain a significant capability over simply compressing the RTP header.

Under RFC 2508's compression scheme the IP/UDP/RTP 20-byte header is reduced to 2 or 4 bytes, with 2 bytes when no UDP checksums are sent while 4 bytes results when UDP checksums are used. To achieve these outstanding compression ratios of 20/2 or 20/4, the

ISP	Latency (ms)	Minimum Packet Delivery (Percent)	Network Availability (Percent)
UUNET	65	99	100.00
Genuity	65	99	99.97
Cables & Wireless	70	99	100.00

Table 7-4. Comparing ISP SLAs

compression algorithm takes advantage of the protocol being used. For example, half the bytes in the IP header remain constant over the life of a connection. Thus, after sending the uncompressed header once, the fields that remain constant can be removed from the compressed headers that follow. The remaining compression results from differential coding on the changing fields to reduce their size and by eliminating the changes for certain fields by calculating the changes from the length of the packet.

For RTP header compression RFC 2508 specifies similar techniques used for IP and UDP compression. However, a majority of compression results from the fact that although several fields in the header change from packet to packet, the difference between packets is often constant, resulting in a second-order difference of zero. By maintaining both the uncompressed header and first-order differences in the session state shared between the compressor and decompressor, all that needs to be communicated is an indication that the second order difference was zero, allowing the decompression to reconstruct the original header.

In a Cisco environment you can enable compressed RTP, referred to as CRTP, on a serial interface through the use of the **ip rtp header-compression** command. The format of this command is shown below:

```
ip rtp header-compression [passive]
```

If you include the keyword passive the software compresses outgoing RTP packets only if incoming RTP packets on the same interface are compressed. When the keyword passive is omitted all RTP traffic is compressed. By default 16 RTP header compression connections are supported on an interface. You can change the default through the use of the **ip rtp compression connections** command whose format is:

```
ip rtp compression connections number
```

where the number represents the total number of RTP header compression connections to be supported on an interface. The following example illustrates the enabling of RTP header compression on the serial 0 interface of a router.

```
interface serial 0
ip rtp header-compression
encapsulation ppp
ip rtp compression-connections 30
```

In this example the Point-to-Point Protocol (PPP) will be used to support up to 30 RTP header compression connections.

Although we will examine the operation of several gateways in detail in Chapter 8, our focus on gateways in this chapter is primarily to obtain an understanding of their economics of use. Thus, readers are referred to Chapter 8 for an in-depth examination of the configuration and operation of several representative products.

Equipment Examination

Now that we have an appreciation for the black-box approach to telephony over the Internet, let's turn to the operation of equipment that provides this capability. In doing so, we will examine five specific vendor products: the Internet PhoneJACK from Quicknet Technologies, the Pingtel xpressa phone, the Voice over IP (V/IP) series of products from Micom Communications Corporation, the Vocal Tech telephony gateway, and the MultiVOIP voice-over-IP gateway from MultiTech Systems.

Internet PhoneJACK

The Internet PhoneJACK was developed by Quicknet Technologies, Inc., of San Francisco, California, as a mechanism to allow a standard telephone to be used for making and receiving Internet telephone calls. In doing so, this product provides compatibility with a number of Internet telephony applications, including IDT's Net2Phone, Microsoft's Net Meeting, and Vocal Tec's Internet Phone. Thus, the Internet PhoneJACK is considered a hybrid mechanism, as it allows a standard telephone to be used with Internet telephony applications.

Voice Support   The Internet PhoneJACK is a sophisticated conversion device that uses a digital signal processor to perform voice compression and decompression in hardware. A user can select the voice-compression method to obtain compatibility with one of several Internet telephony applications programs. Currently the Internet PhoneJACK supports PCM 8 and 16-bit A- and m-law encoding as well as ITU G.711 and G.723.1 and TrueSpeech speech-encoding methods.

Fabrication   From a physical perspective, the PhoneJACK was originally fabricated on a half-size Industry Standard Architecture (ISA) adapter card, which is inserted into a system expansion slot in the system unit of an IBM PC or compatible computer. The card contains three physical connectors for the connection of a conventional telephone. The second connector is an RJ-12, which accepts a handset or headset. The third connector is actually a pair of 3.5-mm stereo jacks, which provide support for the use of a microphone and speaker.

A newer version of the Quicknet Internet PhoneJACK, referred to as the Internet PhoneJACK-PCI, features a higher-performance Digital Signal Processor mounted on a 32-bit PCI adapter card. Through the use of a PCI adapter the card functions as a Plug and Play device that is easily installed under Windows 95/98 and Windows NT. Because the Internet PhoneJACK appears as a Windows-compatible audio device, you can use your existing sound card for music or games while communicating via the Internet PhoneJACK.

Through the use of the PhoneJACK, you can keep your Internet telephony calls private. Incoming calls can be answered via your telephone instead of by talking into a microphone and hearing the other party on speakers. However, the Internet PhoneJACK also supports the use of speakers and a microphone for those who prefer hands-free communications.

OS Support   Currently, the Internet PhoneJACK operates under Windows 95, Windows 98, Windows NT, and Linux. However, it is expected that Quicknet Technologies will broaden support to include Windows ME and Windows 2000, which could result in the ability of its product to work with PCs using those operating systems by the time you read this book.

Operation   The Internet PhoneJACK is a full-featured device that includes full-duplex audio support, a speakerphone, and hardware based echo-cancellation. You can use this product by simply connecting a standard phone to the RJ-11 analog phone jack on the rear of the card. As alternatives, you can connect a headset to an RJ-12 headset port on the rear of the card or a microphone and speaker to individual circular 3.5-mm connectors also fabricated on the rear of the card.

Once you connect your preferred device and assuming your PhoneJACK is installed in your computer, you can make long-distance and international calls over the Internet to any regular telephone. Currently, the Internet PhoneJACK includes software to use the services of Net2Phone which, as described earlier in this chapter, provides both free PC-to-phone and PC-to-PC calling in the United States and discount international calling. While you could use a microphone and speakers connected to a sound card and bypass the need for the Internet PhoneJACK, its use may eliminate annoying echos as well as provide a higher level of voice quality since DSPs on the card are used instead of requiring computer processing cycles to effect voice compression.

Figure 7-17 illustrates the Quicknet Technologies PhoneJACK-PCI, the fourth generation version of the vendor's Internet PhoneJACK.

Pingtel xpressa

Although the Pingtel Corporation xpressa telephone may appear to represent a stylish telephone, it is far more than that. xpressa represents the world's first Java VoIP telephone that, in addition to making and receiving calls, supports call-hold and call-transfer capabilities. Unlike a regular phone, xpressa is similar to a PC, containing memory and a microprocessor that enables applications to run on your phone.

OS Support   Each xpressa phone represents an interface to the Internet. As such, it has an identifying URL and can be assigned a telephone or extension number. The xpressa phone uses the Session Initiation Protocol (SIP) to set up calls with other parties; how-

Figure 7-17.   Quicknet Technologies PhoneJACK-PCI, the fourth generation version of the Internet PhoneJACK (*Photograph courtesy of Quicknet Technologies*)

ever, the phone can support calls with non-SIP phones but may not then support some basic and enhanced phone services.

Figure 7-18 illustrates the xpressa phone. Note that the phone includes an LCD display, screen display buttons that correspond to actions on the display, a message lamp in the upper-right corner that illuminates when a message is waiting, a dial pad consisting of the standard 12 buttons, and fixed function buttons that control volume, transfer, conferencing, speaker, mute, and similar operations.

The scroll wheel on the right of the phone is the primary tool for navigation. To view different portions of a list or a text description on the screen, you would simply turn the scroll wheel.

Currently, the Pingtel xpressa phone supports G.71 and G.729 codecs. Because xpressa supports SIP, the address for a call must conform to the SIP standard, which does not use basic digits for call addresses. Instead, SIP requires call addresses to consist of SIP URLs, entered either as a dotted decimal number or as an e-mail address. For example, the author's SIP address could be entered in either of the following two formats:

```
sip:198.78.46.8
sip:gil_held@myco.com
```

Micom V/IP

In 1996, Micom Communications Corporation of Simi Valley, California, introduced its V/IP (Voice over IP) telephone and fax IP gateway product, which provides the potential to integrate voice over an IP network on an enterprise-wide basis. In actuality, Micom's V/IP represents a family of analog and digital voice interface cards that enable digital and analog PBXs to be connected to the card. Each card, which services a number of voice inputs, is inserted into an IBM PC or compatible computer, which functions as a voice

Figure 7-18.   Pingtel Corporation xpressa telephone supports SIP and represents the world's first Java VoIP telephone. (*Photograph courtesy of Pingtel Corporation*)

gateway. The gateway, which is connected to a LAN, compresses each voice conversation into an 8-Kbps data stream, packetizes the data stream, and uses a local database to map the destination office telephone number to a remote V/IP gateway's IP address. Although the Micom V/IP gateway was popular during the late 1990s, the introduction of PCI bus-based PCs limited the market for its EISA adapter cards. Thus, we will primarily focus our attention upon its cost a few years ago and how we can use the cost of a gateway to examine the economics associated with the use of a gateway.

Basic Configuration Figure 7-19 illustrates the basic configuration of a V/IP phone/fax gateway, based on the assumption that the organization using the gateway operates on an Ethernet LAN. In examining the use of the Micom V/IP gateway shown in Figure 7-19, several items require a bit of elaboration. First, since each V/IP card contains onboard processors that perform required compression and signaling operations, the platform for the gateway requires a minimal amount of processing power. This means you can use an Intel 486 or any type of Pentium processor-based computer as the platform for the gateway. Second, the onboard digital signal processors used on each V/IP card compress and convert both voice and fax signals into IP packets. This enables a PBX to be programmed so that both voice and fax calls can contend for the same V/IP connection between the PBX and the V/IP gateway. To better understand the capability of the V/IP gateway, let's turn to an example and discuss its operation.

Operation When configuring a V/IP gateway, each gateway in a network would be connected to a LAN, resulting in an IP address being assigned to each gateway. Since calls received from a directly connected PBX require routing to a different V/IP gateway, you must associate destination office numbers with the IP address of different V/IP gateways

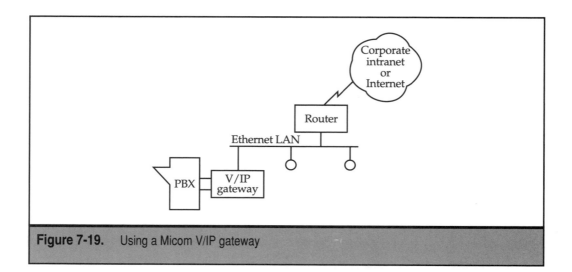

Figure 7-19. Using a Micom V/IP gateway

that serve different organizational locations. Thus, you would configure each gateway with a list of intracompany destination office numbers and the IP address of each.

Once the V/IP configuration process is completed and your organization's PBX is connected to one or more V/IP cards, you will also have to configure your PBX to recognize the V/IP card or cards and route calls to ports on those cards. In configuring your PBX, you would set the signaling method so that both the PBX and V/IP card ports are compatible. For example, setting both to a specific type of E&M signaling would ensure signaling compatibility. Then you would program the PBX to recognize a prefix code, such as the dialing digit 5 or 7, as a signal to route outbound calls through the PBX to the V/IP gateway. Let's assume an employee wants to dial the telephone number 555-1234 in the Chicago branch office where another V/IP gateway is installed and connected to the PBX in that office. By dialing 5-555-1234 or 7-555-1234, the call is first routed to the local V/IP gateway. That gateway looks up the destination telephone number 555-1234 in the phone directory database and extracts the IP address of the V/IP gateway in Chicago. The local V/IP gateway uses that IP address as the destination address for forming packets consisting of voice-compressed data, using a combination of the ITU's G.729 voice-compression standard with silence suppression to reduce the data stream associated with a voice conversation to approximately 4 Kbps. Due to the overhead associated with packet headers, the actual bandwidth required for a voice conversation commonly averages 6 Kbps, which is still a significant improvement over PCM's 64-Kbps operating rate.

Overcoming Predictability Problems   Both the Internet and private IP networks are essentially unpredictable with respect to bandwidth allocation. This problem results from the fact that IP networks were developed primarily to interconnect local area networks, and both inter- and intra-LAN communications are random and bursty in nature. Without a priority mechanism, it becomes possible for packets transporting voice to be delayed behind long packets transporting data, resulting in a degree of distortion to reconstructed voice. Therefore, a mechanism to provide a degree of packet-arrival predictability is required to ensure reconstructed voice is not adversely distorted because of delays in the routing of packets through an IP network.

The Micom V/IP gateway attacks the predictability problem in two ways. First, it supports existing router-priority protocols based on the use of router queues that enable relatively short packets transporting voice to obtain priority over longer packets transporting data. Although this support enables voice-encoded packets to be prioritized for routing onto a wide area network transmission facility, it does not guarantee that the priority will be maintained on public networks such as the Internet, where intermediate routers may not support a specific router vendor's priority mechanism. Thus, Micom added support for the ReSerVation Protocol (RSVP) that can guarantee end-to-end bandwidth if all routers between source and destination support RSVP.

Although the support of RSVP provides an excellent mechanism to overcome the predictability problem associated with the transmission of voice on IP networks, it is currently better suited for private rather than public network use. This is because an organization can control the upgrade of its networking equipment to support RSVP. In comparison, the ability to establish an RSVP connection through the Internet depends on the support of the protocol by all equipment between source and destination. This means that anyone linked through an ISP or NSP that does not support RSVP renders the ability to reserve bandwidth inoperative.

Economics   To illustrate the economics associated with the use of Micom's V/IP product, we will first consider expenses associated with constructing a small V/IP gateway at two corporate locations, assuming an IP network already exists and sufficient bandwidth is available to support several compressed voice calls without requiring an upgrade to existing transmission facilities.

Gateway Cost   When the first edition of this book was written, the cost of a V/IP card was approximately $770 per voice/fax channel. Since the establishment of a V/IP gateway requires the installation of one or more V/IP cards in a PC, we must also consider the cost of the personal computer. That computer, as previously mentioned in this chapter, does not require the latest in processing technology since each V/IP card contains its own digital signal processors. Thus, a minimal Pentium processor-based PC or even an older 486-based PC would be sufficient. If we assume we have to purchase a Pentium-based PC, we can do so for $1500 and obtain a sufficient platform for building a V/IP gateway. If we add $100 for an appropriate network card and $1540 for a two-port V/IP card, the total cost for a two-port voice-over-IP gateway becomes $3140, or $6280 for two identical V/IP gateway locations.

Computing Return on Investment   Determining the possible return on the investment required to support two V/IP gateways requires an additional set of assumptions. First, let's assume that at each corporate location employees work an average of 22 days per month. Next, let's assume that calls originating at each location destined to the other location average 30 minutes per day, which is probably not unreasonable over an 8-hour business day. Since calls are bi-directional, with an assumed 30 minutes per direction, this results in the V/IP gateways eliminating a total of 60 minutes of public switched telephone network activity per day. If we assume calls are between two offices located in the United States, our organization might be billed $.15 per minute for calls made between 8:00 a.m. and 5:00 p.m. Thus, on a daily basis our cost avoidance would be 60 minutes × $.15 per minute, or $9.00 per day. Using 22 working days per month, your monthly PSTN cost avoidance would be $9.00/day × 22 days/month, or $198.00. Thus, on an annual basis, the one-time expenditure of $6280 would result in savings of $198/month × 12 months, or $2376. This is a 37.8 percent return on your investment in voice-over-IP equipment, assuming a minimal amount of communications is routed via your organization's internal IP network instead of over the PSTN.

Examining Increased Usage   To illustrate how potential savings can substantially increase as usage of the V/IP cards increases, let's compute the economic savings associated with different usage scenarios. Table 7-5 lists the potential monthly and annual savings associated with eliminating between 60 and 480 call minutes per day between the two locations. In examining Table 7-5, the column labeled "Dual-Port Business Day V/IP Occupancy"

Calling Minutes/Day	Dual-Port Business Day V/IP Occupancy (%)	Monthly Cost Avoidance	Annual Cost Avoidance
60	6.25	198.00	2,376.00
120	12.50	396.00	4,752.00
180	18.75	594.00	7,128.00
240	25.00	792.00	9,540.00
300	31.25	990.00	11,880.00
360	37.50	1188.00	14,256.00
420	43.75	1386.00	16,632.00
480	50.00	1584.00	19,008.00

Table 7-5. Potential PSTN Cost Avoidance Based on a $.15-per-Minute Calling Rate

indicates the percentage of usage of a dual-port V/IP gateway during an eight-hour business day. Note that 60 calling minutes per day represents only 6.25 percent of the capacity of the dual-port V/IP gateway, yet it results in an annual cost savings of $2376. Also note that an increase in usage of the V/IP gateways between the two locations to 120 minutes per day results in an annual cost avoidance of $4752, or approximately a 75 percent (4752/6280) return on the one-time expenditure required to implement this small two-site V/IP network.

If calling increases to 180 minutes per day, which represents only 18.75 percent occupancy or utilization of the V/IP gateway, the annual savings associated with the use of the dual-port V/IP gateway system increases to $7128. This level of savings pays for the equipment in less than 11 months! As you can see from Table 7-5, as usage in terms of calling minutes per day increases, the annual cost avoidance increases, resulting in a greater return on investment and a quicker payback period. Since the typical corporate return on investment of 15 to 20 percent is considered quite good, the potential return on investment from the use of Micom's V/IP gateways can be expected to warm the heart of the organization's chief financial officer. Although the preceding economic analysis was based on a cost avoidance of $.15 per minute, which may be high for large organizations with contracts for bulk-rate usage of a communications carrier's transmission facilities, such organizations may be able to use a V/IP gateway at a significantly higher level of utilization. The end result would still be an excellent return on investment. For example, assume your organization pays only $.05 per minute for calls within the United States. If your organization could use the V/IP gateway configuration for 480 calling minutes per day, it would save $24 per day, or $528 for a 22-working-day month. On an annual basis, this would result in a cost avoidance of $6336, which, while substantially less than the

$19,008 amount listed in Table 7-5 (based on a calling rate of $.15 per minute), still provides for the full payback of the organization's investment in less than one year.

International Calling   When used for the routing of international calls, the higher tariffs associated with those calls can provide a substantial rate of return, even when usage between locations fills only a small fraction of the capacity of a V/IP gateway during normal business hours. For example, assume a bank installs V/IP gateways in branches located in Chicago and London. A discount PSTN rate for a call between the two locations of $.50 per minute would represent an attractive rate for calling between the two locations. Assuming 60 calling minutes per day between the two locations, the use of two V/IP gateways would save $660 per month, or $7920 on an annual basis. In this example, the return on investment occurs in less than one year even though the actual occupancy of the V/IP gateways is less than 6.25 percent over an eight-hour business day. Although your organization's actual return will vary based on the projected amount of calling and the cost per minute of the call, you can normally expect to obtain significant economic savings by transmitting voice on a private IP network.

MultiTech MultiVOIP   In concluding this section, we will turn our attention to the MultiTech MultiVOIP voice-over-IP gateway. This relatively recent product, initially marketed in early 1999, has a suggested retail price of $1749, which represents an extremely low cost for a gateway. While we will defer a detailed examination of the configuration and operation of MultiVOIP gateway until Chapter 8, in concluding this section we will discuss its basic operation and use its cost to examine its utilization economics.

The MultiVOIP gateway, illustrated in Figure 7-20, is manufactured as a stand-alone device. Each gateway has five connectors, including an RJ-45 Ethernet port, an Ethernet-to-serial port connection to connect a controlling PC, an AC power connector, and two ports for connecting analog devices. The two analog connection ports support two- or four-wire Types I, II, IV, and V E&M signaling as well as foreign exchange (FX) and loop-start two-wire connections.

There are a large number of voice-over-IP network options to consider; however, it is important to first determine how your organization plans to use this capability. This information will enable you to select a solution commensurate with satisfying those requirements.

Features   Although the MultiTech MultiVOIP voice-over-IP gateway weighs only 2 lbs. and is housed in a relatively small 6″ × 1.6″ × 9″ package, it supports a full range of vocoders. Vocoder families supported include ITU G.711, G.723, G.726, and G.727. In addition, the MultiVOIP gateway also supports a series of proprietary NetCoder voice-compression methods that results in users having 22 vocoders from which to select.

Figure 7-20.   The MultiTech MultiVOIP voice-over-IP gateway. *(Photograph courtesy of MultiTech Systems)*

In addition to a wide choice of vocoders, the MultiVOIP gateway is scheduled to add a gatekeeper/billing service option in a pending update. Although this can be used for charge-back purposes, you should note that while the current product is ideal for small-office utilization, its lack of scalability makes it difficult for large-scale implementations, as doing so would require the setup and management of one gateway for every two analog ports.

Utilization Economics   Figure 7-21 illustrates the use of a pair of MultiVOIP gateways to support the transport of voice over the Internet or a corporate intranet. Because of the relatively low cost of a MultiVOIP gateway, for an approximate one-time cost of $3,500 you can obtain the ability to continuously transmit two-voice calls. If your organization can configure PBXs to first attempt to use the two ports and keep them busy 8 hours per day, 22 days per month, you would obtain:

```
8 hours/day × 22 days/mo × 60 min/hr
× 2 ports = 21,120 call minutes/month
```

At a cost avoidance of just $.10 per minute, a pair of MultiVOIP gateways would pay for themselves in under two months. Even at half occupancy during the business day, the payback period is approximately three months, while two hours of voice activity per port per day results in a still very attractive payback period of six months.

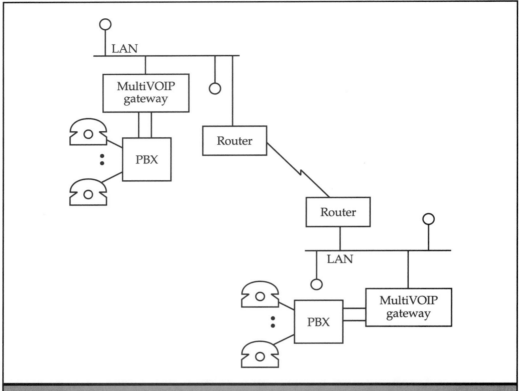

Figure 7-21.   Using the MultiTech MultiVOIP gateway

CHAPTER 8

Working with Voice-over-IP Gateways

In this chapter we will turn our attention to specific information concerning the options you may have to consider when installing and configuring a voice-over-IP (VoIP) gateway. While we briefly examined the generation operation and utilization of several VoIP gateways in the preceding chapter of this book, we will now revisit certain products as well as examine the operation and configuration of other products. Specifically, we will first turn our attention to the Multitech MVP200, the first in a series of VoIP gateways designed for use in a small office environment. This gateway is similar to other gateways in the Multitech series in that it represents an analog interface device. The second gateway we will examine is the Cisco AS 5300, a VoIP gateway that is constructed on a router modular chassis. As we will note later in this chapter, the AS 5300 can scale to support up to four T1 digital interfaces, which provides support for a maximum of 96 channels when connected to a PBX. This VoIP gateway connects to a PBX via the use of one or more T1 digital interfaces and can be considered to represent a digital VoIP gateway.

WORKING WITH THE MULTITECH MVP200

In this section we will examine the configuration of the MultiTech MVP 200 for operation in a small-office environment. As previously mentioned, the MVP 200 only supports the connection of analog devices. Thus, this type of VoIP gateway is commonly referred to as an *analog gateway*.

Overview

The MultiTech MVP200 represents the first in a series of VoIP gateways introduced by MultiTech System of Mounds View, Minnesota (www.multitech.com). Perhaps because MultiTech Systems has long been known for its series of analog modems, the MVP200 resembles a modem because its housing is very similar in size and shape to its line of high-speed external analog modems. Readers are referred to Chapter 7 for a picture of the MultiTech VoIP gateway. If you compare the picture of that gateway to a stand-alone modem you will note a high degree of similarity between the two.

Similar to an external modem, the front panel LEDs are used to indicate the status of key device elements. On the model 200 the LEDs display the status of the Ethernet connection and each voice/fax channel, as well as information about the boot status of the device and its power status. The model 200 includes a 10BASE-T connector for cabling to an Ethernet hub, a power connector, a command connector for configuring the MultiVoIP, and three connectors per channel. Concerning the latter, the channel connections permit you to use each device channel in one of three ways. If you use the E&M connector you can connect a channel directly to an E&M trunk on a PBX. Although the MultiVoIP supports E&M signaling types I through V, the default is type II and you must physically change a jumper block position to support a different type of E&M trunk. The second connector for each channel is an FXS, which supports a common station device, such as an analog telephone or a fax machine. The third connector is an FXO, supporting the connection of the channel to the station side of a PBX. Although you can obviously use only one connector at a time per channel, you can mix or match connectors across channels, which provides a considerable amount of flexibility. For example, you could connect both channels on a model 200 to separate PBX trunks or connect a telephone to one channel and a fax machine to the other channel.

Network Utilization

Figure 8-1 illustrates the general method by which a MultiTech MVP200 can be used within an IP networking environment. In examining Figure 8-1, note that three possible connection methods can be used for each MVP200 voice/fax channel. That is, you can directly connect an analog telephone or fax machine to an MVP200 channel via loop start or ground start signaling, you can connect the analog extension side of a PBX (FXO), or you can connect the analog trunk side of a PBX (E&M). Thus, you must carefully consider the manner by which members of your organization will access the gateway and configure the gateway to reflect the connection method or methods to be used.

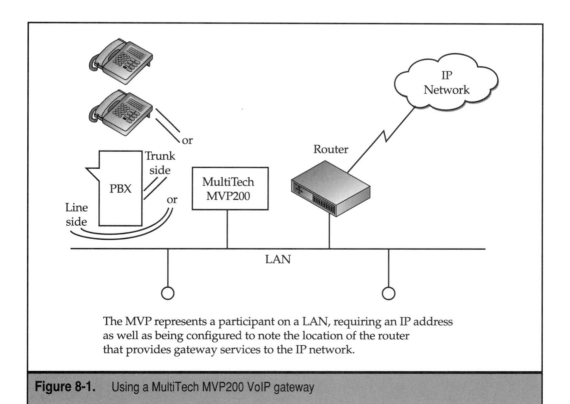

Figure 8-1. Using a MultiTech MVP200 VoIP gateway

Because the MultiTech MVP200 currently is limited to supporting 10 Mbps Ethernet, you cannot use it directly with a Fast Ethernet network. However, you can use it through a 10/100 Mbps switch port so the physical operating constraint can be overcome with applicable hardware.

Returning our attention to Figure 8-1, it is important to note how packets transporting voice flow with respect to the LAN as this will also explain why certain configuration options must be set manually. Thus, let's turn our attention to this topic.

Packet Flow

Once a connection method is configured for each channel you must also consider the manner by which voice coding will occur. That is, you must select a voice-coding algorithm. While this may appear to be a relatively simple process, in actuality it requires some careful thought. This is because the lower the voice digitization rate the higher the coding delay since they are inversely proportional to one another. While a low voice digitization rate is preferable in order to reduce bandwidth requirements, if the coding delay is too long the application may not provide the quality of reconstructed voice necessary to have employees use the application.

Most voice-over-IP applications can withstand a total one-way latency of approximately 250 ms. If delay begins to exceed that value on a sustained basis, it becomes difficult for one party to a conversation to understand precisely when the other party is finished talking. As a result of this, the parties begin to "clash," with the conversation needing the old CB radio "over" statement to ensure we speak at an appropriate time. Obviously, this situation does not bode well for a conversation. Because the difference in delay between one voice coder and another can be as high as 20 or more ms, it may be possible to obtain a better operating voice-over-IP application by experimenting with the use of different voice coders.

One additional item in Figure 8-1 warrants attention prior to moving on to some specifics concerning the MultiTech MVP200. That item is addressing. If you examine Figure 8-1 you will note that the MVP200 is a participant on the LAN. This means that in an IP environment the gateway must be assigned an IP address as well as be configured to direct packets that do not reside on the local network to the router. Thus, similar to other TCP/IP devices, you must configure the VoIP gateway with a default gateway (router) address. Because you must use multiple MVP200 systems to construct a voice-over-IP network you will also need to define an internal telephone exchange of telephone numbers. Doing so allows a call originated into one MVP200 to be routed to another MVP200 device.

Although the MVP200 only has two channels, MultiTech manufactures other VoIP gateways with additional voice/fax channel capacities. At the time this chapter was written MultiTech offered two-, four-, and eight-channel products. In fact, the software provided with the two-channel MVP200 is used throughout the vendor's product line since the configuration options remain the same and the only difference between products is the number of channels supported.

Initial Configuration

In a MultiTech voice-over-IP network you will designate one gateway as the master while all other gateways become slaves. The master gateway will contain a phone directory. The latter will contain a phone number entry that will correspond to each active channel on each gateway, with gateways differentiated from one another by their IP address. Once the phone directory is configured on the master gateway, each slave gateway can download the directory via the IP network. MultiTech includes a configuration disk with each MVP200 that operates under all modern versions of Windows. Once you install the software, the program allows you to select the method to be used to configure the MVP200, either via a COM port cabled to the console port or via the LAN.

The Setup Screen

Figure 8-2 illustrates the MVP200 setup screen. Note that in this example the selected port is shown as COM1 since this author decided to configure the MVP200 via a directly-cabled laptop. Also note that the MVP200 comes configured with a default IP address of 200.2.9.1. While it might be possible to configure the MVP200 directly via the LAN, unless a router is configured to accept multiple network addresses on a port it can be rather difficult to configure a device with a 200.2.9.1 address on the 198.78.46.0 network used by the author. Due

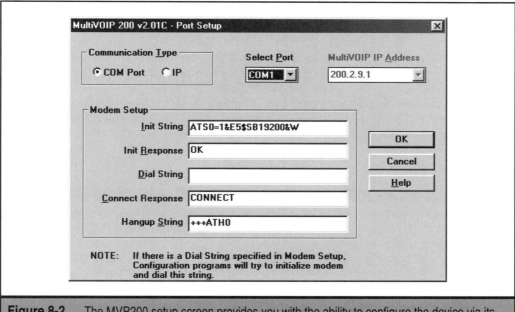

Figure 8-2. The MVP200 setup screen provides you with the ability to configure the device via its serial console port.

to this it was easier to connect the COM port of the author's laptop to the MVP200 for its configuration.

Address Configuration

Once you click on the OK button shown in Figure 8-2, the setup process will allow you to begin configuration of several key gateway parameters. These parameters concern the IP protocol setup and include the frame type, IP address, subnet mask, and gateway address.

Figure 8-3 illustrates an example of the IP protocol setup screen. Note that by default the MVP200 supports Ethernet Type II frames. However, this support can be changed to SubNetwork Access Protocol (SNAP) as the MVP200 also supports SNAP frames. In Figure 8-3 entries are shown for the IP address of the MVP200, its subnet mask and the default gateway address, with the latter referencing the address of the router on the network. As an aside to the address configuration process readers should not get confused by configuring the VoIP gateway with a gateway address. This is because the term "gateway" used by MultiTech and other vendors is a carryover from the era when devices that routed information between networks were referred to as gateways. While the term "router" is now preferred, all TCP/IP configuration screens including MultiTech's still refer to the router as a gateway, which can result in a bit of confusion as some people assume their voice gateway address should be entered each time they view the label "gateway address" on a configuration screen. After the previous explanation you will hopefully not fall into this trap.

Channel Setup

One of the most important screens in the MVP200 configuration process is the Channel Setup dialog box. As mentioned earlier in this section, the MVP200, like other members of the MultiTech family of gateway products, supports three types of connections to each channel. Thus, this dialog box and its tabbed screens must reflect entries commensurate with the method used to connect voice-producing traffic to the gateway for the gateway to operate properly. In addition, as we will shortly note, through this Channel Setup screen we obtain the ability to configure a specific voice-coding method that can have a considerable effect upon the end-to-end latency of the application.

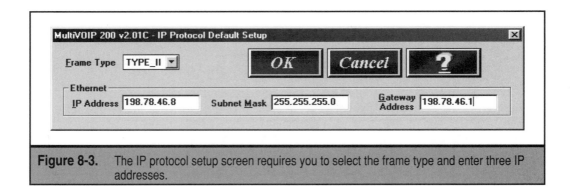

Figure 8-3.   The IP protocol setup screen requires you to select the frame type and enter three IP addresses.

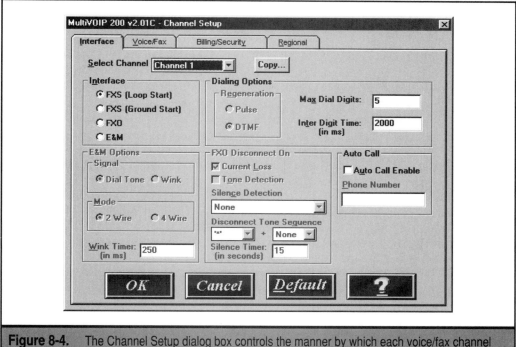

Figure 8-4.   The Channel Setup dialog box controls the manner by which each voice/fax channel operates.

Figure 8-4 illustrates the initial display of the Channel Setup dialog box with its Interface tab selected. Note that by default the Interface group for all channels is predefined as FXS (Loop Start). While this is the correct setting if you intend to plug an analog telephone or fax machine that supports loop start signaling directly into a channel on the MVP200, this setting will not work if you intend to connect the MVP200 to your PBX. Neither will this setting work if the device you intend to connect uses ground start signaling instead of loop start signaling.

If you are connecting an analog trunk on your PBX to a channel on the MVP200 you would then select the E&M option. When you do so the other E&M options shown in light gray would become available for selection. Similarly, if you select the FXO interface to support a connection from the line side of the PBX to a channel on the MVP200, specific options applicable to the FXO options that are shown in gray in Figure 8-4 would become available for selection. Because you can configure each channel differently your actual end-to-end voice path can take one of several configurations. While some of these pairs, such as E&M to E&M, are well known and in common use, other pair options, such as FXO to FXO, may not be as well known. Concerning this option, you must select a specific type of FXO disconnect—either current loss, tone detection, or silence detection—when FXO is configured. If you select silence detection you will be faced with some additional options, including setting one-way or two-way and the number of seconds of silence before disconnect. While these options are numerous it is important to realize that you do not have to configure all options at one time. However, you do need to take note of the different options and coordinate, if applicable, information from the PBX administrators or operators as to the type of signaling used if you are not familiar with voice operations. At the very least you may wish to cycle through all configuration windows and dialog boxes for any VoIP gateway you are installing prior to actually activating the product. This way you can prepare a list of configuration options as well as any questions that require resolution. These actions will facilitate the correct configuration of the gateway.

While the interfaces on voice gateways do not lend themselves to a plug-and-play environment, with some careful reading of applicable vendor manuals and coordination with the PBX side of your organization you should be able to obtain information necessary to correctly configure your gateway. While this environment is certainly not plug-and-play, your effort will ensure your gateway does not become a plug-and-pray product.

Prior to moving forward in our examination of the various options associated with the Channel Setup dialog box tabs, a few words are in order concerning the Select Channel pull-down menu which is shown set to Channel 1 in Figure 8-4. After configuring a given channel (1 or 2 on the MVP200), you can spy that channel's configuration to the other channel by clicking on the Copy button. When you do so all settings on the Interface tab are automatically copied to the other channel, facilitating the configuration process.

Configuring Voice and Fax

While the Channel Setup Interface tab governs the signaling method supported by a MultiTech MVP200 channel, you must also consider the voice-coding method that will be used to compress analog voice. In addition, if you intend to route fax over your IP network you will also have to turn to the Voice/Fax tab in the Channel Setup dialog box.

Figure 8-5 illustrates the Voice/Fax tab in the Channel Setup dialog box. In this example the voice coder drop-down scrollable menu was displayed to illustrate the fact that the MVP200, like most VoIP gateways, supports multiple voice coders. Although the MultiTech display indicates the operating rate of each voice coder, it does not indicate the latency associated with each coding method nor does it indicate the Mean Optimum Score (MOS), a value between 5 (best) and 1 (worst) that indicates the general clarity of reconstructed voice based upon the use of a specific voice coder. The reason we say MOS provides a general indication of the clarity of human voice is that reconstructed voice is graded subjectively. That is, under an MOS test persons listen to a speaker and write down what they hear, which is then graded.

Based upon the preceding, the selection of an appropriate voice coder can require a bit of research. First, you might consider using PING to determine the end-to-end delay between two networks where you intend to locate VoIP gateways. If the latency is near or above 200 ms you will then want to consider selecting a high bit rate coder since the coding rate and coding algorithm delay are inversely proportional to one another. Thus, this action will result in a lower coding delay. Unfortunately, MultiTech is no better nor

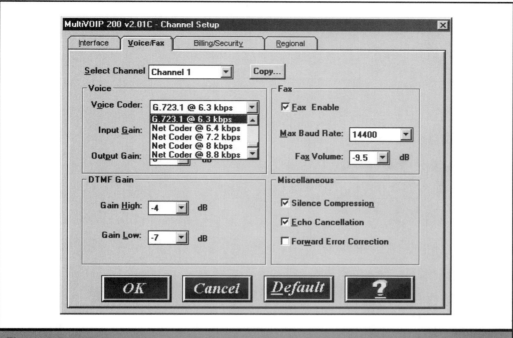

Figure 8-5.   The Voice/Fax tab in the Channel Setup dialog box provides the ability to select a voice coder as well as enable fax for use on a selected channel.

worse than other vendors, leaving it to the user to research information concerning MOS scores, latency, and the bit rate of coders. Readers who are interested in these areas, which are essential for selecting an appropriate coder when delays produce marginal reconstructed voice, should use the index or table of contents of this book to locate information on the delays associated with approximately a dozen voice coders. However, readers may also wish to contact MultiTech as any coder described as "Net Coder" with an operating rate behind the name references a proprietary voice coder for which this author was not able to obtain coding delay information when he revised this book.

Returning to Figure 8-5, you can both enable the use of fax over a specific channel as well as fix the maximum transmission rate for fax on a particular channel. Other options included on the Voice/Fax tab allow you to change voice and DTMF gain, enable or disable silence compression, echo cancellation, and forward error detection. Because it is very easy to set up a DTMF gain that can disturb the ability of touch-tone dialing, the MultiTech manual recommends that certain changes on the screen should only be done in conjunction with MultiTech support personnel.

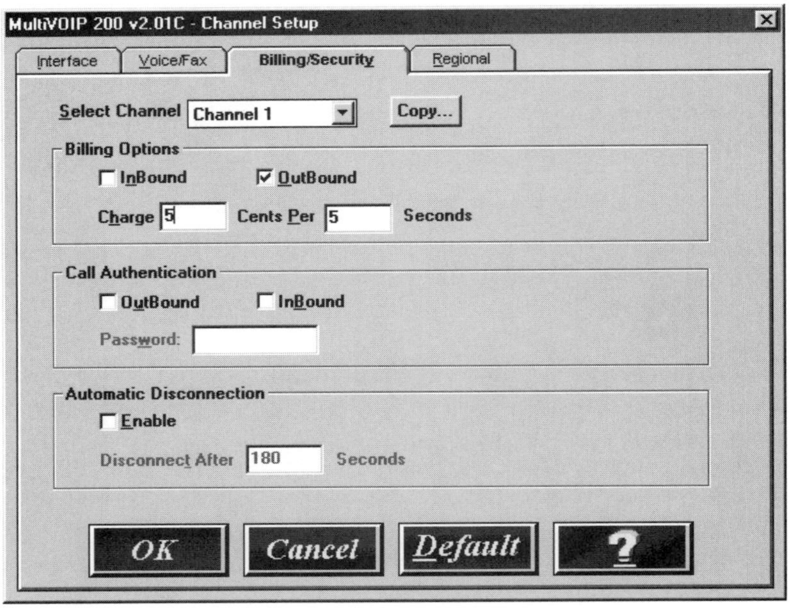

Figure 8-6.   The Billing/Security tab provides you with the ability to charge users for their use of the gateway as well as to assign a password for call authentication.

Billing and Security

The Billing/Security tab shown in Figure 8-6 provides users with the ability to perform elementary chargeback to recover the cost of the gateway. In addition, you can use this screen to assign a password for inbound or outbound call authentication.

The default charge of $.05 per 5 seconds is more suitable for an era when AT&T was the only long-distance company. In today's competitive environment you would have to charge below $.05 per 60 seconds to be competitive with the numerous "1010" carriers that have been formed to capture a few percent of the long-distance pie. If you want to restrict access to the voice gateway, you can do so through the call authentication option. If you enable password authentication on inbound or outbound calls, you must then enter a password of up to 14 numeric characters in the Password field.

The bottom of the Billing/Security tab has an automatic disconnection option that by default is not enabled. This option is easy to misconstrue due to its label. While this option does result in an automatic disconnect, it also serves to limit a call to the duration set when enabled. Other gateways use an automatic disconnection option only when a predefined period of inactivity has occurred. It should again be noted that this option on the MVP200 does not work in this manner. Instead, it restricts all calls to the number of seconds entered regardless of whether or not a conversation is ongoing. Thus, you must carefully consider the meaning of each option prior to enabling or disabling its use.

Regional Considerations

The Regional tab in the Channel Setup dialog box is illustrated in Figure 8-7 and represents a considerable recognition by MultiTech that the world is more than North America, Europe, and Japan. Because there are many different telephone systems located around the world that use different frequencies and cadences for common signals, MultiTech went beyond what is available on many other vendor gateway products. Instead of limiting the selection to the U.S., Japan, or the U.K., MultiTech includes a "custom" option that, when selected, allows the user to customize the gateway to support different dial, busy, and ring tones that can considerably vary from country to country. Of course, MultiTech leaves it for the user to figure out the tone pairs and cadence to enter if you plan to ship a slave gateway to Upper Volta or another exotic location.

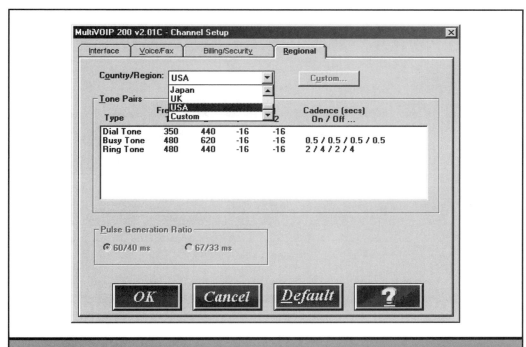

Figure 8-7. The Regional tab allows users to configure the MVP200 to support different dial, busy, and ring tones.

The Phone Book

Earlier in this section it was mentioned that you must associate each channel with a telephone number for your internal dialing plan. However, until now we left out the details concerning how to accomplish this.

The MultiTech setup program includes a main screen display that provides access to the Phone Book. Previously, we went directly to the voice channel configuration and bypassed mention of the main setup window. For those of us from Missouri that may wish to verify it exists, Figure 8-8 illustrates the main setup window.

The main setup window permits you to directly configure one or more voice channels, observe the progress of calls, examine gateway usage statistics, and configure the Phone Book. In concluding our examination of the MultiTech MVP200 gateway we will turn our attention to the Phone Book as it represents the key for insuring that calls are correctly routed within the voice-over-IP network you are establishing through the use of two or more gateways.

Figure 8-9 illustrates the display of two phone directory screens superimposed on a Windows desktop. If we assume that one gateway will be installed in our organization's

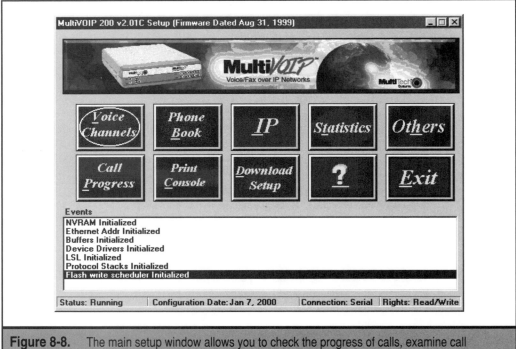

Figure 8-8.   The main setup window allows you to check the progress of calls, examine call statistics, and work with the Phone Book.

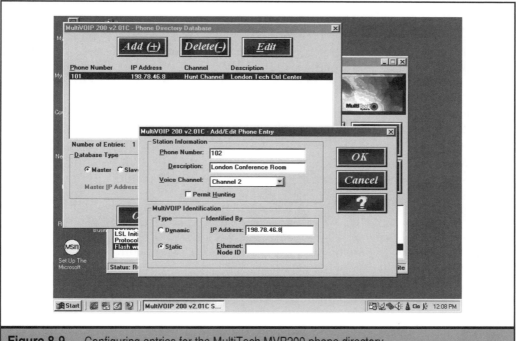

Figure 8-9. Configuring entries for the MultiTech MVP200 phone directory

London, Ohio, office we would assign a phone number to each channel on that gateway. The top window in Figure 8-9 illustrates the initial entry into the phone directory. Note that IP address 198.78.46.8, which was previously used in our setup process, continues to be used. The foreground window in Figure 8-9 illustrates the addition of a second entry to the phone directory. In the foreground window we entered the number 102 to represent the second phone number, which will be assigned to Channel 2. If we check the box labeled Permit Hunting the channel identifier in the phone directory will be labeled Hunt Channel. This means that if a call destined to one channel encounters a busy signal, the next channel will be used to deliver the call.

By assigning a phone number to each channel it becomes possible to implement a dialing plan. For example, let's assume there are several MVP200 gateways in our network, with the first having the IP address of 198.78.46.8. Let's also assume that channel 1 on the voice gateway whose IP address is 198.78.46.8 has the phone number 101 while channel 2 has the phone number 102. With these phone number assignments any person that obtains a connection to any MVP200 in the network only has to dial 101 or 102 to be connected to either channel 1 or channel 2 on the gateway whose IP address is 198.78.46.8, without having to know about the IP address of any gateway.

Other Factors to Consider

If you focus your attention upon the lower-left portion of Figure 8-9, you will note that the MultiTech MVP200 supports both dynamic and static addresses. While slave MVP200s can be configured for either static or dynamic IP addresses, the master MVP200 must have a static IP address. This is necessary to ensure that each remote MVP200 can directly locate the master to obtain a copy of the master phone directory. Two additional factors you must consider prior to being able to use your MultiTech MVP200 or another VoIP gateway include the configuration of your router's access list and any firewall used in your network. The MVP200 uses ISDN Q.931 signaling for call control between gateways. Call control signals are transferred commencing on UDP port 900 using even numbered ports, with the number of ports used based upon the number of voice channels supported. That is, a two-channel device uses UDP ports 900 and 902 while a four-channel gateway uses UDP ports 900, 902, 904, and 906 and so on. Similarly, the MVP200 uses the Real-Time Protocol (RTP) and the Real-Time Control Protocol (RTCP) to convey time-stamping and synchronization information. The MVP200 uses pairs of UDP ports commencing at 5004 (RTP) and 5005 (RTCP) for each channel built into a specific gateway. Because it is critical to permit signaling and synchronization via RTP and RTCP to flow between gateways, it is also critical to insure that your router access lists and firewalls are configured to permit traffic to flow through those ports. If you are using a different vendor product, it is important to note that there are no standards that govern the use of UDP and TCP ports for VoIP applications. In fact, this author has worked with several different vendor products and could note that the only similarity between products was the non-standard use of ports for different call control and signaling functions.

WORKING WITH THE CISCO AS5300

Currently, Cisco Systems markets a series of VoIP products to include routers that can support installed voice cards in an expansion slot and VoIP gateways. Most router products that Cisco markets with the capability to support voice modules, such as the Cisco 2600 and 3600 routers, are designed for use in small offices that require the ability to integrate a small number of voice calls for transport over an IP network. Both the Cisco 2600 and 3600 telephony-enabled modules can be configured to support analog and ISDN Basic Rate Interfaces (BRI). The Cisco 2600 supports up to 4 voice ports, while the Cisco 3600 supports up to 12 voice ports. While both products enable a router to provide the functionality of a voice gateway, neither product can economically scale beyond support for a small office environment as the voice modules require installation in a router. Perhaps recognizing the need for a more scalable VoIP gateway, Cisco introduced its AS5300, which is the focus of this section.

Overview

The Cisco AS5300 VoIP gateway represents a features-packed gateway that can be used at a corporate home office or another major enterprise location, by an ISP, or even by a communications carrier. The AS5300 supports the direct connection of up to four T1 or E1 digital interfaces, permitting support for up to 96 or 120 voice calls. When T1 lines are interfaced to the AS5300, up to 96 calls can be supported, while the use of E1 circuits increases support to 120 calls. By providing compatibility with voice modules used in the Cisco 2600 and 3600 platforms, the AS5300 can be used as the center-point of an organization, providing voice services at a major headquarters to and from branch offices. The AS5300 is H.323-compliant and supports both authentication and billing servers, which enables this VoIP gateway to be used by alternate long-distance communications carriers.

Figure 8-10 illustrates how the AS5300 could be used in a large city, such as New York or Los Angeles, and provide economical long-distance calling to other locations. In this example the communications carrier is shown using an AS5300 in New York and one in Los Angeles, while a Cisco 2600 with four voice ports is shown serving Macon, Georgia. Note that only one authentication server and one H.323 gatekeeper is shown, as we will

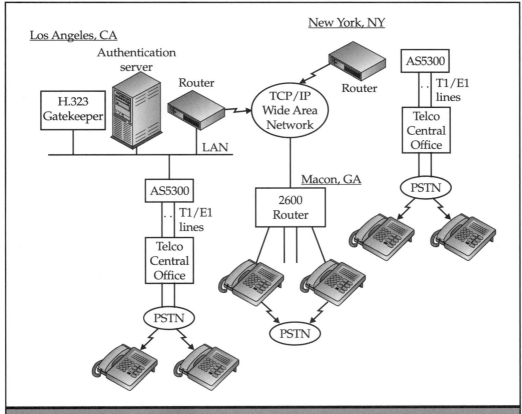

Figure 8-10. Using the AS5300 to provide long-distance calling via a TCP/IP network

assume that the carrier configured the VoIP network as a single H.323 zone and authentication is supported from a central location.

Assuming the communications carrier is selling long-distance voice at a discount, subscribers would dial a "1010" prefix to access the carrier's network via the telco central office. Calls would be routed via a series of T1 (or in Europe, E1) lines, with each line supporting 24 calls per T1 line and 30 calls per E1 line to the AS5300. After authentication and in the event of the use of a commercial carrier billing initiation, the digitized call will flow over the TCP/IP network. While calls from subscribers via the PSTN are routed into the AS5300 in the form of 64 Kbps PCM-encoded data, the AS5300 will normally be configured to convert such calls via the use of a low-speed codec, such as the dual-speed G.723.1 codec that can digitize voice at a data rate of 5.3 Kbps. When you compare the PSTN's dedicated use of 64 Kbps of bandwidth to support PCM versus the ability of the AS5300 to digitize voice at 5.3 Kbps it becomes obvious that the use of low bit rate coders is economically more efficient. When you consider the fact that the AS5300 also supports silence suppression and only forms packets when people speak, the efficiency associated with transporting voice over a TCP/IP network in comparison to the use of the PSTN becomes even more pronounced.

Features

The Cisco AS5300 uses voice/fax coprocessor cards, supporting two cards per chassis. Each card supports two T1 or E1 connections, resulting in the AS5300 obtaining the capability to scale up to 96 (T1) or 120 (E1) voice connections within a single chassis. Because the AS5300 can be installed in a stack configuration, Cisco offers a cabinet-style rack referred to as the AccessPath-VS3 into which approximately 20 AS5300 devices can be installed. Thus, when combined with one or more AccessPath-VS3 rack mounts, it becomes possible for the AS5300 to support VoIP at a very large scale, such as for a carrier class VoIP operation.

QoS Support

Similar to many Cisco data products the AS5300 supports a wide range of QoS methods. Cisco QoS supported by the AS5300 includes recognition of the Service Type/DiffServ byte in the IPv4 header, RSVP, and such queuing methods as weighted fair queuing (WFQ) and weighted random early detection (WRED). In addition, the AS5300 supports Multiclass Multilink Point-to-Point Protocol (MP) fragmentation, which results in large data packets being fragmented into smaller packets. By interleaving the fragments with packets transporting voice, this technique will considerably reduce voice latency since packets transporting digitized voice cannot get stuck behind lengthy packets transporting data.

Codec Support

Currently, the AS5300 is limited to supporting five codecs. Those codecs include G.711, G.729, G.726, G.279a, and G.723.1, the latter the dual-speed codec while G.729a represents a lower-complexity but fully interoperable version of G.729. In actuality, support for specific codecs depends upon the codec feature set selected under Cisco VCWare. At the time this book revision was prepared VCWare software supported three codec feature sets. One feature set supports all codecs plus fax relay. A second codec feature set is designed for use with an early generation of the AS5300, while the third feature set is limited to supporting G.711 and G.729a codecs and fax relay.

The digital signal processors used by Cisco include support for the ITU E.165 echo cancellation method into the circuit-switched network. In addition, the DSPs also support voice activity detection, which permits periods between voice activity to be suppressed.

Other Features

As mentioned earlier in this section, the Cisco AS5300 supports a large number of features. Table 8-1 lists a portion of those features, some of which we previously described and a few that deserve a bit of elaboration.

The adaptive jitter buffer permits compensation for variations in delay resulting from packets flowing through a dynamic TCP/IP network. Although packets may flow into the network with a uniform period between each, as packets flow through routers on their way to their destination they encounter random delays based upon the processing power of each router and traffic arriving at the router. By routing packets into an adaptive jitter buffer it becomes possible to remove interpacket jitter.

Another feature that deserves mentioning is the fact that the AS5300 provides a dial plan mapping capability. This feature enables the mapping of dialed telephone numbers to IP addresses. Mapping can occur via the direct configuration of the AS5300 or maintaining mappings in H.323 gatekeepers that communicate to multiple gateways through the use of H.323 Registration, Admission, and Service (RAS) messages.

Another useful feature of the AS5300 is its direct inward dial support. This feature enables the direct dialing of users located behind a PBX, alleviating the necessity for a person to first dial the main telephone number of an organization and then have to enter an extension.

Two additional AS5300 features include its international calling support and tone-generation capability. Concerning the former, the AS5300 supports both µ-law and A-law en-

Scalable to 96 or 120 voice ports	Uses an adaptive jitter buffer
Supports interconnection of digital T1/E1 lines	Maps dialed phone numbers to IP addresses
Supports G.711, G.729, G.729a, and G.723.1 codecs	Allows direct inward dialing to users behind a PBX
ITU E.165 echo cancellation	Provides international calling capability
Voice activity detection and silence suppression	Generates secondary dial tone and call progress tones

Table 8-1. Major Cisco AS5300 Features

coding on any channel. Thus, the AS5300 can support connections to North America as well as to telephone systems outside North America. Concerning tone generation, the AS5300 supports a secondary dial tone that enables access to a VoIP network implemented in a two-stage calling process, where the first dial tone is generated by the local telephone company while the second dial tone serves as a prompt to let the subscriber or employee know they reached the VoIP equipment or equipment operator. The AS5300 also supports call progress tones, such as dialing, ring-back, busy, and other tones, many of which can be varied based upon the use of the VoIP gateway in a specific country.

Equipment Restrictions

The use of the AS5300 involves certain tradeoffs. Because the level of CPU utilization affects the voice processing capability of the gateway, one key tradeoff is latency versus CPU utilization. This results from the fact that as the CPU load increases, a delay in packet processing occurs, resulting in additional latency per packet.

To maintain a high level of CPU performance, which in turn can be expected to reduce packet latency, Cisco recommends AS5300 operators should use a maximum of two serial ports, with each port limited to an operating rate of 2 Mbps; employ a 2:1 ratio of PCM calls to encapsulated VoIP bandwidth; and disable compressed RTP. Concerning the latter, if you turn to Chapter 7 in this book and refocus your attention upon the description of compressed RTP, you will realize that it is very processor-intensive. Based upon Cisco's recommendation, we can reasonably expect that the load placed by CRTP on the AS5300 negates the advantages of compressing the IP/UDP/RTP headers. If you read a caution note in Cisco documentation you will note that CRTP should only be enabled in limited cases where 10 to 24 or less total calls are to be processed. In such situations the call rate should be limited to .5 calls per second or one call per two-second period. When CRTP is disabled the AS5300 can support a sustained call rate up to two calls per second. While the AS5300 cannot directly prevent a higher calling rate from begin processed, CPU utilization can be adversely affected beyond two calls per second, which will adversely affect packet latency. However, as we will soon note, we can control calls via the use of a gatekeeper. Thus, let's examine the relationship between these two key devices.

Gateway vs. Gatekeeper

To better understand the relationship between the VoIP gateway and the need for an H3.23 gatekeeper, we will look at the functions performed by each device and how one complements the other to provide a VoIP networking capability.

Gateway

The key function of the AS5300 gateway is to accept calls originated from the PSTN and convert 64 Kbps PCM digitized voice from those calls to a low–bit-rate voice-encoding method, directing calls to their destination. In the opposite direction, the gateway is responsible for terminating a call received from another gateway by converting it to 64

Kbps PCM and routing it onto the PSTN. In addition to the preceding functions, the AS5300 can be configured to work with a debit card billing server, which enables the gateway to be used in a commercial environment. In fact, at the time this book was revised several Internet Telephony Service Providers (ITSPs) were using the AS5300 in their networks. As a reminder, the basic operation of credit card and debit card calling was described in Chapter 7.

Gatekeeper

A key function of one or more H.323 gatekeepers is to complement gateways. Gatekeepers are used to manage an H.323 zone and H.323 end points in a consistent manner, provide bandwidth management through the use of the RAS protocol, and provide address resolution services to gateways. Because the H.323 gatekeeper controls admission, it is possible to obtain a QoS capability on a private IP network without requiring the use of RSVP, DiffServ, or other traffic-expediting techniques. For example, assume your organization operates a T1 line between two LANs for transporting voice traffic directed from a PBX or telco office in each location to the other location. In the wonderful world of PCM you could route 24 calls at a time in each direction over a T1 line. Because you will more than likely use a low-bit-rate codec to reduce each conversation to 16, 8, or even 5.3 Kbps, you can support several times the PCM limit. By examining the actual bandwidth to include packet header overhead per call as well as the effect of silence suppression, you can determine the number of simultaneous calls a gateway should allow to occupy the T1 bandwidth. Then, you can configure the gatekeeper to restrict admission to that call level, in effect guaranteeing QoS over the T1 line.

Addressing   A gatekeeper recognizes one of two types of terminal names—H.323 identifiers or ITU E.164 addresses. H.323 identifiers represent arbitrary, case-sensitive text strings, with Cisco recommending the use of fully qualified e-mail names as the H.323 ID when an H.323 network involves interzone communications. In comparison, the ITU E.164 address represents the international public telecommunications numbering plan and would be preferable when calls are routed to a gateway for delivery to a destination on the PSTN. Now that we have an appreciation for the difference in functionality between the gateway and gatekeeper, we will conclude our examination of the AS5300 by focusing our attention upon its interactive voice response (IVR) capability.

Interactive Voice Response

One interesting features of the AS5300 is its IVR capability. If you consider the fact that the AS5300 is scalable as well as stackable to provide support for hundreds to thousands of simultaneous voice calls that require a billing mechanism, it is obvious that subscribers require prompting. Thus, the key function of the IVR system is to provide audio prompts to collect subscriber or employee identification codes, passwords, or PIN numbers, and the destination phone number the subscriber or employee wishes to call.

The AS5300 includes several IVR scripts that are embedded in IOS software. While operators cannot modify an embedded script they can use IVR commands to control what scripts are executed and when they are executed. In addition, audio files used for IVR prompts can be modified by the AS5300 operator.

To support IVR functionality Cisco added several commands to IOS. For example, the *audio-prompt load* command is used to initiate the loading from Flash memory into RAM of a specified audio file. Through the use of the IVR AS5300, operators can create a sequence of scripts that play prompts, collect account number and associated PINs, deny access, and re-prompt a subscriber if their initial authentication fails, as well as collect destination telephone numbers for use. While Cisco does not currently provide AS5300 operators with the ability to modify embedded scripts, you can use them as building blocks to tailor interactive voice response to your specific operational requirements.

Summary

As indicated in the brief overview of the Cisco AS5300, this VoIP gateway represents a very scalable product that works in conjunction with an H.323 gatekeeper. As such, it provides a standardized mechanism for routing calls through a TCP/IP network. Because the AS5300 is compatible with Cisco 2600 and 3600 series routers, the AS5300 provides a mechanism for integrating small to large offices within an integrated TCP/IP network or it allows a carrier to provide service to locations where the subscriber base is not sufficient to justify the use of an AS5300.

CHAPTER 9

Voice over Frame Relay

In previous chapters in this book, we examined the operation of frame relay and the effect of using different methods on the bandwidth required to digitize voice. In this chapter, we will put that knowledge to use as we investigate the transmission of voice over frame relay networks. First, we will examine some key technical issues associated with the transmission of voice over a frame relay network. Then we'll turn our attention to the operation and utilization of several types of vendor equipment developed to provide users with the capability to transmit voice over their frame relay network connections. As we cover the operation and utilization of vendor equipment, we will also look at the development of economic models you can use to estimate the cost of adding a voice transmission capability to your organization's public or private frame relay network. Finally, because the Frame Relay Forum is so important, we will conclude this chapter with a discussion of the FRF.11 and FRF 12 Implementation Agreements (IA), which set the standard for the transmission of voice over frame relay.

Although just about all equipment vendors were offering proprietary solutions for transmitting voice over frame relay as this book was written, the reason for the proprietary nature of products was the lack of standards in this area. Now that the voice over frame relay IAs are a fact, we can expect that vendors will announce compliant products in the near future, assuming that they have not done so by the time this book is published. In fact, based on telephone conversations with several vendor representatives while preparing this book, this author found a willingness of vendor personnel in both engineering and product marketing divisions at several firms to predict that they would eventually modify their existing products to become fully compliant with the FRF.11 and FRF 12 Implementation Agreements.

TECHNOLOGICAL ISSUES

Frame relay was developed as a fast packet switching technology designed to eliminate the delays associated with the use of an X.25 packet network. In doing so, frame relay became a suitable transport for interactive query-response applications in which a client was located on one network and the server was located on a geographically separated network, with both networks linked together through the use of a frame relay network. The fact that CRC checking was employed only at switches to determine whether or not to drop a frame (instead of using full error detection and correction by retransmission) enabled data to flow end to end with minimal delay. This capability was extremely important to support query-response applications in which client-station users would enter a variety of short queries that would otherwise be significantly delayed through a conventional packet switching network and would degrade the ability of the user to type in a normal manner, adversely affecting productivity.

Although the use of a frame relay network enables access to SNA mainframes as well as LAN-to-LAN connectivity that otherwise might be awkward when using an X.25 network, frame relay uses a variable-length information field. Thus, from a technical perspective, the variable frame length can adversely affect the arrival and interpretation of the digitized-voice frame when its contents are examined and used to reconstruct voice. In addition to the variable-length frame presenting problems with respect to the transportation of digitized voice, a variety of other technical issues face equipment developers. Table 9-1 lists ten of the key issues faced by equipment developers and voice-over-frame relay implementers. In the remainder of this section, we will examine each issue. Not only will this give you an appreciation for the obstacles equipment developers and implementers had to overcome, it will also help to clarify the technological features so you can compare equipment manufactured by different vendors.

Frame Length Handling

As noted earlier in this chapter and in Chapter 4, the information field in the frame relay frame is variable in length. To illustrate the problem this can cause, consider the FRAD or router in Figure 9-1, which provides a connection to a frame relay network for a transmission to and from a LAN and a PBX. In this example, there is one serial port connection used to link the FRAD or router, which services the LAN and the PBX, to the frame relay network. This means that the transfer of a file, a long response from the server to a client query from another location connected to the LAN (as shown in Figure 9-1) via the frame relay network, or another lengthy data transfer will tend to fill the information field in the frame relay frame.

Effect of Variable-Length Frames

Although an Ethernet LAN limits the information field length to 1500 bytes, a Token Ring LAN operating at 16 Mbps can have an information field that can be up to approximately 18 Kbytes in length. If a LAN station obtains access to the serial interface of the router or FRAD between frames transporting digitized speech, the delay will depend on the operating rate of the serial connection to the network and the length of the frame. Concerning the former, there are significant delay differences between accessing a frame relay network at 56/64 Kbps and a T1 operating rate of 1.544 Mbps. For example, at 64 Kbps, one

Frame length handling	Frame prioritization
CIR selection	Frame loss handling
Echo cancellation	Frame delay handling
Silence suppression	Voice-compression method
Frame overhead	Telephony signaling
Multiplexing technique	Service-level agreements

Table 9-1.

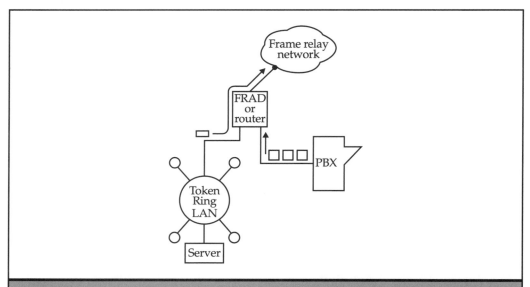

Figure 9-1.   A lengthy frame transporting data that is serviced between frames transporting digitized voice can adversely delay the reconstruction of speech at its destination, making it sound awkward.

8-bit byte carried in the information field of a frame relay frame requires 125×10^{-6} seconds (8/64,000) for transmission. Thus, a 1500-byte information field in a frame relay frame that results from the transportation of data in the maximum-length Ethernet information field to the FRAD or router would require 0.1875 seconds ($125 \times 10^{-6} \times 1500$) to be transmitted to the network, without considering the overhead associated with control fields. This means that, excluding frame overhead and processing time associated with the FRAD or router forming a frame for placement onto the access line connected to the frame relay network, a lengthy Ethernet LAN packet could delay a frame transporting voice by almost 0.2 second.

APPLICATION NOTE: FRADs or frame relay-compliant routers must be configured to place a limit on the length of frames transporting data to minimize their effect on frames transporting digitized voice.

Considering End-to-End Delay

It should be noted that the previously described delay represents only the delay associated with accessing the frame relay network. The actual end-to-end delay can be significantly higher, since slight delays are introduced at each node in the frame relay network, and additional delay may result if a frame transporting data arrives at a network node prior to a frame transporting digitized voice. The latter situation can easily occur if your organization uses a frame relay network to interconnect a number of LANs, with PVCs set up to provide an any-LAN-to-any-LAN communications capability. In this situation, there is a higher probability that a frame transporting data will arrive between a sequence of frames transporting digitized speech at a network node for delivery to a FRAD or router via an access line from the network to a subscriber's location. In fact, a station on a LAN at the originating site could be transmitting to a different destination than the destination of the frame carrying digitized voice, while a station on a LAN at a third location could be transmitting a sequence of frames carrying data to the destination location of the digitized-voice frame. If any frame carrying data arrives before the frame carrying digitized voice, then the delay in effect doubles. Thus, the use of 64-Kbps access lines could result in an end-to-end delay of 0.3750 seconds, a gap sufficient in duration to result in the distortion of reconstructed speech.

Worst-Case Delay

Although a delay of 0.375 seconds is significant, that delay assumes a worst-case situation when frames transporting digitized speech are delayed by frames transporting data in an information field 1500 bytes in length. Now let's assume that the LAN used is a Token Ring network operating at 16 Mbps. As previously discussed, this network can transport frames with an information field of up to 18,000 bytes in length. Since the maximum length of the information field of a frame relay frame is 8192 characters, this means that the FRAD or router would fragment the contents of a lengthy Token Ring information field that exceeds 8192 bytes into two or more frame relay frames, with the first frame always having a maximum-length information field of 8192 bytes. Again, returning to an access line operating rate of 64 Kbps, this means that the worst-case delay resulting from a frame transporting data will be 125×10^{-6} seconds/byte $\times$ 8192 bytes, or 1.024 seconds, clearly a most unsuitable situation. Once again, it becomes possible for a long frame transporting data from a different location to arrive at the destination FRAD slightly ahead of the frame transporting digitized voice. This action doubles the 1.024-second delay between frames transporting voice to 2.048 seconds, making a bad situation intolerable.

Varying the Access Line Operating Rate

Now let's assume our organization installs a T1 circuit as the local access line to a frame relay network operator. Although this transmission facility operates at 1.544 Mbps, the operating rate includes an 8-Kbps sequence of framing bits that cannot be used for the transmission of the contents of frame relay frames. This means that the data transmission capacity of the access line will be 1.536 Mbps, or 24 times the capacity of a 64-Kbps circuit. This also means that the latency or delay resulting from a frame transporting data being processed by a router or FRAD just prior to a frame transporting digitized voice will adversely affect the digitized voice frame by 1/24th the time indicated by our prior computations for the same situation, resulting from the use of a 64-Kbps access line. Table 9-2 summarizes the one-way delays resulting from 1500- and 8192-byte frames carrying data being processed by a FRAD or router prior to a frame transporting digitized voice via access lines operating at data rates from 56 Kbps to 1.544 Mbps, including fractional T1 operating rates of 128 Kbps, 256 Kbps, 384 Kbps, 512 Kbps, and 768 Kbps.

	Frame Length	
Access Line Operating Rate	**1500 Bytes**	**8192 Bytes**
56 Kbps	0.21428571	1.17028571
64 Kbps	0.18750000	1.02400000
128 Kbps	0.09375000	0.51200000
256 Kbps	0.04687500	0.25600000
384 Kbps	0.0312500	0.17066667
512 Kbps	0.02343750	0.12800000
768 Kbps	0.01562500	0.08533333
1.544 Mbps	0.00781250	0.04266667

Table 9-1. Data Frame Delay Times (in Seconds) as a Function of Frame Length and the Access Line Operating Rate

In examining the entries in Table 9-2, your first instinct to solve the delay problem caused by a variable-length information field may be to significantly increase the operating rate of the access line to the frame relay network. Although this would undoubtedly reduce the delay time caused by frames that transport data arriving at a FRAD or router prior to frames that transport digitized voice, it is also important to note that very rarely does a frame transporting data arrive at a FRAD or router as a single entity. That is, if a workstation initiates a file transfer, there is a high probability that a flow or sequence of LAN frames will arrive at the FRAD or router, and in fact may fill its buffer memory unless there is a mechanism that subdivides buffer memory into independent queues and uses a priority mechanism to service data placed in each queue. Even then, separate queues and a priority scheme may not be sufficient, because at certain access line operating rates, a lengthy frame transporting data can adversely delay a frame transporting digitized voice. Thus, a better solution to this problem is both fragmentation and prioritization of data.

Fragmentation

Fragmentation provides a mechanism to subdivide relatively long frames into a series of shorter, less-delay-creating frames. Through a fragmentation process, an even flow of minimized-length frames will be created, such that frames transporting digitized voice do not have to wait too long behind frames transporting data. This concept is illustrated in Figure 9-2, which shows how a long data frame is subdivided into a sequence of shorter-length frames by a FRAD or router. In this example, it was assumed that the FRAD or router was programmed to interleave frames transporting voice and data.

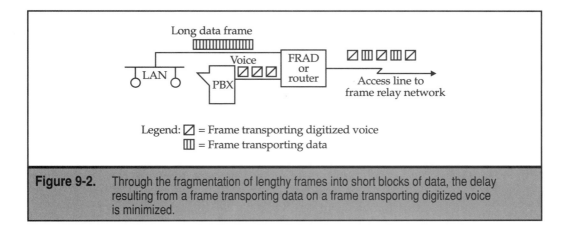

Figure 9-2.   Through the fragmentation of lengthy frames into short blocks of data, the delay resulting from a frame transporting data on a frame transporting digitized voice is minimized.

However, in many operational situations, fragmentation by itself or with simple interleaving is not sufficient. What is then needed is a frame-prioritization scheme that works in conjunction with frame fragmentation.

One important factor you must recognize is the effect of fragmentation. Although the fragmentation of packets transporting voice is a necessity to minimize delays to packets transporting digitized voice, the fragmentation process increases network overhead. Simply put, prior to implementing a voice-over-frame relay application it is extremely important to determine the potential effect of fragmentation on your data transportation.

APPLICATION NOTE: Although frame fragmentation is vital for the success of mixed voice and data applications flowing over a common serial port, the fragmentation process introduces additional overhead that adversely affects the transfer of data.

As previously mentioned at the beginning of this chapter, the Frame Relay Forum recognized the need for a standard method of fragmenting data. In December 1997, approximately seven months after publishing their Voice over Frame Relay Implementation Agreement the Frame Relay Forum published its Frame Relay Fragmentation Implementation Agreement. In the last two sections in this chapter we will examine each IA in detail.

Frame Prioritization

Prioritization is a technique in which frames are processed based on predefined criteria. When a FRAD or router is servicing both data and digitized voice, it logically makes sense to prioritize delay-sensitive traffic, such as digitized voice, ahead of non-delay-sensitive traffic or less-delay-sensitive traffic.

Memory Partitioning and Priority Queues

Prioritization must be used with fragmentation to be effective. Many equipment vendors will partition the memory of their FRAD or router into priority queues. Frames containing data that exceed a predefined length are first fragmented. Then, all frames transporting data, including fragmented and nonfragmented frames, are placed into a low-priority queue. Similarly, frames transporting digitized speech that exceed a predefined length are also fragmented. However, unlike frames transporting data, frames carrying digitized voice are placed into a high-priority queue. The primary reason frames transporting digitized voice are fragmented is that the sampling source produces a lengthy frame, and fragmentation will produce a more regular flow of voice information. It also reduces the potential that the loss or delay of a packet as it flows through the network will adversely affect the reconstruction of the voice signal at its destination.

Prioritization Techniques

The method used to prioritize traffic will obviously provide a preference to voice. However, depending on the type of data presented to the FRAD or router, there may be certain types of frames transporting different types of data that also require prioritization. For example, the transmission of SNA traffic can result in session timeouts if frames carrying such data are adversely delayed. Thus, prioritization techniques must consider the type of data being transmitted as well as the fact that frames transporting digitized speech should receive a higher priority than most frames transporting data.

APPLICATION NOTE: It is important to examine the manner by which queue servicing occurs and whether such servicing is adjustable.

In most voice-enabled FRADs the prioritization of voice and fax ahead of frames transporting data is done automatically. However, some equipment permits a degree of adjustment based upon the coder used as well as total end-to-end delay and other traffic. For example, you cannot delay SNA traffic to the degree that you can delay an FTP file transfer.

CIR Selection

In Chapter 4 we examined the role of the Committed Information Rate (CIR). In that chapter we noted that the frame relay standard allows the network service provider to control congestion by discarding any frames which exceed a subscriber's CIR.

We can obtain an appreciation for the role of the CIR upon voice transport by examining its relationship to the line access rate, the committed burst size (Bc) and the excess burst size (Be). As a refresher, Bc represents the amount of data in bits that a frame relay network agrees to transfer under normal network conditions during a predefined time interval (T_c). Most frame relay network operators set T_c at 1 second, resulting in the following relationship:

$$CIR = \frac{Bc}{T_c} = Bc$$

In comparison to the Bc, the Be represents the minimum amount of uncommitted data in bits above Bc that the network will attempt to deliver during the time interval T_c, which as previously explained is set to a value of 1 second by most network operators. Based upon the relationships between the CIR, Bc, Be, and the line access rate, we can revise Figure 4-5 to be a bit more meaningful concerning frames that are eligible for being discarded and those that are discarded.

Figure 9-3 illustrates the relationship between the previously mentioned frame relay metrics. In examining Figure 9-3 note that only frames within the CIR are guaranteed delivery. In comparison, when your transmission rate exceeds the CIR your frames will either have their discard eligible (DE) bit set or may actually be discarded when frames exceed the sum of the committed burst size and the excess burst size. For example, assume your organization has a contract with a frame relay service provider for a CIR of 128 Kbps. However, you use a T1 line access rate of 1.544 Mbps that permits bursts well above the CIR. As your FRAD or router bursts above the 128 Kbps CIR each frame that exceeds the 128 Kbps rate will have its DE bit set. Because there is no way to tell whether a DE bit setting represents the discard region or the discard eligible region after a frame flows through the first network switch, that switch is the only switch that will first discard frames arriving at a rate greater than the sum of Bc + Be prior to discarding frames in the discard eligible region. Thereafter, any switch under congestion in the network will drop frames with their DE bit set regardless of the frame rate into the network.

While an occasional dropped frame will not adversely affect a voice conversation, it is important to remember that unlike the transport of data where higher layers in the protocol stack compensate for frame loss via retransmission this is not possible when real-time voice is transported. This means that the transport of data where subscribers commonly burst above the CIR knowing that higher layers in the protocol stack compensate for frame loss can result in poor results when voice is transported. To obtain reliable delivery sufficient for good voice quality, you have to have a high enough CIR to cover voice

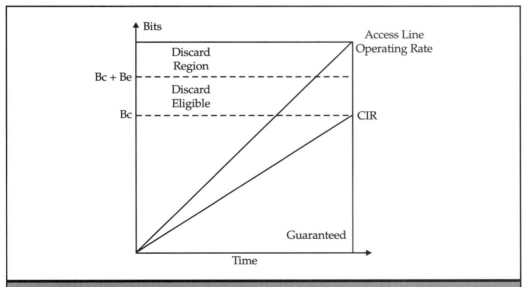

Figure 9-3. A sufficient CIR is critical to maintain voice quality

usage. Another way to achieve a similar result is to consider the service-level agreement (SLA) guarantees with respect to all frames and DE marked frames, a topic we will cover later in this section.

APPLICATION NOTE: One way to ensure frames transporting digitized voice reach their destination and are not dropped by the network is to have a sufficient CIR for your application.

Frame Loss Handling

When data is transmitted over a frame relay network, the loss of a frame due to congestion or error results in higher layers at the end points performing a retransmission of the lost frame. Although the retransmission process is commonly used to correct a previously lost frame, it introduces a delay that is not suitable for handling digitized speech. That is, if a frame transporting digitized speech is for some reason dropped by the network, the retransmission of the frame results in an arrival delay that distorts the reconstruction of the digitized voice signal. Recognizing this fact, equipment vendors handle frame loss with respect to the transportation of digitized speech in one of two ways. Some vendors simply generate a period of silence, while other vendors use the contents of a sequence of previously arrived frames to generate speech for the missing interval. This is accomplished by means of an interpolation technique based on the contents of the last or a few previously arrived frames. For both techniques, vendor equipment will not retransmit lost frames, as their arrival at the destination would result in a sufficient degree of delay that would render their use impractical.

It is important to note that the second technique actually represents a combination of techniques. A short period of time can transpire prior to a frame either being marked as missing or delayed and an interpolation of the prior frame being used. That period or gap is commonly compensated for by the generation of noise; however, because the gap is a very short period of time that is commonly unnoticeable to the human ear, some vendors may elect to use a period of silence. Thus, interpolation may be combined either with noise or silence.

Echo Cancellation

Echo is a phenomenon resulting from the connection of two-wire subscriber access lines to the four-wire infrastructure used to form the long-distance trunks that interconnect telephone company central offices to one another. The actual connection of two-wire to four-wire circuits is performed by a device referred to as a *hybrid*. When energy is transmitted across the hybrid, a portion is reflected back, and this reflection is called an *echo*. The echo can be an annoyance to the speaker, especially when the reflection of his or her voice is delayed sufficiently to become highly noticeable.

There are actually two types of echoes in voice conversations: near-end and far-end. The *near-end echo* results from the reflection of energy at the hybrid in the caller's central serving office. The *far-end echo* is caused by the hybrid located in the central office serving the called party. From the perspective of degree of disturbance, the far-end echo travels a much longer distance and would therefore usually arrive at the talker's location well after he or she spoke, causing an echo similar to what you might encounter in a canyon. To suppress the effect of echoes, communications carriers use *echo suppressors*. An echo suppressor is an electronic device inserted into a four-wire circuit to function as a blocking mechanism with respect to reflected energy.

Frame relay networks do not use echo-suppression equipment in their networks. This is because those networks were constructed to support the transfer of digital data. However, since the annoyance factor of an echo is a function of its delay, the transmission of digitized speech over a frame relay network can result in disturbing echoes when two sites separated by a sufficient distance are interconnected via a frame relay network. Since the network operator does not use echo-suppression equipment, this becomes the responsibility of equipment vendors.

Although most equipment vendors who market products for the voice-over-frame relay market include an echo-suppression or echo-cancellation capability, not all do. However, prior to using this feature as an evaluation discriminator, it is important to note that only when the average round-trip propagation time exceeds 25 to 50 ms is the use of echo suppression recommended. Otherwise, the use of a frame relay network to connect a few locations in close proximity to one another may not be necessary.

Frame Delay Handling

As frames are routed through a network, they will encounter a variety of nonuniform processing delays. These delays, which depend on such factors as the activity level of a switch, the length of a frame processed ahead of another frame, and the processing power of the switch, result in a uniform data stream by the time it is transmitted into a frame relay network being received at its destination with random delays between frames. This is illustrated in Figure 9-4, which shows how precise time intervals between a sequence of frames presented to a frame relay network could be altered during their flow through the network, resulting in random delays between received frames.

Jitter and Jitter Compensation

The delay between received frames is commonly referred to as *jitter* and can result in awkward-sounding regenerated speech. Recognizing this problem, many voice-over-frame relay equipment vendors incorporate a buffer area in their products. Buffering the frames that transport digitized speech removes the variable delays, in effect facilitating the process by which voice is reconstructed to prevent annoying periods of random delays.

The buffer area used to compensate for jitter is referred to as a jitter buffer. Most jitter buffers can be selected to a value between 0 (no buffering) to 255 ms of hold time or delay. The setting of the jitter buffer governs the amount of hold time. That is, a 50-ms jitter buffer setting means up to 50 ms of speech can be held and extracted according to the RTP time-stamp to provide a smoothed output.

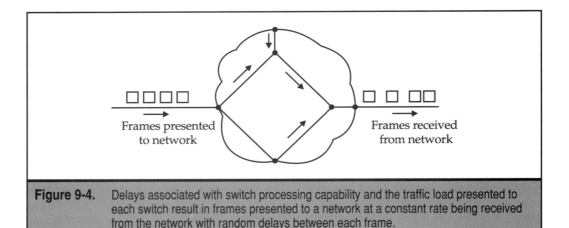

Figure 9-4. Delays associated with switch processing capability and the traffic load presented to each switch result in frames presented to a network at a constant rate being received from the network with random delays between each frame.

It is important to note that the jitter buffer hold time adds to the overall end-to-end delay. Thus, under most situations you more than likely will set the jitter buffer to a relatively low value. For example, if the end-to-end delay is 120 ms on average, with a low of 90 ms and a high of 130 ms, the variability is 40 ms. In this example you would set the jitter buffer to 40 ms or perhaps even beyond since you would still be below a total of 250 ms, which represents the point where a voice conversation begins to resemble a half-duplex transmission.

APPLICATION NOTE*:* Many equipment vendors have selectable-delay jitter buffers, typically supporting 0 to 255 ms of delay. Although a longer delay facilitates the reconstruction of voice, it also adds to the overall latency. Too high a setting can force users into a CB mode of having to say "over" to alert the other party to proceed.

Silence Suppression

Although many frame relay network providers do not bill for the number of bytes transmitted, from a practical standpoint it makes no sense to transmit digitized voice samples that contain only a period of silence. Frames containing a period of silence add to network traffic and can adversely affect the transmission and delivery of other frames carrying digitized voice samples from the same or different callers. Recognizing this problem, some equipment vendors include a silence-suppression capability in their equipment.

The effective implementation of silence suppression requires the use of a time tag for each frame transporting a digitized speech sample. This time tag enables the receiver to reconstruct a person's voice, including gaps and pauses between words and sentences, which results in the ability to maintain the natural quality of a person's speech. Unfortunately, the methods used to suppress periods of silence currently vary between some vendor products. However, the adoption of the Frame Relay Forum Voice over Frame Relay Implementation Agreement (FRF.11 IA) facilitates the interoperability between vendor equipment.

Voice-Compression Method

Although many technical issues can affect the quality of reconstructed voice, the most important long-term factor is the method of voice compression used. As previously discussed in Chapter 5, voice-compression methods can be generally grouped into three distinct categories: waveform coding, vocoding, and hybrid coding, with the latter representing a combination of the first two categories. Although such waveform-encoding methods as PCM and ADPCM result in very high quality reconstructed speech, they also require a large amount of bandwidth to transmit a digitized conversation, with PCM consuming 64 Kbps and ADPCM requiring 32 Kbps to provide a toll-quality conversation. Waveform coding provides such a high level of reconstructed voice that it serves as a benchmark for users to compare the quality of other compression methods, but it does not maximize the utilization of bandwidth. In fact, on certain access lines, such as a 64-Kbps connection to a frame relay network, the use of PCM or ADPCM would either preclude the use of the connection for data or consume all or half of the bandwidth of the access line—neither of which is a pleasant situation. Thus, a more practical employment of voice on a frame relay network requires the use of low-bit-rate compression algorithms, either resulting from vocoding or hybrid coding techniques.

Low-Bit-Rate Coding

Currently, equipment vendors offer products that support a variety of voice-compression algorithms, with many vendors using algorithms in the Code Excited Linear Prediction (CELP) family, whose employment can result in voice digitization rates from a high of approximately 16 Kbps down to 2.4 Kbps. Table 9-3 lists three examples of low-bit-rate standardized voice-compression algorithms commonly supported by frame relay equipment vendors.

In examining the low-bit-rate voice-compression methods listed in Table 9-3, you should consider the ability of equipment to support multiple methods. Doing so provides

ITU Standard	Compression Method	
G.723.1	ACELP	5.3/6.3 Kbps
G.728	LD-CELP	16 Kbps
G.729	CS-ACELP	8 Kbps

Table 9-2.   Low-Bit-Rate Standards-Based Voice-Compression Methods

you with the ability to select an alternative if for some reason the primary method results in reconstructed voice that is not to your liking.

Frame Overhead

One of the most overlooked facts associated with the transmission of voice over a data network is the overhead associated with transporting voice. While many publications indicate that an active conversation occurring over a frame relay network may only require 4 Kbps, that is only part of the story. To obtain an appreciation for the full story let's begin at the beginning and examine how many trade publications compute the 4 Kbps bandwidth consumption and, unfortunately, stop there.

If we assume a speech coder operating at 8 Kbps is used, we would start our bandwidth consumption computation at 8 Kbps. Next, we must consider the overhead associated with the header and trailer of the frame relay and RTP timing and sequencing information with respect to the actual data transported. Because small intervals of speech are transported in a frame the actual overhead can be approximately 25 percent or 2 Kbps. Thus, the gross frame relay bandwidth becomes 8 Kbps + 2 Kbps, or 10 Kbps.

If we assume silence suppression is supported, approximately 60 percent of a conversation's period consists of silence. This is because most conversations are half-duplex, which results in a 50 percent period of silence. After adding time for thinking, pausing for air, and perhaps scratching behind an ear, we can easily add an additional 10 percent period of silence. Thus, if we subtract 60 percent from 10 Kbps, we obtain a net bandwidth of 4 Kbps. Table 9-4 summarizes the previously discussed computations.

While the previously described computations are true, they only tell part of the story. If your organization is transmitting data frames typically longer than 64 bytes they will be fragmented. To illustrate the effect this has upon the efficiency of transmitting data, let's assume a frame transporting 1500 bytes requires fragmentation. Without fragmentation and assuming a 2-byte control field the overhead is 6 bytes to transport 1500 bytes of information. If fragmentation becomes necessary due to the need to minimize delay the 1500 bytes will be transported through the use of 24 frames. The first 23 frames would

Function	Bandwidth
8 Kbps Codec	8 Kbps
Frame overhead	2 Kbps
Gross bandwidth	10 Kbps
Less 60% silence suppression–6 Kbps	–6 Kbps
Net bandwidth	**4 Kbps**

Table 9-3. Computing Voice over Frame Relay Net Bandwidth Requirements

each transport 64 data bytes, while the 24th frame would transport 28 data bytes. Because 24 frames are now required overhead increases to 6 bytes/frame × 24 frames, or 144 bytes. Thus, overhead increased from 6/1500 or .04 percent to 9.6 percent for this example. This is the other part of the story most publications ignore and which, in reality, needs careful consideration.

APPLICATION NOTE: It is important to note that fragmentation is a dual-edged sword. While it is required to minimized delay when a frame transporting data arrives at a serial interface ahead of a frame transporting digitized voice, it also adds to the overhead associated with transporting data.

Telephony Signaling

The effective transmission of voice over a frame relay network requires the ability to alert the destination to an incoming call and to inform the originator of the progress of the call. Such activities are associated with telephony signaling, and a FRAD or voice-compliant router must be able to transfer appropriate telephony signaling. In doing so, it is important to note that the meaning of a telephone signal can be represented by a transition as well as by time between pulses and the sequence of pulses. This makes the appropriate coding and transfer of telephony signaling much more challenging than the simple passage of data signals that are represented by a transition from a high to low pulse or a low to high pulse. Since PBXs may use different signaling methods, it is important to ensure compatibility between the signaling method supported by the FRAD or router and the signaling method used by the PBX.

Multiplexing Techniques Used

In their quest to enhance the performance of voice over frame relay and to develop a mechanism that can differentiate their products from those of others, vendors offering voice-capable FRADs have incorporated different bandwidth-optimization methods into their products. Two such techniques that are incompatible with one another are *logical link multiplexing* and *subchannel multiplexing*.

Logical Link Multiplexing

Logical link multiplexing (LLM) enables frames transporting voice and data to share the same PVC. This technique is suitable for the situation in which your organization uses a voice server on a LAN to digitize voice. It requires both digitized voice-encoded frames as well as frames transporting LAN data to be carried via a common frame relay connection to the LAN. By using logical link multiplexing, you can more than likely reduce your organization's cost to use a frame relay network, since most network operators include a cost component based on the number of PVCs used.

Subchannel Multiplexing

A second multiplexing technique used by some FRAD manufacturers is subchannel multiplexing (SM). Under subchannel multiplexing, portions of multiple voice conversations are combined within one frame. By transmitting samples of several voice conversations within one frame, the overhead of the frame in comparison to its payload is reduced. This can be an especially important consideration when transmitting multiple digitized voice conversations over a low-speed frame relay access circuit, such as a 56- or 64-Kbps access line.

Service-Level Agreements

Beginning in 1997 several frame relay network operators began to offer different service-level agreements (SLAs) as a mechanism to differentiate one vendor from another. Most SLAs were originally focused upon throughput, response time, and availability, with refunds provided to subscribers if one or more SLA areas were not satisfied within the vendor level of guarantee.

While different SLA options will appeal to different categories of users, in a voice-over-frame relay environment there are several key options that warrant consideration. Those SLA options include throughput, data delivery rate, and response time or latency.

Throughput in a frame relay environment is normally expressed as the CIR range supported by the service provider. Another related service option is the maximum throughput rate, which normally permits bursts up to the access line operating rate.

The data delivery rate is normally expressed in terms of the CIR and bursting above the CIR. For example, a network service provider might offer a 99.99 percent data delivery rate for frames within the CIR and a 99.0 percent data delivery rate for frames that exceed the CIR.

As discussed earlier in this chapter, you can ignore the effect of bursting above the CIR if your service provider provides a relatively high data delivery rate. For example, an SLA that provides a data delivery rate of 99.0 percent for frames with their DE bit set means you can expect 99 out of 100 frames bursting above the CIR to reach the other end. If each frame transports 20 ms of voice, this means that you can expect to lose 20 ms out of every 2 seconds of speech, a small enough amount that will not be noted by the human ear.

Another key service option is response time or one-way network delay. When examining this SLA metric it is important to note that there are two methods service providers use to denote network delay. Some network providers only guarantee delay from the point where a frame enters the service provider's network to when it exits the network. In comparison, some service providers guarantee an SLA delay on an end-to-end basis. For both types of delays service providers commonly differentiate between different types or classes of traffic, such as LAN versus SNA. For example, an SLA for network delay might be 64 ms for LAN traffic and 50 ms for SNA traffic. Another method of differentiation concerning end-to-end delay is based upon the line access rate and traffic class. For LAN traffic, when this book revision was prepared Sprint's SLA was 130 ms at 56 Kbps, 85 ms with an access line operating rate of 256 Kbps, and 70 ms when the access line was a T1 circuit. In comparison, for an SNA class of traffic the SLA end-to-end delay was 115 ms at 56 Kbps, 70 ms at 256 Kbps, and 55 ms at a T1 operating rate.

APPLICATION NOTE: By carefully selecting an applicable SLA data delivery rate and end-to-end delay you can obtain a Quality of Service for your voice-over-frame relay application.

Now that we have a general appreciation of the major technologically related issues concerning the transmission of voice over a frame relay network, let's turn our attention to the operation and utilization of vendor equipment developed to provide this capability. We will also discuss the general cost of acquiring certain types of equipment, and we'll combine that with the cost of using a frame relay network to develop economic models for voice over frame relay transmission under several networking scenarios.

EQUIPMENT OPERATION AND UTILIZATION

Since 1995, a number of vendors have introduced equipment that supports the transmission of voice over a frame relay network. These vendors include ACT Networks (which was acquired by Clarent Communications during 2000), Fast Comm Communications Corporation, Memotec Corporation, Motorola Corporation, Nuera Communications, and Castleton Network Systems Corporation (which was in the process of being acquired by Newbridge Networks when this book revision was written). In addition, due to the potential economic savings associated with the use of this technology, additional vendors were introducing new products as this book was written. Thus, a viable market of voice-over-frame relay equipment providers is being established that will competitively offer equipment to satisfy organizational networking requirements. In addition, the Frame Relay Forum's Voice over Frame Relay and Frame Relay Fragmentation Implementation Agreements provide the foundation for interoperability between different vendor products. Thus, an expanding and eventually interoperable series of products should be available for consideration by the time you read this book.

In this section, we will first examine the operation and utilization of several commercially available products. We'll learn how they facilitate the transmission of voice over a frame relay network, and we'll weigh the economics associated with using certain types of vendor products. This in-depth economic analysis will include both the cost of hardware and the cost of frame relay network services. Once this is accomplished we will turn our attention to the ACT Networks NetPerformer series of products that can be considered to represent a second generation of multiservice access products that allow voice, fax, SNA, and LAN traffic to be transported over both frame relay and IP networks.

ACT Networks

ACT Networks is one of several vendors of communications products with a relatively long history in the development of frame relay products. In fact, ACT was the first company to commercially market frame relay access devices that could be used to transmit voice, fax, and data over public and private frame relay networks. In addition to manufacturing customer-provided equipment (CPE) for frame relay operations, ACT also manufactures a carrier class central office voice FRAD for use by communications carriers or for use on the backbone used to form high-speed internal networks. In this section, we will examine the operation and utilization of two ACT Networks products—its Integrated FRAD and a specific model of its IFRAD called the SN-8800 IFRAD. We will also examine the economics associated with using the SN-8800 to transmit voice over a frame relay network.

The Integrated FRAD

As previously noted in this chapter, there are a large number of technical issues associated with obtaining the ability to effectively and efficiently transport digitized voice over a frame relay network. Recognizing those technical issues, ACT Networks developed a family of FRADs designed from the ground up as a mechanism to transmit voice, data, and fax via public or private frame relay networks. Since these FRADs provide an integrated transmission capability, ACT Networks refers to them as Integrated Frame Relay Access Devices, or IFRADs. Such IFRADs are designed to do the following: fragment data frames to alleviate the potential adverse effect of those frames on frames transporting digitized speech; use buffers to minimize the effect of jitter for frames transporting speech; employ a priority scheme to prioritize voice and fax over data; and use a combination of silence detection and digital speech interpolation to take advantage of the half-duplex nature of voice communications.

Operational Features   The best way to obtain an appreciation of the capability and functionality of ACT Networks' IFRADs is to examine their operating features. We'll discuss seven features that enable Integrated FRADs to transport voice, data, and fax over a frame relay network in an effective and efficient manner, resulting in the reconstruction of speech almost identical to its original quality. These features are as follows:

1. Signal prioritization
2. Frame fragmentation
3. Predictive congestion management
4. The use of jitter buffers to alleviate signal sampling delays
5. The implementation of silence detection and the application of Digital Speech Interpolation to take advantage of the half-duplex nature of human conversations
6. The support of four speech compression methods
7. Signaling and telephone operational support features necessary to convey telephone calls across a frame relay network

Signal Prioritization   An ACT IFRAD supports the use of three priority queues. By assigning delay-sensitive traffic to high-priority queues and delay-insensitive traffic to lower-priority queues, you can facilitate the transfer of delay-sensitive information through the IFRAD. In doing so, ACT Networks recommends assigning faxes to the highest-priority queue, as such signals are more sensitive to delay than is voice when transmitted on a real-time basis. It is important to note that if you are using a store-and-forward fax system through a frame relay network, fax files would be handled in a manner similar to conventional data files. That is, you would assign the transfer of data files, including those containing digitized fax, to a low-priority queue.

In addition to providing a signal-prioritizing method via the assignment of different categories of information to different queues, the IFRAD supports the selling of the discard eligibility (DE) bit on selected channels. This additional feature enables users to specify to the network which data sources the network can selectively drop during periods of congestion.

Although queuing is an important mechanism for servicing high-priority traffic over delay-insensitive traffic, by itself it only ensures that information in high-priority queues is serviced before information in lower-priority queues. This means that a relatively long frame transporting data that arrives at an IFRAD slightly before a frame containing digitized speech will be serviced first, which can adversely affect the arrival of digitized speech at its destination. This in turn could result in an awkward portion of reconstructed voice. To prevent this situation from occurring, ACT Networks' IFRADs include a frame fragmentation feature.

Frame Fragmentation   ACT's IFRADs attack the previously mentioned problem through the use of a frame fragmentation process. An ACT Networks' IFRAD limits the length of frames based upon the type of information they transport. Frames transporting voice are limited to a maximum of 83 bytes. Frames transporting data are limited to a maximum length of 71 bytes for asynchronous data and 72 bytes for synchronous data. In comparison, frames transporting fax are limited to a maximum length of 58 bytes.

By limiting the length of packets the ACT Networks' IFRAD fragmentation process ensures that those frames transporting voice and fax are not significantly delayed by the prior servicing of data packets. Thus, prioritization and fragmentation go hand in hand to facilitate the transfer of frames carrying time-sensitive information through an IFRAD while minimizing the potential delay effect when a packet transporting data is serviced by the device.

Predictive Congestion Management   The use of FECN and BECN bits by a frame relay network for congestion control can be viewed as a reactive, after-the-fact method of congestion control. This is because network congestion has already occurred when the network sets those bits. Recognizing this problem and its potential effect on frames transporting time-sensitive information, ACT Networks included a feature called predictive congestion management in its IFRADs. Under predictive congestion management, an IFRAD responds to variances in traffic loading by varying the length of queues before congestion occurs. This enables frames transporting time-insensitive data to be further delayed, reducing the load on the network and in effect reducing the probability of congestion adversely affecting frames presented to the network. Predictive congestion management supplements the network's use of FECN and BECN bits but does not replace it. This means that an IFRAD, on receiving frames with those bits set by a network switch, will reduce traffic to the network on the data link control identifier(s) associated with the set bit or bits.

Jitter Buffers   Instead of accepting received packets and immediately delivering them to their destination, an IFRAD employs a jitter buffer to remove random time delays between frames transporting digitized voice. Those time delays result from different processing requirements occurring at each switch as frames flow through a network. Thus, although a sequence of frames transporting digitized speech may be presented at a uniform rate to a frame relay network, it will more than likely be received from the network with variable delays between frames. Since variable delays would result in awkward-sounding reconstructed speech, ACT Networks' IFRADs incorporate a configuration jitter buffer. This buffer can be set to hold a maximum of 255 ms of speech and serves as a mechanism to remove the variable time gaps between frames received from the network. That is, after frames are temporarily stored in the IFRAD's jitter buffer, they are dumped from the buffer at a constant rate.

Silence Detection   Most human conversations are fairly civil. When we speak, there are gaps between words, pauses between sentences, and slight delays as we form responses mentally prior to vocalizing them. Thus, human conversations are inherently half-duplex in nature and contain frequent short pauses between sentences, and even between words in a sentence. To take advantage of these characteristics of human speech, the ACT Networks' IFRAD uses the gaps in speech from one ongoing conversation in another conversation. This technique is more formally referred to as Digital Speech Interpolation. When conversations are truly half-duplex, this technique can result in the ability to improve bandwidth utilization by up to 50 percent.

Voice-Compression Support   IFRADs are modular in design, obtaining support for voice and fax via the installation of a voice/fax card. Each voice channel on the vendor's voice/fax card supports four voice compression methods, ranging in bandwidth from PCM's 64-Kbps operating rate to Algebraic Code Excited Linear Prediction's (ACELP's) 4.8-Kbps operating rate. This extended voice-compression support enables users to select an appropriate voice-compression algorithm based on a particular application and the level of background noise (or lack thereof). For example, if a connection to a voice/fax card occurs from a telephone in an industrial area, background noise commonly has a significant effect on vocoding-based voice-compression techniques. In comparison, the low background noise associated with a normal office environment would have a negligible effect.

Telephone Signaling Support   As we discussed in Chapter 6, there are several methods by which telephone signaling can be conveyed across a network. ACT Networks' IFRADs support four types of signaling, and they perform adaptive echo cancellation to ensure delays in the reflection of speech energy do not adversely affect a voice conversation.

The SN-8800 NetPerformer

In concluding our examination of ACT Networks' IFRADs, we will turn our attention to the use of this vendor's SN-8800 NetPerformer, which represents a small branch office IFRAD. We will use this IFRAD for our economic analysis because of its proven suitability for connecting small branch offices and its relatively limited expansion capability. Unlike other IFRADs that have the ability to be expanded in a manner that provides an almost unlimited number of networking configurations, the SN-8800 contains only two expansion slots and supports only three types of cards that can be used in those slots. This makes the use of the SN-8800 relatively straightforward and easy to price. In addition, the SN-8800 comes equipped with two built-in voice channels in its base unit, which facilitates its use for voice over frame relay in small branch offices that might have a limited voice-transmission requirement to other branch offices or to headquarters.

The SN-8800 base unit consists of four serial data ports and two voice channels. Two expansion slots are available on the base unit for the installation of Ethernet or Token Ring LAN adapters or a T1/E1 CSU1/DSU. At the time the first edition of this book was written, the retail price of the SN-8800 was $3995, with an Ethernet LAN card priced at $395, a Token Ring LAN card priced at $895, and a T1/E1 CSU/DSU card priced at $795.

Although the SN-8800 and other ACT Network IFRADs were replaced by their NetPerformer series of products during 2000, we can obtain an appreciation for the economics associated with voice over frame relay by examining the cost of that product and its use. This will provide an indication of economic parameters you need to consider regardless of the cost and type of product used in your frame relay networking environment.

Figure 9-5 illustrates the use of an ACT Networks' SN-8800 IFRAD to support the communications requirements of a typical branch office into a frame relay network. In this example, the SN-8800 is used to support the connection of an IBM 3174 control unit that supports an SNA connection for a group of legacy SNA terminals accessing a mainframe, as well as for LAN-to-LAN connectivity via the use of an Ethernet adapter that connects the IFRAD to an Ethernet network. Since the SN-8800 has two built-in voice channels, we will assume that we connected two ports on a PBX at the branch office to the SN-8800 and programmed the PBX to route calls to the IFRAD when employees dial a 6 prefix.

The actual cost of an ACT Networks SN-8800 based on its list price would be $5185 per IFRAD to support two channels of voice along with the transmission of Ethernet and SNA traffic via a common frame relay network access line. Since the voice ports are built into the SN-8800, we cannot directly break out the cost associated with transporting voice. However, from a network perspective, the addition of voice would require the addition of only two PVCs, each at a very low CIR. Assuming the cost of a PVC at a 16-Kbps CIR is $10 per month, which represents an average of several frame relay network provider billings obtained during 2000, then the cost for adding four PVCs to enable communications between two branch offices would be $40 per month. Assuming you were calling only at night (which is unreasonable for a business), and you obtained the Sprint

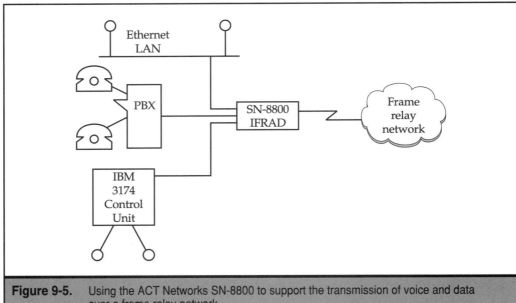

Figure 9-5. Using the ACT Networks SN-8800 to support the transmission of voice and data over a frame relay network

"dime lady" rate of $.10 per minute, you would need to use the two voice channels a total of only 400 minutes per month, or approximately 18 minutes per day (based upon 22 working days per month), to recover the added cost of the PVCs. Thus, the use of the SN-8800 IFRAD provides a very economical method for adding a voice communications transmission capability to a branch office to bypass the PSTN. If the branch office is located overseas, where the per-minute cost of using the PSTN might be $.30 to $.50 per minute or higher, you would need to use the voice transmission provided by the SN-8800 IFRAD only a few minutes per day to recover the cost of the PVCs. In other words, if it takes only one short call per day to break even, subsequent calls may provide significant savings that can make the person who acquired the SN-8800 the economic champion of the organization.

The SDM-NetPerformers

As previously mentioned in this section, the first generation of ACT Networks IFRADs were replaced by a series of second-generation products from that vendor marketed under the name NetPerformer. At the time this edition was prepared ACT Networks was marketing five NetPerformer products. Those products ranged in scope from the SDM-9350 and SDM-9360 low-cost voice, fax, and data integration devices to the SDM-9380 designed for use at a central site and the SDM-9400 regional site and SDM-9500 central site expandable device. The latter product can be expanded and is more suitable for use in a large regional office or central office location. To obtain an appreciation for the capability of this second-generation series of products, we will examine how NetPerformer products can be networked together as well as their functionality.

Networking NetPerformers   Figure 9-6 illustrates the manner by which ACT Networks NetPerformer 9360/9380 products can be used at branch office locations to provide communications with NetPerformer 9400 and 9500 products located at regional and corporate headquarters locations.

In examining Figure 9-6 you can obtain an appreciation for the networking capability of the ACT Networks series of NetPerformer products. Because all products in the series support SNA, frame relay, IP, and according to the vendor's Web site will soon support ATM, they can be used in almost any type of networking environment.

The SDM-9350 is designed for use in branch offices and represents the entry model in the NetPerformer product line. This device has two analog voice ports that can be configured for E&M, FXS, or FX0 support. The 9350 is expandable to four voice ports and supports both

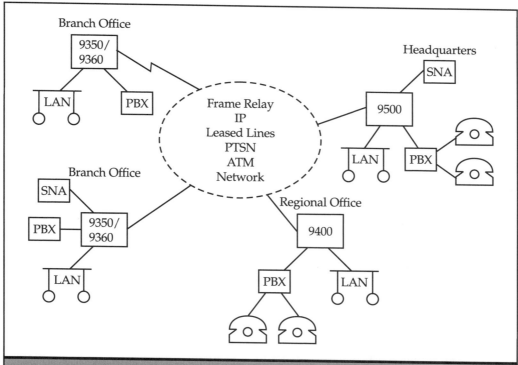

Figure 9-6.   Networking ACT Networks NetPerformer products

Ethernet and Token Ring LAN connections. The NetPerformer 9350 includes four serial ports that enable the device to be connected to different types of networks as well as to aggregate SNA and asynchronous traffic as well as the input from other routers for transmission with digitized voice over a common network connection.

To expedite traffic the 9350 supports 8 classes of service using 16 priority weights. Voice compression algorithms supported include ACELP at 8 Kbps, 5.8 Kbps, 4.8 Kbps; ADPCM G.726; and PCM G.711. The NetPerformer also supports Group III fax at 2.4, 4.8, 7.2, 9.6, 12.0, and 14.4 Kbps. Unlike some products that cannot support modem modulation, the NetPerformer 9350, as well as other members in the series, will support V.32bis modulation up to 14.4 Kbps.

The NetPerformer 9360 maintains the core features of the 9350 but provides support for up to 16 analog or 30 digital telephony channels. Thus, the NetPerformer 9360 is more suitable for a larger branch office or a regional office environment. In addition to supporting the same analog interfaces as the 9350, the NetPerformer 9360 adds support for digital telephony by providing the capability for connecting to a PBX via a 24 channel lT1 or 30 channel E1 port.

Although you might expect the NetPerformer 9400 to have more capacity than the 9360 since the former is designed for use in a regional office, this is not the case. The 9400 unit can only support up to 8 analog or 30 digital voice channels, although the device includes support for all other features supported by other members of the NetPerformer product line. At the top of that product line is the NetPerformer 9500. This is a modular, chassis-based central-site product that has the capability to be considerably expanded by adding additional modules into the chassis.

Now that we have an appreciation for the general features of one vendor's voice-over-frame relay products, let's look at a few other products and the economics associated with using a pair of FRADs to transport voice over a frame relay network.

Memotec

Memotec Corporation, which is headquartered in Montreal, Canada, is a well-known manufacturer of communications products, including a series of FRADs developed to provide support for the transmission of voice, data, and legacy communications protocols over a frame relay network. In April 1996, Memotec announced its CX900 "flex-FRAD," which provides a modular solution for users needing to support a variety of traffic over a frame relay network. The vendor's CX900 flex-FRAD was followed by the introduction of its CX900e frame relay access switch, which is based on a chassis design that permits the addition of various types of modules so that you can custom-tailor a configuration to satisfy your requirements.

The CX900e was followed with the Memotec high performance CX950. The CX950, according to the vendor, was the first multiservice platform to integrate voice, data, and video over frame relay, IP, ISDN and even X.25 and ATM. The best way to understand the versatility of a FRAD is to study its features. In doing so, we will focus our attention on the CX900e and the CX950, the latter a recent addition to the Memotec product line.

CX900e FRAD

As previously discussed, the CX900e is a modular FRAD that is fabricated as a chassis into which different modules are installed so that device functionality can be customized to a specific set of user requirements. The CX900e supports six types of modules, including LAN, Voice, Serial Port, ISDN, DSU/CSU, and V.34 Dial Modem modules. The ISDN Modules can be used as a primary method for network access or as backup to the frame relay network, while the V.34 Dial Modem module is used only for dial backup in the event of a primary-link failure.

Currently, Memotec markets both Ethernet and Token Ring LAN modules, enabling the FRAD to function as a LAN bridge/router supporting either type of LAN network over a frame relay network. The serial port module provides device connectivity to a wide area network via V.24, V.35, or X.21 interfaces, while the DSU/CSU module represents an integrated DSU/CSU required for transmission onto a digital access line. The ability to integrate support for voice and fax via the Memotec FRAD is accomplished through the use of one or more voice modules. Each voice module supports one voice/fax connection, and a CX900e can support up to four voice modules. The voice/fax capability of the CX900e is based on the functionality of the vendor's voice modules, so let's turn our attention to the characteristics of the voice modules supported by the FRAD.

Voice Module Support   Voice/fax support is accomplished through the installation of one or more voice I/O cards into the FRAD chassis, with a maximum of four I/O cards supported per FRAD. Each voice I/O card contains an analog voice/fax port that is controlled by a specific type of line interface driver. Memotec offers foreign exchange drivers as well as a two-wire and four-wire E&M driver support for each voice card.

Operation   The Memotec CX900e is similar to the ACT Networks' IFRAD with respect to the manner in which it supports the transmission of time-sensitive information and its removal of random spacing between received frames carrying digitized speech. That is, the CX900e prioritizes frames, using fragmentation to facilitate the transfer of frames carrying voice and fax, while employing an elastic buffer to alleviate the effect of jitter caused by the flow of frames through the network. Concerning fragmentation, although the Memotec FRAD uses default parameters to define the maximum-length default of frames transporting voice, fax, and asynchronous and synchronous data, a network manager or administrator can change those values. Similar to the IFRAD, the CX900e permits a user to set the discard eligibility (DE) bit for individual channels. Also similar to the ACT Networks' IFRAD, the Memotec CX900e supports the use of Digital Speech Interpolation to take advantage of the half-duplex nature of human conversation and the pauses that occur during speech. One area of significant differences between ACT and Memotec products is in their support of different voice-digitization methods. Unlike the ACT Networks' IFRAD, which supports the use of four distinct methods of voice compression, including toll-quality PCM, Memotec's voice-digitization support is limited to 8- and 5.8-Kbps versions of the ITU G.729 Algebraic Code Excited Linear Prediction (ACELP), with the method used software selectable. The actual voice-digitization method supported by Memotec is referred to as ACELP II, which is considered to represent a superset of the ITU G.729 standard.

CX9500 Access Switch

As previously mentioned, the CX950 represents a multiservice platform that supports the integration of voice and data over all popular wide area network technologies. In addition to providing a voice-over-frame relay capability, the CX950 also provides voice-over-IP and voice-over-ATM support.

The CX950 is a modular device that obtains its functionality via the installation of different modules. In the area of voice support the CX950 supports up to seven single port analog voice modules or up to four dual port analog modules. The single port analog modules support ACELP 5.8 Kbps and 8 Kbps compression, while the dual port analog modules support the G.729B compression method. In addition to supporting analog voice, the CX950 can support T1, E1, and ISDN Basic Rate Interface (BRI) and Primary Rate Interface (PRI) digital voice interfaces. When used with T1, E1, or PRI lines the CX950 can support a maximum of 62 timeslots with up to 30 supporting voice.

Other modules supported by the CX960 include single and five-port serial I/O, LAN modules, CSU/DSU modules, an ISDN module, and a V.34 modem module. The serial I/O modules support the V.24, V.35, and X.21 interface standards. There are several types of LAN modules supported by the CX950. Those modules include Ethernet, Fast Ethernet, and Token Ring. CSU/DSU modules include a 56 Kbps/64 Kbps module as well as a high-speed T1/E1 CSU/DSU module. The ISDN module supports BRI with two B-channels and one D-channel. The CX950 supports up to seven ISDN BRI modules, enabling transmission and reception of data and voice over an ISDN network. Last, but not least, the V.34 modem module provides the CX950 with a dial backup capability and an alternate connection to the network in the event the primary link fails.

Similar to the previously described ACT Networks NetPerformer products, you can use Memotec CX900e and CX950. That is, you can install the CX900e in branch offices while the modular CX950 is more suitable for installation in regional and central offices. In fact, you can replace NetPerformer equipment shown in Figure 9-6 by Memotec equipment to obtain the same general networking capability. However, as with any network project, you need to compare the functionality and cost of vendor products against your actual organizational requirements.

Nuera Communications

Another interesting vendor product located by this author during the preparation of this book was the Access Plus F200ip from Nuera Communications of San Diego, California. Similar to some previously mentioned products that support either voice over frame relay or voice over IP, the F200ip supports both. Thus, it provides a degree of flexibility similar to that obtainable with other products. Since this chapter is focused on voice over frame relay, we will limit our coverage of the Nuera Communications F200ip to its frame relay capability, which is essentially equivalent to the vendor's F100 FRAD, whose operation was rated number one in quality for voice over frame relay by listeners during a test performed by *Data Communications Magazine*. Thus, in this section we will review the functional capability of the Nuera Communications F100 FRAD and then examine the economics associated with its use.

The F100 FRAD

The Nuera Communications F100 FRAD is based on an eight-slot chassis. The chassis accepts the installation of one-port analog or four-port digital/fax cards, with each card containing its own digital signal processor to perform all required voice-coding operations. Included in the chassis is a dedicated frame relay serial port and three data ports for connecting the FRAD to routers and other network devices. This means that you can easily use the F100 to compress 24 voice channels from a PBX T1 connection or up to 30 from an E1 connection to a PBX. The use of built-in dedicated serial ports frees the chassis for a full installation of eight voice I/O cards. Users need not worry about the support of a wide area connection degrading from the functional capacity of the FRAD to support certain voice communications requirements.

Within the F100 chassis is a network processor that performs such functions as echo cancellation and frame prioritization. A separate packet processor is employed to perform call-processing functions, including the ability to switch each call independently of the others. This enables the connection of a T1 or E1 line from a PBX to have up to 24 or 30 calls routed to appropriate digital/fax cards, unlike some other FRAD designs that require a sequence of individual PBX connections from a PBX to individual ports on voice/fax cards installed in a FRAD. Thus, the design of the F100, which is illustrated in Figure 9-7, facilitates cabling between a PBX and the FRAD. However, unlike some other FRADs that support the installation of LAN adapter cards to enable the FRAD to become a direct participant on a local area network, the F100 connects to a LAN via a serial interface to a router. Although this requires the use of a router instead of a network adapter card, it frees slots in the chassis for use by other types of cards.

The Nuera Communications F100 FRAD includes features such as silence suppression, frame fragmentation, prioritization, and the use of a jitter buffer that functions in a manner similar to voice-capable FRADs manufactured by other vendors. However, the F100 also includes several unique features that provide an additional margin of performance to facilitate the transmission of voice over a frame relay network. Those features include the use of asymmetric fax channels, adaptable echo cancellation, the use of digital voice/fax cards that support five types of voice compression, and the capability to support the use of eight four-port voice cards. These features can be an important consideration when evaluating equipment, so let's briefly discuss them.

Asymmetric Fax Channels   Fax transmission results in the majority of data flowing in one direction, while a limited amount of control information flows in the reverse direction. This transmission characteristic enables the use of asymmetric fax channels to considerably reduce the use of bandwidth, which can then be used by other traffic.

Adaptable Echo Cancellation   The F100 includes an adaptable echo canceler. This canceler measures reflected energy and adapts from 0 to 49 ms to provide a consistent voice quality for all calls serviced by the FRAD.

Digital Voice/Fax Card Support   The F100 uses digital voice/fax cards that support up to four voice channels per card and up to 30 channels per system, providing among the

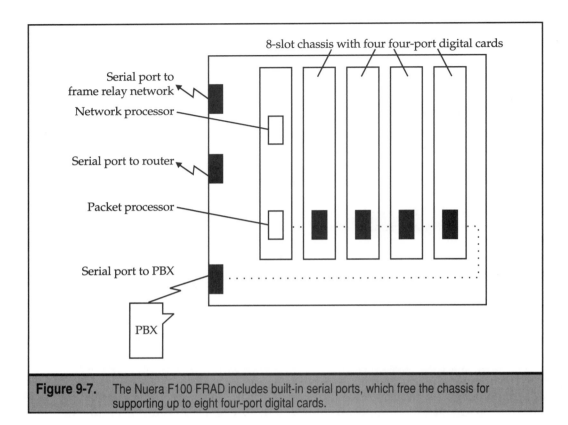

Figure 9-7. The Nuera F100 FRAD includes built-in serial ports, which free the chassis for supporting up to eight four-port digital cards.

highest voice support of products currently marketed for transmitting voice over a frame relay network. Each voice/fax card supports five methods of voice compression:

▼ ATC, from 7.4 to 32 Kbps

■ G.729 CS-ACELP at 8 Kbps

■ G.728 LD-CELP at 16 Kbps

■ G.726 ADPCM at 32 Kbps

▲ E-CELP at 4.8, 7.47, and 9.6 Kbps

Between the voice-call capacity and the ability to select from five voice-compression methods, the F100 gives users a significant capability to "test the concept" and, once the concept proves justifiable, to significantly expand the capacity of the FRAD.

Economics of Use   To obtain an appreciation for the economics associated with the use of the Nuera Communications F100 FRAD, let's examine the use of a pair of FRADs, each containing four voice ports and several low-speed ports to enable you to route SNA and perhaps both asynchronous and synchronous data traffic through a common frame relay access line. Although the cost of each F100 obviously varies by configuration, the basic price of the FRAD just mentioned will be between $6000 and $7000 per unit. Let us further assume the use of a T1 access line results in an average cost of $625 per month to obtain a fractional T1 (FT1) access to a frame relay network at a data rate of 128 Kbps. The monthly cost of a 128-Kbps network port will be assumed to be $415, while PVCs with a 16-Kbps CIR will be assumed to cost $10 per month. Based on the use of four PVCs to support voice and data, we can now compute the cost of adding voice to an SNA or a mixture of formerly data-only applications, including both hardware and network operation cost. However, prior to doing so, it is important to note that the one-time cost associated with the purchase of communications equipment should be amortized over its lifetime to determine the monthly cost. This can be added to the monthly fee of the communications carrier to obtain the total monthly cost for both equipment and the use of a transmission facility. Once this is determined, it becomes relatively easy to estimate the cost per minute for the transmission of voice over a frame relay network for different potential levels of voice traffic.

Amortizing One-Time Equipment Cost   Assume the cost of F100s is $7000 each, or $14,000 for a pair that will be used to enable two branch offices to exchange both data and voice over common access lines routed to a frame relay network. If we assume the equipment has a four-year (48-month) life, then the cost per month becomes $14,000/48, or $292 per month.

Adding the Network Cost   Assuming that the monthly cost of the frame relay service is $1080 per site, which includes the access line, port cost, and individual PVCs, the monthly network cost for two sites becomes $2160. Thus, the total monthly cost of equipment plus a frame relay network provider is $2160 plus $292, or $2452.

Computing the Cost Per Minute   Since the cost per minute depends on the volume of voice traffic carried by FRADs between network end-point locations, we will first make some assumptions concerning the potential use of the four voice ports per FRAD and then use those assumptions as a basis to project additional usage of the FRADs to transport varying amounts of voice traffic via a frame relay network. First, let's assume there are 22 working days per month during which the voice capability of the FRADs will be used. Second, assume we will use the port on each voice card 12.5 percent of the business day, or one hour during an eight-hour business day (probably a very low level of utilization for an organization considering the use of voice over frame relay). Based on the preceding assumptions, we anticipate four hours of voice communications per day, occurring collectively on all four voice ports, or 88 hours on a monthly basis. This means that on a per-minute basis, we will begin our economic analysis by assuming the pair of FRADs are used to provide 88 hours × 60 minutes/hour, or a total of 5280 call minutes of voice transmission per month. Based on the previously computed monthly cost of equipment and frame relay service of $2452, this results in a per-minute cost of $2452 divided by 5280 call minutes, or $.464 per minute. At this rate, it would appear that transmitting voice over frame relay would pay off only for expensive international calls. While this may be true for a low volume of call minutes, consider the possibility of adding voice-transmission capacity to an existing FRAD or of increasing the volume of usage. For our example, let's do both, starting with the latter by examining the effect of increased call volume on the cost per minute.

Examining the Effect of Increased Call Volume　Table 9-5 summarizes the cost per minute based on varying the occupancy rates of the four voice ports. Note that you have to keep the ports on the voice/fax card fairly well occupied to obtain a reasonable per-minute cost that is less expensive than normal commercial service via the PSTN. However, Table 9-5 is based on the full purchase cost of a pair of FRADs for voice transmission via frame relay. If we already have an existing pair of FRADs we wish to upgrade to support voice over frame relay, or if we originally intended to acquire a pair of FRADs to support data transmission and added voice and fax transmission capability, we should consider only the cost of a pair of voice/fax cards and the PVC cost for voice. This is because we would still have to purchase the chassis and pay a monthly fee for the access line and network port. Thus, let's again revise our computations.

Examining the Incremental Cost of Voice　We can examine the economics associated with transmitting voice over frame relay by assuming that the basic expense of obtaining a pair of FRADs, a pair of access lines, and a pair of network ports would be incurred even if we were simply transmitting data over the network. Thus, the cost of transmitting voice involves the *incremental* cost associated with adding this transmission capability. To determine the

Full Cost of Equipment and Network Considered			
Hours/Day	Percent Port Occupancy	Monthly Call Hours*	Cost/Min
1	12.5	88	46.4
2	25.0	176	23.2
3	37.5	264	15.5
4	50.0	352	11.6
5	62.5	440	9.3
6	75.0	528	7.7
7	87.5	616	6.6
8	100.0	704	5.8

*Four ports, 22 business days per month.

Table 9-4.　Economics of Increased Call Volume

incremental cost associated with transmitting voice, let's assume that each four-port card can be purchased for $1500, resulting in the equipment expense being reduced to $3,000, or $62.50 per month when amortized over a 48-month period. Instead of considering monthly access line charges and the monthly cost of two network ports, let's assume we simply add a PVC for each PBX-to-PBX transmission, which adds $20 per month to our frame relay network carrier bill. Thus, our total monthly cost now becomes $82.50.

If we use the same network usage assumptions as before, then a 12.5 percent occupancy rate generates 5280 call minutes of voice transmission per month. This results in a per-minute cost of $82.50/5280, or approximately $.016 per minute, which is super-competitive. Table 9-6 indicates the economics associated with an increased call volume for the situation where we amortize only the additional cost associated with adding a voice and fax transmission capability to FRADs that we would have had to purchase anyway to satisfy a data transmission requirement. As indicated in Table 9-6, the cost per minute can get so low that you may consider becoming a communications carrier in competition with the big firms!

General Observations   In comparing the costs listed in Tables 9-5 and 9-6 for increased call volumes, we can make some general observations. First, unless you anticipate an extremely high level of voice communications, the transmission of voice over a frame relay network may be hard to justify from an economic perspective. Second, if you have an existing frame relay connection using FRADs that can be upgraded to support voice, or if

	Only Incremental Cost of Voice Capability Considered		
Hours/Day	**Percent Port Occupancy**	**Monthly Call Hours***	**Cost/Min**
1	12.5	88	1.56
2	25.0	176	0.78
3	37.5	264	0.52
4	50.0	352	0.39
5	62.5	440	0.31
6	75.0	528	0.26
7	87.5	616	0.02
8	100.0	704	0.19

*Four ports, 22 business days per month.

Table 9-5.   Economics of Increased Call Volume

you need to acquire hardware and establish a connection to a frame relay network to support data transmission, then the incremental cost associated with adding a voice transmission capability will normally result in a very low level of call volume being sufficient to justify the additional cost of supporting voice. Now that we have an appreciation for the economics associated with transporting voice over a frame relay network, let's conclude this section with a feature checklist to facilitate our potential equipment-acquisition process.

Feature Checklist

In concluding this section, Table 9-7 provides you with a comprehensive checklist of voice/fax-related features to consider when evaluating different vendor products. It includes a column for listing your requirements with respect to a particular voice FRAD feature as well as two columns labeled "Vendor A" and "Vendor B." Use these to compare products from two vendors against your specific requirements. You can add more vendors if you wish.

Feature	Requirement	Vendor A	Vendor B
DE bit setting	_____	_____	_____
Echo cancellation adjustability	_____	_____	_____
Fax support			
V.21 300 bps	_____	_____	_____
V.27ter 2.4/4.8 Kbps	_____	_____	_____
V.29 9.6 Kbps	_____	_____	_____
V.17 14.4 Kbps	_____	_____	_____
Fragmentation			
Fixed	_____	_____	_____
Adjustable	_____	_____	_____
Logical link multiplexing	_____	_____	_____
Jitter buffer			
Adjustable	_____	_____	_____

Table 9-6. FRAD Voice-Related Features to Consider

Feature	Requirement	Vendor A	Vendor B
Prioritization			
Fixed queues	_____	_____	_____
Selectable	_____	_____	_____
Subchannel multiplexing	_____	_____	_____
Voice/fax card			
Type			
Analog	_____	_____	_____
Trunk Interface			
FXS	_____	_____	_____
FXO	_____	_____	_____
E&M	_____	_____	_____
	_____	_____	_____
Digital			
T1 interface	_____	_____	_____
E1 interface	_____	_____	_____
ISDN interface	_____	_____	_____
Ports/card			
Number	_____	_____	_____
Maximum supported	_____	_____	_____
on FRAD	_____	_____	_____
Compression method(s)	_____	_____	_____
Voice signaling support			
Dial tone	_____	_____	_____
Busy tone	_____	_____	_____
Trunk busy tone	_____	_____	_____
Ringing	_____	_____	_____
Compression method(s)	_____	_____	_____

Table 9-7.   FRAD Voice-Related Features to Consider *(continued)*

Feature	Requirement	Vendor A	Vendor B
LAN Connectivity			
Ethernet	_____	_____	_____
Fast Ethernet	_____	_____	_____
Token Ring	_____	_____	_____
Management			
SNMP	_____	_____	_____

Table 9-7.   FRAD Voice-Related Features to Consider *(continued)*

THE FRAME RELAY FORUM VOFR IA

In this section we will focus our attention on the Frame Relay Forum Voice over Frame Relay (VoFR) Implementation Agreement. Frame relay IAs are based on relevant ANSI, ITU, or other standards and highlight areas where the agreements differ from existing standards. IAs are developed to provide unambiguous agreements between switch equipment vendors and network operators on how options within a standard may be implemented to facilitate true worldwide multivendor interoperability. Thus, IAs are meant to be read along with relevant standards.

Overview

The Frame Relay Forum FRF.11 IA consists of four sections and eight annexes. The first section is introductory in nature and covers the purpose of the IA, includes an overview of its contents, and provides a list of abbreviations used in the document and the relevant standards. The second section, titled "Reference Model and Service Description," illustrates how a Voice Frame Relay Access Device (VFRAD) would exchange voice and signaling information with another VFRAD via the use of a network reference model. In doing so, this section refers to the use of different voice compression schemes that are contained in Annex A of the IA and the packaging of data frames according to rules specified in Annex C of the IA. Other portions of Section 2 discuss the use of encoded fax, how periods of silence can be encoded, the use of signaling bits, and similar information, with detailed information provided by references to appropriate annexes in the IA.

With the exception of some of the annexes, Section 3, "Frame Formats," is probably the most important section with respect to interoperability between vendor equipment. It provides a detailed description of how voice and data payloads are multiplexed within a frame. This section uses the term *subframe* to refer to a payload package that is placed in a frame's information field.

The fourth and final section in the document, entitled "Minimum Requirements for Conformance," defines the support vendors must provide to ensure that their products can interoperate with other vendor products that support the VoFR IA.

We will now turn our attention to the formation of subframes and the basic compliance requirements a FRAD must support to conform to the VoFR IA.

Subframes

The use of subframes is optional; however, they serve as a mechanism to enable more efficient processing and transmission of frames containing digitized speech. Each subframe contains its own header and payload, with the header identifying the voice/data subchannel that originated the payload and, optionally, the type of payload and its length.

Figure 9-8 illustrates the relationship between frames and subframes. In this example, a single DLCI is used to support two voice channels and one data channel. This is accomplished by two voice payloads being packaged in the first frame, while a data payload is shown being placed into the second frame.

In examining Figure 9-8, it is important to note that payloads can transport voice, fax, data, or signaling information. To provide a mechanism for interoperability, the format of the subframe is defined in the IA. Figure 9-9 illustrates its format.

The Subframe Format

In examining Figure 9-9, note that 6 bits in the first byte and 2 bits in the second byte can be used to define the subchannel identification, providing the ability to multiplex up to 256 channels within a common DLCI. The actual structure of the first two octets and the inclusion of octets 1a and 1b depend on the settings of the extension indication (EI) and length indication (LI) bits in the subframe. Those settings are indicated in the lower portion of Figure 9-9. For example, when the EI bit is set to a value of binary 1, it serves as a flag to indicate the presence of octet 1a. This extends the subchannel identification to 8 bits. Otherwise, when the EI bit is set to a value of 0, the CID uses 6 bits, and the maximum value of the subchannel identification is limited to 63. Returning to the setting of the EI bit

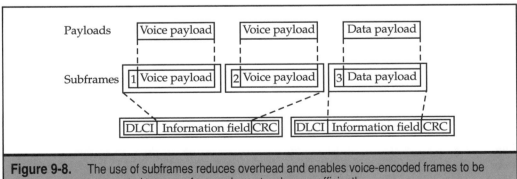

Figure 9-8.   The use of subframes reduces overhead and enables voice-encoded frames to be transported across a frame relay network more efficiently.

Bit position	8	7	6	5	4	3	2	1
Octet 1	EI	LI	Subchannel identifier (CID)					
Octet 1a	CID		Spare	Spare	Payload type			
Octet 1b	Payload length							
Octet *n*	Payload							

Notes:

1. When the EI bit = 1, the structure of octet 1a's payload type field applies as follows:

Bits	4	3	2	1	
	0	0	0	0	Primary payload transfer
	0	0	0	1	Dialed digit transfer
	0	0	1	0	Signaling bit transfer
	0	0	1	1	Fax relay transfer
	0	1	0	0	Silence information description

2. When the LI bit = 1, the structure of octet 1b applies.

Figure 9-9.   The subframe format

to 1, as indicated in Figure 9-9, this setting in conjunction with the Payload Type field values defines how the entries in the Payload Type field are interpreted. Note that although 4 bits are used to define the payload type, only 5 out of 16 possible values are presently defined.

Through the use of the setting of the EI and LI bits and the values assigned to the Payload Type field, there are a number of possible combinations of subframes as well as variances in the information transported in those frames. Figure 9-10 illustrates two examples of the formation and composition of subframes. In Figure 9-10a, setting the EI and LI bits to 0 results in the elimination of the need to include octets 1a and 1b in Figure 9-9. Thus, the payload directly follows the CID.

In the example illustrated in Figure 9-10b, both the E1 and LI bits are set to values of binary 1. The setting of the EI bit to a value of 1 means that octet 1a will be included in the header. That byte includes a Payload Type (PT) field whose value is shown set to 1. As indicated in Figure 9-9, this signifies that dialed digits are being transferred in the payload. The setting of the LI bit informs the receiver that octet 1b, which defines the payload length, follows, and it is simply labeled "octet 1b" in Figure 9-10. Once the payload in the form of dial digits is used to complete the subframe, a second subframe is included in the frame. The following subframe is transmitted on a different CID and has its LI bit set to 0, which indicates that no payload-length octet follows. Instead, the use of a Payload Type field value of 0 indicates to the receiver that the information directly following the Payload Type field represents the primary payload transfer in a standard voice Payload field.

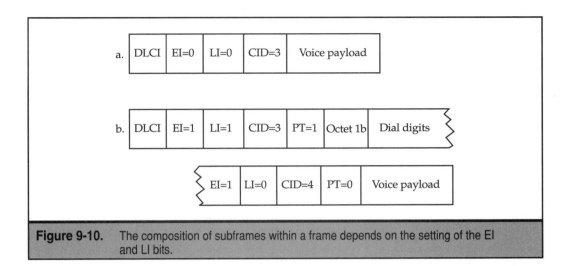

Figure 9-10.   The composition of subframes within a frame depends on the setting of the EI and LI bits.

Conformance Issues

The key to equipment interoperability is the ability of different vendor products to support a minimum number of common functions. Under the VoFR IA, there are two classes of compliance requirements. Vendor products must satisfy one or the other or both to provide interoperability with other vendor products. Class 1 compliance is for products that support high-bit-rate voice digitization at 32, 24, or 16 Kbps, while Class 2 compliance is for products that support low-bit-rate voice-digitization methods. Thus, a vendor product can be Class 1-compliant, Class 2-compliant, or both Class 1- and Class 2-compliant.

For both Class 1 and Class 2 compliance, equipment must support the frame structure defined in the IA, including the subframe format. For Class 1 compliance, equipment at a minimum must support the ITU G.727 ADPCM voice-compression method for transmission at 32 Kbps and for the receiver at 32, 24, and 16 Kbps. The support of all other payload transfer methods, including fax, is optional; however, the IA conformance section defines the support for signaling bits that enables voice calling information to be passed between vendor products. Class 2 compliance is similar to Class 1, with the key difference being that CS-ACELP voice digitization is mandatory in place of G.727 voice digitization. Under Class 2 compliance, a vendor product must support either the ITU G.729 or the G.729 Annex A method of voice encoding.

For Further Reference

The Voice over Frame Relay Implementation Agreement uses a series of annexes to the 16-page, four-section document to specify the exact details necessary for vendors to develop products that can interoperate with one another. Annex A covers the syntax of dialed digit transfers, which provide a uniform method for transmitting and interpreting both dialed digits and the power level of frequencies used for DTMF signaling. Annex B specifies the structure and procedure for the transfer of signaling bits across a frame relay network. In Annex C, the structure of the subframe payload is defined, while Annex D provides detailed information concerning the transport of fax data. Annexes E through I focus on the coding of voice digitized by different encoding methods. Thus, the annexes provide detailed information necessary for vendors to design and manufacture VoFR-compliant products. A complete copy of the Voice over Frame Relay Implementation Agreement FRF.12 is included in the appendix of this book courtesy of the Frame Relay Forum. Readers should also access the Frame Relay Forum's World Wide Web site at **http://www.frforum.com**. This site provides a wealth of material, from frame relay background to technical information to copies of various Implementation Agreements. Visit this Web site to explore additional aspects of frame relay, such as data compression, that are beyond the scope of this book.

THE FRAME RELAY FORUM FRAGMENTATION IA

A second Frame Relay Forum Implementation Agreement that warrants attention is FRF IA 12, which covers fragmentation. The purpose of this IA was to define a mechanism vendors could follow that permits a common approach to the fragmentation of long frames and their reassembly. FRF IA-12 supports fragmentation locally across a frame relay user-to-network interface (UNI) between Data Terminal Equipment (DTE) to Data Communications Equipment (DCE) peers, locally across a frame relay network to network interface (NNI) between DCE peers, and end-to-end between two frame relay DTEs interconnected by one or more frame relay networks.

Overview

The promulgation of FRF IA 12 was based upon the recognition that long frames transporting data must be fragmented when they share the same UNI or NNI link with frames transporting voice. This IA defines a fragmentation procedure and three fragmentation models—locally across a UNI, locally across an NWI, and on an end-to-end basis. Each model shares a common fragmentation procedure defined by the IA. That procedure commences by taking a frame, removing the leading flag field and its address field, original FCS field and trailing flag. This results in the creation of an information field that is then broken into smaller pieces and encapsulated by the appropriate frame relay fields to form a sequence of frames, each carrying a fragment. To provide a mechanism for the correct reassembly of data, the first fragment in the series will have its B bit set, while the final fragment will have its E bit set.

Recognizing that different vendors approach fragmentation based upon different operational considerations FRF IA 12 does not recommend any specific fragment size. Instead, this IA defines how the information field is subdivided and the resulting fragments are put back together. In doing so this provides a mechanism for different vendor equipment to become interoperable as the only frame length restriction is that receivers must be capable of reassembly of complete frames up to 1600 bytes in length.

For Further Reference

Readers are referred to Appendix A, which contains the complete description FRF IA 12, courtesy of the Frame Relay Forum. Readers should also access the Frame Relay Forum's World Wide Web site at **http://www.frforum.com**.

CHAPTER 10

Voice over ATM

I n this chapter we will turn our attention to a transmission technique that for many organizations can make both Internet telephony and telephony over the Internet a practical reality. That transmission technique involves taking the advantage of the use of Asynchronous Transfer Mode (ATM) in a wide area network infrastructure. Because ATM supports multiple service types with varying levels of service guarantees, it becomes possible with the applicable hardware and software to take advantage of appropriate ATM classes of service to obtain the predictability and reliability required to transport voice end to end and obtain a high quality of reconstructed voice at the destination.

In this chapter we will first turn our attention to the operation of ATM. In doing so we will examine the format of ATM cells, the manner by which ATM switches operate, ATM transmission services, metrics, admission control into an ATM network, the ATM protocol stack and the different ATM classes of service that are tailored to support different applications. Concerning the latter, we will note how certain classes of service are more suitable for transporting real-time voice than other classes of service. Once the preceding is accomplished we will then examine how this technology can provide us with the ability to obtain a true Quality of Service when linking geographically separated IP and frame relay networks.

OVERVIEW

ATM represents a fast packet-oriented, switch-based transfer technology based upon asynchronous time division multiplexing. Instead of variable packets ATM uses fixed-length cells. Each ATM cell is 53 bytes in length consisting of a 48-byte payload and a five-byte cell header.

The use of a fixed-length cell and a header where routing information is contained enables ATM switching to be performed in hardware. While this fact may appear to be trivial, it permits extremely fast cell switching to occur and allows ATM to be scaled from a T1 access line rate of 1.544 Mbps to an OC-48 rate of 2.5 Gbps.

Cell Formats

There are two ATM cell formats, with a slight difference between the two. Once cell format is used at the user-to-network interface (UNI). The second cell format occurs at the network-to-network interface (NNI).

The UNI represents the interface between a subscriber and an ATM network. In comparison, the NNI represents the interface between two ATM networks.

Figure 10-1a indicates the ATM cell format at the UNI, while Figure 10-1b indicates the NNI cell format. As we can note by comparing the two cell formats, the only difference between the two concerns the use of the first byte in each cell. We will compare and contrast the fields of both cell formats by discussing them in their order of placement in each cell.

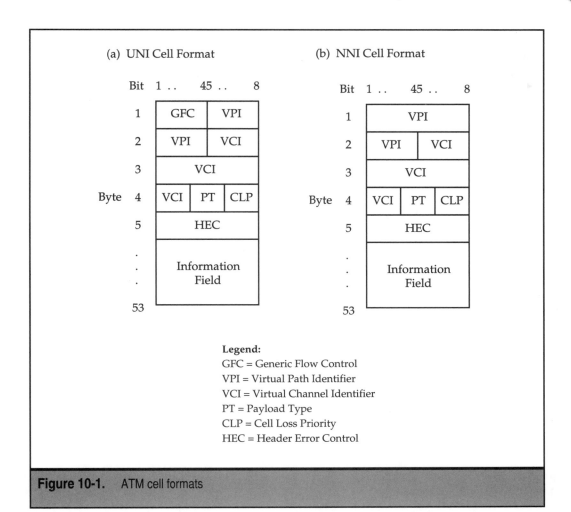

Figure 10-1. ATM cell formats

Generic Flow Control

In the UNI cell header the first four bits represent the Generic Flow Control (GFC) field. This field can be used to define the parameters for flow control from a subscriber into an ATM network. Because flow control is not applicable when connecting ATM networks, the FGC field is not used in cells at the NNI.

Virtual Path Identifier

The Virtual Path Identifier (VPI) represents a unique identifier that denotes a virtual path on a physical circuit. The UNI cell format uses 8 bits, which provide 256 unique values.

In comparison, the NNI ATM cell format is extended to 12 bits through the use of the GFC field. This results in the NNI cell format providing 4096 possible values.

The relationship between a virtual path and a virtual circuit is shown in Figure 10-2. Note that multiple virtual circuits can be assigned to a virtual path. This relationship provides a mechanism for bundling a group of communications to expedite routing.

If you examine Figure 10-1a and 10-1b, you will note that the VPI field precedes the VCI field. This structure enables ATM switches to rapidly route cells based upon their VIP field value within an ATM network. Once we complete our examination of the fields in an ATM cell we will note an example of cell routing in an ATM network, which will hopefully clarify the use of VIPs and VCIs.

Payload Type Identifier

The Payload Type Identifier (PTI) is used to indicate the type of data transported in the payload portion of the cell. Table 10-1 indicates the values associated with the 3-bit PTI field.

Cell Loss Priority

The Cell Loss Priority (CLP) one-bit field is used to prioritize cells. When this field is set to a value of binary 1 it indicates a low priority, while when set to a value of binary zero it means the cell has a high priority.

When a switch is under congestion the CLP value is used to determine the sequence of cells to discard. That is, the switch will first attempt to discard cells with the CLP set to 1 before discarding those with a value of zero.

Header Error Control

The Header Error Control (HEC) is 8 bits in length. This field is used to provide an error correction capability that can correct one-bit error in the cell header. The HEC field is only applicable to the ATM header and does not provide protection to the payload.

Services and Connections

ATM provides three types of transmission services. Those services include permanent virtual connections (PVCs), switched virtual connections (SVCs), and a connectionless service.

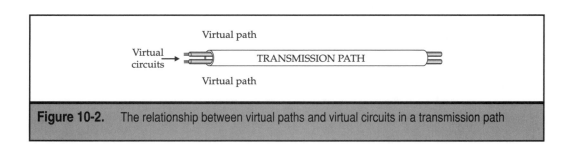

Figure 10-2. The relationship between virtual paths and virtual circuits in a transmission path

Payload Value	Meaning
000	User data cell, no congestion, cell type 0
001	User data cell, no congestion, cell type 1
010	User data cell, congestion experienced, cell type 0
011	User data cell, congestion experienced, cell type 1
100	Maintenance information between adjacent switches
101	Maintenance information between source and destination
110	Resource management cell
111	Reserved for future function

Table 10-1. Values of the ATM Cell PTI field

PVC

A permanent virtual connection is similar to a leased line. That is, a PVC emulates a leased-line type of service, although it does not represent a physical circuit. Instead, a PVC is created by using shared resources with other ATM users. Because the path from source to destination is dedicated, no call setup signaling is required.

SVC

A switched virtual connection represents a temporary path established through an ATM network. This results in a need for signaling information to establish a call as well as to remove the virtual channel from service once the call is completed.

Connectionless

A third type of ATM service is connectionless service. Connectionless service is used for supporting LAN-to-LAN communications via an ATM network. Now that we have an appreciation for ATM services and connections, let's turn our attention to the flow of data through an ATM network.

Data Flow

Figure 10-3 illustrates a three-switch ATM network as a mechanism for noting the flow of data through the network. In this example let's assume three data sources are assigned VCI values of 2, 3 and 4. Let's further assume VCI2 and 3 are to be transported via the ATM network from location A to location B while VCI 4 is to be forwarded to location C. Because they are being routed to a common location they can be transported via the use of a common VPI. In Figure 10-3 it was assumed that VPI 8 was used to bundle the three VCIs from A to B. At location B we will assume VCI 50 is to be routed to location C. Because VCI4 will be switched at location B to location C, a new bundle would be created. In this example VPI 50 is shown as the path between locations B and C, which transports VCI 4 and VCI 50 from B to C. This miniature example of ATM data flow also indicates that the VPI changes at each connection point in an ATM network.

ATM Metrics

There are several metrics that define the performance of ATM. Those metrics include the peak cell rate (PCR), sustained cell rate (SCR), maximum burst size (MBS), and variable bit rate (VBR).

Peak Cell Rate

The peak cell rate defines the upper boundary of an ATM connection or the maximum cell rate. The PCR is the inverse of the minimum inter-arrival time between cells.

Sustainable Cell Rate

The sustainable cell rate (SCR) defines the amount of traffic an end point can burst into a network. For those persons familiar with frame relay, the ATM SCR is similar to frame relay's Committed Information Rate (CIR).

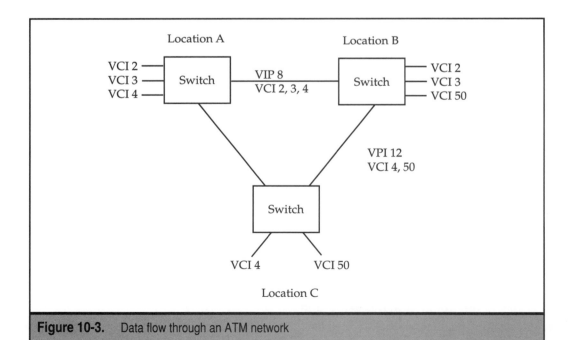

Figure 10-3.   Data flow through an ATM network

Maximum Burst Size

The maximum burst size (MBS) represents the maximum number of cells accepted over a period of time. When the cell rate exceeds the MBS, cells can be dropped.

Variable Bit Rate

There are two types of variable bit rates supported by ATM—variable bit rate real time (VBR-rt) and variable bit rate non-real time (VBR-nrt). The VBR represents traffic guaranteed for delivery under the SCR. However, traffic above the SCR may be discarded if it exceeds the MBS. The VBR cannot exceed the PCR as the latter represents the port speed on an ATM switch.

ATM Connection and Admission

There are two classes of parameters that govern connection admission control into an ATM network. First, source traffic characteristics in the form of the peak cell rate, average cell rate, maximum cell burst, and peak duration permitted are exchanged. Next, information concerning the required Quality of Service, such as cell transfer delay, delay jitter, cell loss ratio, and maximum cell burst and cell loss are exchanged. In addition, the cell loss priority bit in the cell header allows users to generate two different priority traffic flows. The resulting two classes are treated separately by the network connection admission control. Also note that the cell transfer delay (CTD) represents another important metric. The CTD represents the elapsed time between a cell exiting one measurement point and entering another measurement point.

Switch Performance

In addition to network performance parameters there are five parameters that characterize ATM switching performance. Those parameters include throughput, connection blocking probability, cell loss probability, switching delay, and cell delay variation.

The throughput is the rate at which cells depart a switch per unit time. The connection blocking probability represents the probability that demand exceeds resources available between ingress and egress ports on a switch. At this time cells will be dropped based upon the setting of the CLP bit in the cell header.

The cell loss probability represents the probability that a cell will be lost. This occurs when a switch queue is filled and additional cells arrive.

The switching delay represents the time required to route an ATM cell through a switch while the cell delay variation represents the probability that the delay of a switch exceeds a certain value. This metric is also referred to as jitter delay. While switch metrics are important when purchasing hardware it is the previously mentioned two classes of parameters that govern connection admission control and primarily govern the flow of traffic through an ATM network.

THE ATM PROTOCOL STACK

The ATM protocol stack is relatively simple, consisting of either two or three layers based upon the location of data flowing into, through, and out of an ATM network. At end points where data enters or exits an ATM network, the protocol stack has three layers. Those layers are an ATM Adaption Layer (AAL) responsible for adapting different classes of traffic to the ATM layer, in effect taking protocol data units (PDUs) and breaking them into cells; an ATM Layer that is responsible for relaying cells between the AAL and the physical layer; and the physical layer. Because ATM switches only work upon cells, there is no need for a mechanism to convert PDUs into cells. Figure 10-4 illustrates the ATM protocol stack within an ATM network. Thus, the AAL is only applicable for ATM end points. Because the AAL is critical for providing a specific class of service, we will focus our attention upon the top layer in the ATM protocol stack.

Types of AAL

Originally, five types of AALs were recommended by the Consultative Committee for International Telephone and Telegraph (CCITT), the predecessor to the ITU Telecommunications Standardization (TS) body. Those AAL layers were referenced as AAL1 through AAL5. However, layers 3 and 4 were merged into a common joint layer referred to as 3/4.

Sublayers

Each AAL consists of two sublayers, referred to as the Convergence Sublayer (CS) and the Segmentation and Reassembly (SAR) Sublayer.

Convergence Sublayer The Convergence Sublayer (CS) supports end-user applications and is further subdivided into two sublayers—the Common Part Convergence Sublayer (CPCS) that is required and an optional Service Specific Convergence Sublayer (SSCS).

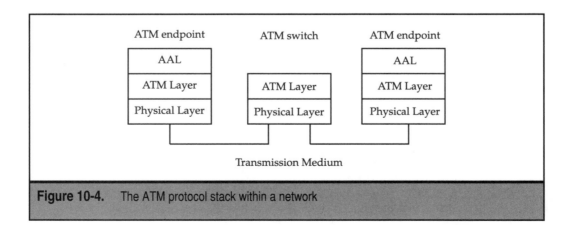

Figure 10-4. The ATM protocol stack within a network

The Convergence Sublayer is responsible for managing multiple data flows between applications and the Segmentation and Reassembly Sublayer.

Segmentation and Reassembly Sublayer   The Segmentation and Reassembly (SAR) Sublayer is responsible for taking a PDU, such as an IP datagram, and breaking it into a sequence of cells, as well as reassembling cells into PDUs. Figure 10-5 illustrates the relationship between the various parts of the ATM Adaption Layer.

In examining Figure 10-5 note that the Convergence Sublayer provides the AAL service at the AAL-SAP (Service Access Point) for higher layers and is service-dependent. The SAP can be considered to represent a mailbox where data is sent to higher layers or received from higher layers. Now that we have an appreciation for how the AAL is subdivided, let's turn our attention to each ATM AAL. Because each AAL employs a specific

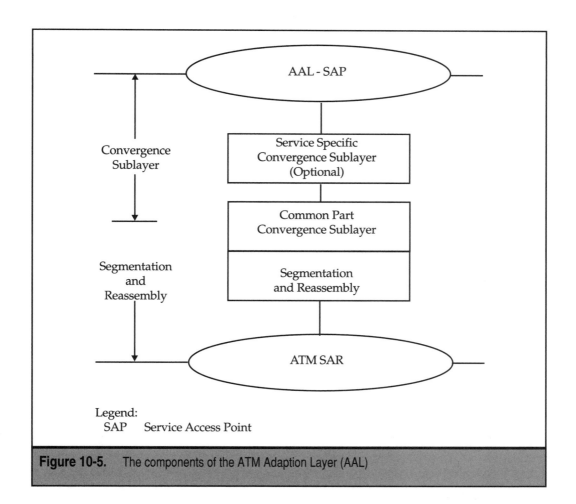

Figure 10-5.   The components of the ATM Adaption Layer (AAL)

SAR and CS, each supports a specific type of traffic, commonly referred to as a Class of Service. Thus, let's examine the four types of AALs and their relationship to different Classes of Service.

There are four ATM AALs, numbered AAL1, AAL2, AAL3/4, and AAL5. As previously explained, two AALs were merged and are now referred to as AAL3/4.

AAL1

AAL1 provides a timing relationship between source and destination. This permits AAL1 to support connection-oriented services that require constant bit rates and have specific timing and delay requirements, such as emulating a DS0 or DS1. The ability to support a constant bit rate provides a true QoS capability. This is because CBR results in the output or egress of a bit stream that has an extract or near-extract relationship to the timing of the bit stream upon entering the ATM network.

AAL2

AAL2 provides more efficient support for voice over ATM. This AAL layer supports variable length packets within the ATM payload. Previously referred to as Composite ATM, AAL2 is now referred to as the ITU-TS I.363.2 standard.

AAL3/4

AAL3/4 supports both connectionless and connection-oriented variable bit rate services. However, AAL3/4 has high overhead and is rarely supported.

AAL5

AAL5 supports connection-oriented variable bit rate services. In comparison to AAL3/4, AAL5 provides a reduction in overhead and is popularly used for variable bit rate applications.

PDU to Cell Conversion

Each AAL has PDUs converted into cells through the use of the Convergence Sublayer and the Segmentation and Reassembly sublayer. However, types 3/4 and 5 have their CS subdivided into Service Specific Convergence Sublayer and Common Part Convergence Sublayer (CPCS), as illustrated in Figure 10-6.

In examining Figure 10-6 note that the SSCS is applicable to variable bit rate applications that require specific services. Now that we have a basic understanding of AALs, let's turn our attention to ATM Classes of Service.

ATM Classes of Service

Under ATM, multiple traffic classes that are commonly referred to as *service types* are supported. Each traffic class or service type has a predefined characteristic as well as a level of service guarantee. Each traffic class definition is based on the use of three attributes: the timing relationship between the source and destination, the variability of the bit rate, and its connection mode.

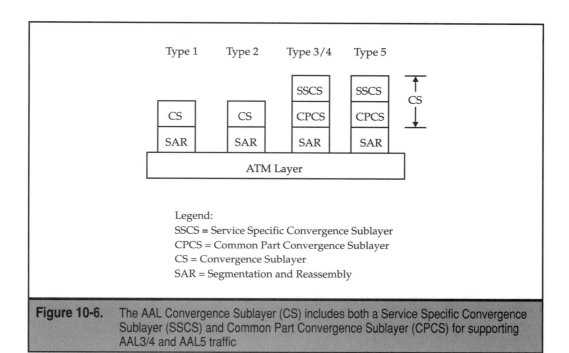

Figure 10-6.   The AAL Convergence Sublayer (CS) includes both a Service Specific Convergence Sublayer (SSCS) and Common Part Convergence Sublayer (CPCS) for supporting AAL3/4 and AAL5 traffic

Timing Relationship

The timing relationship between the source and destination defines the ability of the receiver to receive the original data stream at the same rate at which it was originated. For example, a voice conversation digitized at 64 Kbps via PCM must be "read" by the receiver at that data rate to be correctly interpreted. In comparison, a file transfer occurring via a T1 line into the Internet at a data rate of 1.544 Mbps could be correctly received via an egress access line operating at 56 Kbps. Although the reception of the file requires additional time, different transmission and reception rates do not inhibit the actual data transfer.

Bit Rate Variability

The second attribute governing the class of traffic is the bit rate. Some applications, such as digitized real-time voice, require a constant bit rate. Other applications, such as a file transfer, can occur successfully with either a constant or a variable bit rate.

Connection Mode

A third attribute governing the class of traffic is its connection mode. The connection mode can be either connection-oriented or connectionless. *Connection-oriented* means a connection must be established prior to actual data transfer occurring, while *connectionless* references transmission occurring on a best-effort basis, with an acknowledgment flowing back only after transmission was initiated. Examples of connection-oriented applications include voice calls and IBM SNA data sessions. Examples of connectionless applications include Ethernet transmission and applications that use UDP. Table 10-2 indicates the relationship between key ATM classes of service and their three key attributes.

ATM's Class A is also commonly referred to as a constant bit rate (CBR) class of service. CBR is well suited for transporting digitized voice and in fact was designed primarily for supporting voice communications. CBR allows the amount of bandwidth, end-to-end delay, and the delay variation to be specified during call setup. Class B traffic is commonly referred to as variable bit rate real time (VBR-rt) as it requires a timing relationship. In comparison, because Class C traffic does not require a timing relationship, it is commonly referred to as variable bit rate non-real time (VBR-nrt). Finally, Class D traffic is commonly referred to as unspecified bit rate (UBR).

We can indicate the relationship between ATM classes of service and AALs by comparing the entries in Table 10-2 to our description of ATM Adaption Layers, similar to a famous movie quote, "their way is a better way." That way is to directly illustrate the relationship between the two. This relationship as well as summary of ATM Classes is contained in Table 10-3.

Examining Real-Time Voice Transport AALs

To appreciate how different ATM Classes of Service operate you must have a more detailed understanding of the different AALs. Thus, let's explore the format of the AALs in more detail. However, because only AAL1 and AAL2 is applicable to the transport of real-time voice, we will limit our examination to those two ATM Adaption Layers.

AAL1

AAL1 was standardized by the ITU and ANSI in 1993. This adaption layer is used for Class A cell payloads to provide circuit emulation via a constant bit rate. Thus, AAL1 provides for the transfer of data units with a constant source bit rate and delivers data to the destination at that bit rate. In addition, AAL1 transfers timing and structure information between source and destination.

Class	Timing	Bit	Connection
A	Required	Constant	Connection-oriented
B	Required	Variable	Connection-oriented
C	Not required	Variable	Connection-oriented
D	Not required	Variable	Connectionless

Table 10-2. ATM Classes of Service

Class of Service	A	B	C	D
AAL Service	AAL1	AAL2	AAL5	AAL3/4
Timing (bit rate)	Constant	Variable	Available	Unspecified
Traffic	Delay sensitive	Delay sensitive	Non-delay sensitive	Non-delay sensitive
Connection	Connection-oriented	Connection-oriented	Connection-oriented	Connectionless
Utilization	Circuit	Variable bit	Connection-oriented	Connectionless
Emulation	Rate voice, video	Oriented data	Data	

Table 10-3.   Relationship of ATM Classes of Service and AALs

Figure 10-7 illustrates the AAL1 Class A cell payload format. The Convergence Sublayer Indication (CSI) bit is used to indicate if an 8-bit pointer exists. If this bit is set it indicates the pointer is the first byte in the payload. The purpose of this byte is to allow a cell to be partially filled if required by a user application instead of delaying the cell until the payload is filled. While the CSI byte provides a mechanism to expedite the flow of cells through an ATM network, it also results in the waste of cell transport space. This is because AAL1 is limited to supporting one user per cell. In addition, as we will shortly note, AAL1 is limited to supporting one or more 64 Kbps digitized voice channels and cannot support low bit rate voice compression schemes whose use has gained in popularity during the past few years.

The sequence count field counts sequential ATM cells in modulo-8. This field is used to detect convergence layer gaps due to cell errors or cell loss. The Convergence Sublayer Indication and the Sequence Count form a 4-bit sequence number. This number is used to detect mistakenly inserted or lost cells.

The third field in Figure 10-7 is the Sequence Number Protection field. This 3-bit field contains a CRC computed over the first 4 bits in the AAL1 header. The 1-bit parity field provides simple parity protection over the first 7 bits.

One other item concerning the Convergence Sublayer Indication warrants attention. Bits from four successive cells provide a Synchronous Residual time-stamp for clock recovery. For structured data transfers a value of 1 results in an 8-bit pointer to the first byte of the payload, while a value of 0 results in no pointer for partially filled cells.

Although AAL1 is well defined and very popularly employed to support Class A service, it has several key limitations. First, AAL1 only permits a single user to be supported.

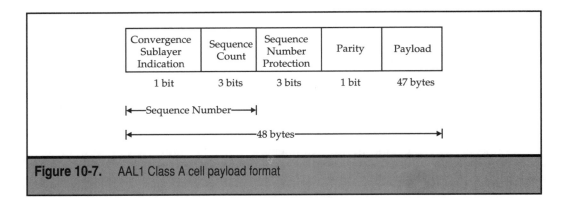

Figure 10-7. AAL1 Class A cell payload format

Second, voice is always transported as 64 Kbps or in bundles of $n \times 64$ Kbps. This means AAL1 does not support compression, silence detection/suppression, idle channel removal, and other features that are associated with a variable bit rate. In addition, bandwidth is always used, even when there is no traffic. Due to this a significant amount of attention has been focused on AAL2, which removes the previously mentioned AAL1 limitations and provides a more efficient method for filling cells.

AAL2

AAL2 was developed for variable bit rate audio and video. This version of AAL includes an expanded CRC used for error detection and correction and has a structure that supports functions not considered possible through the use of AAL1. Those functions include support for low bit rate and short, variable packets with delay sensitivity. In addition, AAL2 supports both variable bit rate and constant bit rate as well as multiple user channels on a single ATM virtual circuit. Concerning the latter, this is a significant improvement over AAL1 as user packets can be split across cells, which can provide more efficiency since digitized voice samples no longer have to be carried in individual cells. AAL2 was defined in late 1997 and has only recently been supported by carrier networks.

CPS Packet Format Figure 10-8 illustrates the format of the AAL2 Common Part Sublayer (CPS) packet format, which is defined in the I.363.2 standard. The CPS permits the identification of users of the ATM Adaption Layer, supports the assembly and disassembly of the payload, and forms a relationship with the Service Specific Convergence Sublayer (SSCS). The CID or Channel Identifier provides a mechanism to identify an individual user within the AAL2. The value of the CID field coding is indicated in Table 10-4.

In examining the entries in Table 10-4 note that the CID field identifies the individual user channels within the AAL2. Because values 8 to 255 are used for the identification of AAL2 user channels, this permits up to 248 individual users within each AAL2 structure.

The Length Indicator (LI) field identifies the length of the packet payload associated with each individual user. The value of this field is one less than the packet payload and has a default value of 45 bytes.

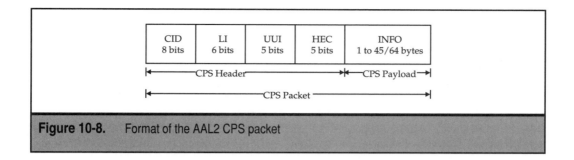

Figure 10-8. Format of the AAL2 CPS packet

The next field in the AAL2 CPS packet header is the User-to-User Indication (UUI) field. The purpose of this 5-bit field is to provide a link between the CPS and the applicable SSCS that satisfies the higher layer application. This field enables different SSCS protocols to be defined to support specific AAL2 user services. Table 10-5 indicates the possible coding of the UUI field.

The CPS-PDU   After the creation of a CPS packet, individual CPS packets are combined to form a CPS-PDU payload. The format of the AAL2 CPS-PDU is shown in Figure 10-9.

In examining Figure 10-9 note that the actual composition of the CPS-PDU can contain data from one or more individual CPS packets. The Offset field (OSF) identifies the start of the next CPS packet within the CPS-PDU. This field functions as a pointer to the payload. The following 1-bit field is the Sequence Number (SN), which alternates between 0 and 1 to promote data integrity. The 1-bit Parity (P) field supplements the SN field for data integrity, while the PAD field can be from 0 to 47 bytes in length.

Adaption Layer Efficiency   To illustrate the key advantage of AAL2, let's assume you wish to use a G.729.A codec, which digitizes voice at 8 Kbps. Under this scenario each frame is 10 bytes in length. Let's further assume that voice is time-stamped using RTP, resulting in a total of 14 bytes required to be transported for each voice sample that is digitized.

Value	Utilization
0	Not used
1	Reserved for layer management
2-7	Reserved
8-255	Identification of AAL2 user channels

Table 10-4. Channel Identifier (CID) Field Coding

Value	Utilization
0-27	Identification of SSCS entries
28-29	Reserved for future standardization
30-31	Reserved for layer management

Table 10-5.   AAL2 CPS UUI Field Values

Because AAL1 does not support variable bit rate, you would have to transport voice as 64 Kbps DSOs using that adaption method. However, when using AAL2 you could either pack digitized voice from multiple sources into each cell or use AAL2 cells to transport multiple digitized voice samples. In either situation you obtain more efficient use of ATM cells.

Figure 10-10 illustrates the manner by which a sequence of two ATM cells can transport multiple G.729.A frames. In this example note that the first cell contains two full G.729.A frames and 10 bytes of a third 14-byte G.729.A frame. The next cell contains the remaining 4 bytes of the third G.729.A frame and two additional G.729.A frames. Because AAL2 provides the ability for multiple digitized voice frames to be carried within a cell it is ideally suited for transmission of low-bit-rate encoded voice over an ATM network. Because the cost associated with using an ATM public network is based upon the quantity of cells transmitted, among other metrics, the ability to pack multiple voice frames into a

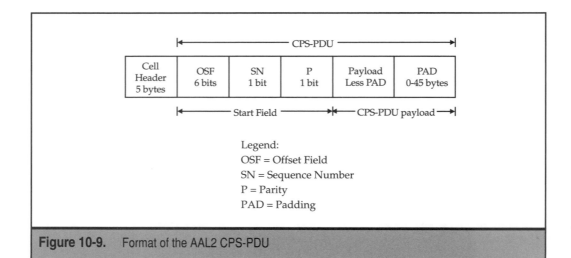

Figure 10-9.   Format of the AAL2 CPS-PDU

cell significantly reduces the number of cells required to transport a conversation. This in turn reduces the cost associated with using ATM as a backbone for linking IP and frame relay devices and provides a financial incentive for using AAL2. Now that we have an appreciation for the role of AAL2 in promoting both efficiency and economy of use of ATM, let's turn our attention to the transmission of IP over ATM.

IP OVER ATM

The key to routing IP over ATM to support the reliable and predictable transport of digitized voice is threefold. First, a mechanism is required to map IP datagrams into an appropriate ATM class of service. Second, an appropriate ATM class of service must be used. Third, equipment must be used that supports the flow of IP traffic and its conversion into ATM cells, provides priority queuing for the conversion, and at the egress from the ATM network performs an appropriate reverse translation from ATM cells into IP datagrams. Because Cisco Systems provides equipment that can be used to obtain this capability, our examination of several aspects of IP over ATM will include a discussion of the features accompanying certain products manufactured by that vendor.

IP Datagram Considerations

Because either digitized voice or data can be transported within an IP datagram, it is important to differentiate between the two. In addition, because your organization may wish to prioritize the flow of certain types of data, such as encapsulated SNA or e-mail, it

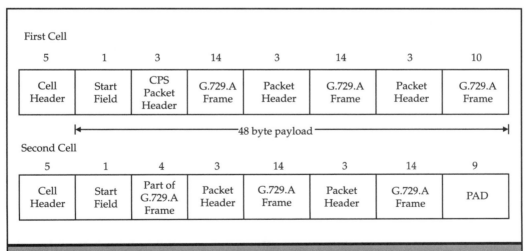

Figure 10-10. Under AAL2 multiple sources of data can be transported via a common cell or AAL2 formed cells can transport multiple digitized voice samples.

becomes important to divide traffic types and users into classes and treat each class differently during peak traffic periods.

When traffic flows on an intranet, it is possible to provide a Quality of Service (QoS) by implementing RSVP or by adding bandwidth. However, when data flows over a wide area network operated by one or more third-party communications carriers, it may be difficult or impossible to control QoS via IP. Fortunately, ATM has a built-in Class-of-Service capability. However, as illustrated in Figure 10-11, a mechanism is required to map IP datagrams into an appropriate ATM class of service at the ingress and from ATM cells to IP datagrams at the egress from the ATM network.

One method that can be used to differentiate IP datagrams is to set appropriate bits in the Service Type byte, which is also referred to as the Type of Service (ToS) byte. RFC 791 defines eight levels of precedence that can be assigned to the Precedence field in the ToS byte. Those settings were previously discussed in Chapter 3. Thus, setting the Precedence field in the ToS byte represents a method to classify traffic. Because the resulting series of datagrams from different locations transporting different applications can be differentiated from one another, Cisco Systems refers to the newly classified traffic as *differentiated service classes*.

As IP datagrams are processed by an internal IP network, the setting of the appropriate ToS bits occurs. Prioritized datagrams that transport digitized voice should be mapped into ATM CBR cells. In doing so, ATM traffic shaping on a CBR ATM permanent virtual circuit (PVC) must occur to provide a constant intercell gap, which is the characteristic of ATM CBR shaping.

In a Cisco Systems router environment, certain vendor router ATM interfaces provide an ATM traffic-shaping capability. In doing so, Cisco equipment supports the configuration of parameters that define the average sustainable cell rate (SCR), peak cell rate (PCR), and maximum burst size (MBS). By configuring the average cell rate and the peak cell rate to the same bandwidth through the use of a Cisco routers **atm pvc** command, you will shape the resulting ATM cell flow with a constant intercell gap. This in turn results in a CBR data flow, because a constant intercell gap represents the characteristic of CBR shaping.

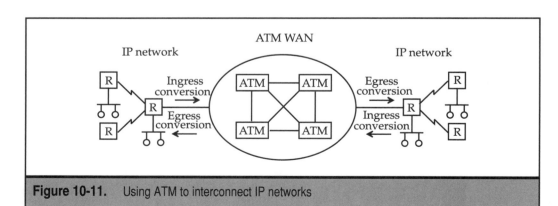

Figure 10-11. Using ATM to interconnect IP networks

Ingress Considerations

In addition to prioritizing IP datagrams and converting traffic to an appropriate ATM Class of Service, it is important to ensure that ingress traffic is prioritized. This is because the router connected to the ATM backbone must place traffic into different virtual circuit queues as well as extract data from those queues based on precedence. To accomplish this, certain Cisco routers support a prioritization technique referred to as Weighted Random Early Detection (WRED).

WRED provides a mechanism to obtain service differentiation across different ATM Classes of Service based on the setting of the ToS byte in an IP datagram. Under WRED, an exponential weight factor can be assigned for a WRED parameter group.

The parameters are used in an equation to define the probability of a packet being dropped. That probability depends on a mapping from the IP precedence value to a WRED parameters value as indicated in Table 10-6.

The following equation defines the WRED average value, with n ranging between 1 and 16.

$$\text{Average} = (\text{old average} \times 1^{1}/2^{\,n}) + \text{current queue size} \times {}^{1}/2^{\,n}$$

Note that the higher the value of n, the more dependent the average is on the previous average. Thus, a large value of n serves to smooth out the peaks and valleys in queue length. Cisco routing packets are marked with a precedence of 6, resulting in a selective discard with a very low loss probability. Thus, you should configure a Cisco router to use precedence 6 and 7 for digitized voice traffic, while data traffic should use precedences 0 to 5.

As a practical matter, the average queue size is unlikely to change very quickly and avoids drastic swings in queue length. However, Cisco recommends that users

IP Precedence	WRED Minimum Threshold Values
0	9
1	10
2	11
3	12
4	13
5	14
6	15
7	16
RSVP	17

Table 10-6.   Default WRED Minimum Threshold Values Based on IP Precedence

start operation or testing with the default value of n, which is 9, because the fine-tuning of WRED to achieve a specific IP service differentiation is a "delicate exercise."

Configuring WRED

In a Cisco router environment WRED is enabled by using the following interface command:

random-detect

The default value for the exponential weighting factor that affects the responsiveness of WRED is 9. You can configure WRED with the range of values from 1 to 16, with 16 resulting in the least level of responsiveness to queue-depth changes while providing the smoothest level of response.

Economics

Although each communications carrier can be expected to price the use of ATM differently, we will conclude this section with an example of the cost associated with the use of an ATM backbone network. In 1999, Quest billed ATM CBR service at $.02 per Mbyte of traffic. If we are not greedy, and we assume a voice-digitization rate of 8 Kbps, then a one-minute conversation results in the transmission of 8 Kbps × 60 seconds/minute × 8 bits/byte, or 60 Kbytes/minute. On an hourly basis this results in 60 Kbytes/minute × 60 minutes/hour, or 3.6 Mbytes/hour. Thus, at a cost of $.02/Mbyte of traffic, a one-hour voice call would cost $.072. While you must still consider the cost of voice-over-IP equipment, the ability to use an ATM backbone for under $.08 per hour to obtain Quality of Service represents an ideal mechanism to lower the cost of voice.

FRAME RELAY OVER ATM

The rapid growth in the use of public frame relay services as well as the migration of several communications carriers to the use of ATM as a backbone resulted in the need to obtain a standardized frame relay-to-ATM internetworking capability. The Frame Relay Forum responded to this internetworking requirement with the creation of two implementation agreements (IAs) that specify the manner by which frame relay and ATM networks and services can be interconnected.

Interconnection Methods

The Frame Relay Forum responded to the requirements of subscribers and network operators to interconnect frame relay and ATM networks with two IAs due to two types of interconnection methods being available for use—one at the network level and one at the service level. FRF IA5, referred to as the Frame Relay/ATM PVC Network Internetworking Implementation Agreement, provides guidance concerning the internetworking of frame relay and ATM terminals and networks. In comparison, FRF IA8 titled Frame Relay/ATM PVC Service for Internetworking Implementation Agreement provides information concerning internetworking between frame relay and ATM services. For both IAs internetworking is specified for PVCs and is applicable for public and private networks. In the wonderful world of ATM networking the ATM Forum standardized two methods for moving frame relay over an ATM network. One method is referred to as Data Exchange Interface (DXI), while the second method is referred to as Frame-Based User-to-Network Interface (FUNI). In this section we will also examine both ATM Forum methods that more correctly represent access protocols.

Networking Relationship

Figure 10-12 illustrates the manner by which ATM is used in the core of a wide area network to provide a transport service for frame relay terminals and frame relay networks. In examining Figure 10-12 note that either frame relay terminals or frame relay networks can be connected to an ATM service. In addition, ATM terminals can obtain connectivity with frame relay terminals and networks. For either situation the ATM service transports frame relay frames transparently to the frame relay user.

Under Frame Relay Forum IA5 the actual point where frame relay and ATM internetworking occurs is left as an implementation detail. However, as a matter of practice the location for the internetworking function is in the ingress switch.

Under FRF IA5 frames are converted to ATM cells based upon the use of AAL5. Under FRF IA8 the router performs the conversion process at the edge of the internetwork.

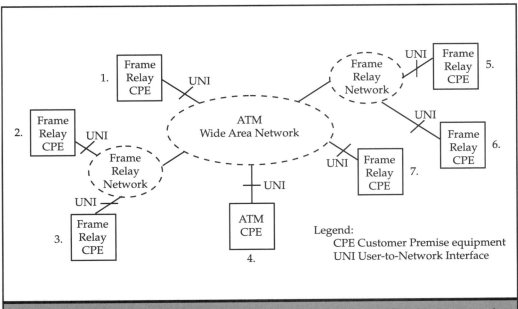

Figure 10-12. Internetworking frame relay terminals and networks via an ATM wide area network

In this section we will focus our attention upon FRF IA5. Because frames are converted into ATM cells based upon the use of ATM's AAL5 we will digress a bit, no pun intended, and examine AAL5. Once we finish our digression and examine FRF IA5 we will then turn our attention to the DXI and FUNI access protocols.

AAL5

The original intent of AAL5 was to provide a more efficient method for data traffic than AAL3/4, which was basically ignored due to its relatively high level of overhead. AAL5 was originally intended to support Class C and Class D traffic. As a quick review, Class C represents no timing, variable bit rate connection-oriented traffic while Class D represents no timing, variable bit rate connectionless traffic.

Figure 10-13 illustrates the ATM Common Part Convergence Sublayer (CPCS)-PDU format. This PDU format has less overhead than the use of AAL3/4 as the payload prefixes an 8-byte trailer contained in the last cell of a PDU. This structure eliminates a per-cell length field and a per-cell CRC associated with AAL3/4. Higher layers in the protocol stack are then responsible for pre-allocating buffers, which eliminates another field used by AAL3/4.

Another technique used by AAL5 that increases its efficiency over AAL3/4 is the assumption of orderly delivery. This assumption eliminates the need for a sequence number, further enhancing AAL5 efficiency.

In examining Figure 10-13 note that the PAD field pads the CPCS-PDU to fit exactly into the ATM cell such that the last 48 bytes created by the Segmentation and Reassembly (SAR) Sublayer are right-justified into the cell. The CPCS-UU (User-to-User Indication) field transfers CPCS user-to-user information. The CPI (Common Part Indicator) field aligns the CPCS-PDU trailer to 64 bites. Finally, the Length Indicator field denotes the payload length. The value of this field must be less or equal to 65535 and when set to 0 indicates an abort.

Figure 10-14 illustrates the manner by which a CPCS-PDU's user data is formed into a sequence of cells and the format of the last AAL5 cell. In examining Figure 10-14 note that actual user data, up to 65535 bytes in length for a PDU, is segmented into a sequence of cells. Then, the last cell, which is shown in the lower portion of Figure 10-14, terminates the PDU.

CPCS-PDU Payload Up to 2 16-1 bytes	PAD 0-47	CPCS-UU 1 byte	CPI 1 byte	Length 2 bytes	CRC 4 bytes

CPCS Common Part Convergence Sublayer
CPI Common Part Indicator

Figure 10-13. The ATM AAL5 CPCS PDU format

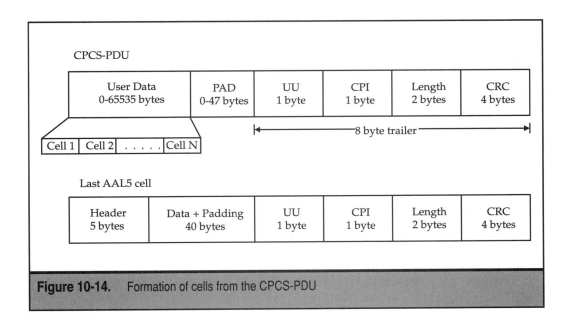

Figure 10-14.   Formation of cells from the CPCS-PDU

In the last AAL5 cell shown in the lower portion of Figure 10-14, the User-to-User (UU) field identifies the payload. Under multiprotocol ATM encapsulation this field is not used. The CPI (Common Part Indicator) field is used to align the CPS-PDU trailer to 64 bits and is then coded to "00" when an alignment occurs. The CRC is computed over the entire AAL5 PDU to include padding and trailer.

While AAL5 was developed to provide efficient connection and connectionless data transmission over ATM, it lends itself for connection-oriented digitized voice when frame relay carries voice. This is because it is possible to place multiple voice conversations into a frame and AAL5's efficiency to include a trailer at the end of the PDU and the lack of sequencing lends itself very well to transporting frames that in turn transport digitized voice.

Internetworking Methods

The method of frame relay-to-ATM internetworking specified under FRR IA5 is standardized as ITU-TI.555. Under this standard there are two internetworking scenarios that are defined. Under the first scenario a frame relay terminal device or end-point communicates with a second frame relay terminal over the ATM network. This method of internetworking is referred to as frame relay transport over ATM. An example of this method of internetworking is represented by the transmission between points 1 and 7 in Figure 10-12.

Under the second scenario a frame relay terminal communicates with an ATM terminal over an ATM network. An example of transmission following this scenario would be from points 1 or 7 to point 4 in Figure 10-12. In this situation the ATM terminal device must support the Frame Relay Service Specific Convergence Layer (FR-SSCS) as part of the ATM protocol stack.

Figure 10-15 illustrates the ATM protocol stack required to support frame relay terminal to ATM terminal communications over an ATM network. Note that the FR-SSCS is defined in the ITU-TS I.365.1 standard.

Through the use of IA5 and ATM's AAL5, ATM becomes capable of supporting variable length PDUs. In addition, such frame relay features as connection multiplexing, frame loss priority indication, and congestion indication in the forward and backward direction are converted to equivalent ATM indicators. To obtain an appreciation of the manner by which frame relay and ATM map key metrics, let's examine the relationship between the frame relay discard eligibility (DE) bit and ATM's cell loss priority (CLP) bit.

DE and CLP Equivalency

As previously noted in Chapter 4, the Frame Relay DE bit identifies frames that can be discarded. In comparison, ATM's CLP bit tells the ATM network which cells can be discarded during periods of congestion.

There are two operational modes that govern the relationship between the DE and CLP bits. In mode 1, for the frame relay-to-ATM flow direction the DE bit is first copied into the FR-SSCS PDU DE bit. Next, the DE bit is mapped to the AAL5 CLP bit for each cell resulting from the segmentation of the FR-SSCS PDU. In the ATM-to-frame relay direction when one or more cells associated with the reassembled FR-SSCS PDDU has its CLP bit set or if the FR-SSCS PDU DE bit is set, then the frame relay DE bit will be set.

Under mode 2 in the frame relay-to-ATM direction the frame relay DE bit setting is directly copied into the FR-SSCS PDU DE bit. Then, the DE bit setting is mapped into the AAL5 CLP bit position for each cell generated by the segmentation of the FR-SSCU PDU as either a 0 or 1 based upon the provisioning of the ATM connection. In the opposite direction under mode 2 there is no provision for mapping between ATM and frame relay

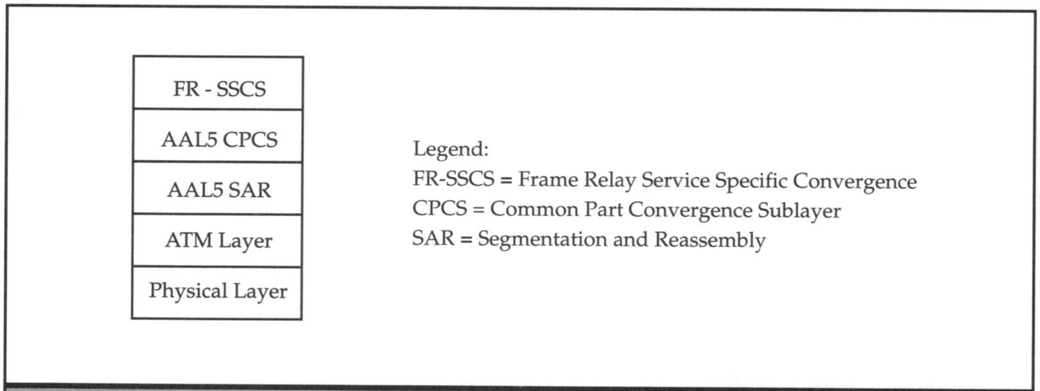

| FR - SSCS |
| AAL5 CPCS |
| AAL5 SAR |
| ATM Layer |
| Physical Layer |

Legend:
FR-SSCS = Frame Relay Service Specific Convergence
CPCS = Common Part Convergence Sublayer
SAR = Segmentation and Reassembly

Figure 10-15.   The ATM protocol stack incorporating FR-SSCS

layers. Instead, the FR-SSCS PDU DE bit setting is copied unchanged to the frame relay DE bit and all AAL5 CLP indications are ignored.

Congestion Mapping

As indicated in Chapter 4 frame relay sets bits to indicate congestion in both the forward and backward directions. Forward congestion is indicated by the setting of the Forward Explicit Congestion Notification (FECN) bit. In comparison, backward congestion is indicated by setting the Backward Explicit Congestion Notification (BECN) bit. At the cell layer ATM is limited to providing a forward congestion indication by setting the Explicit Forward Congestion Indication (EFCI) bit.

In the frame relay-to-ATM direction forward congestion is not mapped to the cell level or EFCI congestion indication. Instead, the FECN bit is copied into the FR-SSCS FECN bit and the EFCI bit in each cell associated with the FR-SSCS is set to "congestion not experienced." In the ATM-to-frame relay direction the forward congestion indication is mapped between services. That is, if the EFCI field in an ATM cell or the FECN of the received FR-SSCS PDU is set to "congestion experienced," then the FECN bit of the resulting frame relay PDU is set to congestion experienced. Otherwise, when congestion has not occurred nothing is set.

In comparison to mapping forward congestion ATM lacks an equivalent to frame relay's BECN. Now that we have an appreciation for FRF IA5 mapping issues let's turn our attention to the two ATM Forum access protocols.

ATM Forum Access Protocols

Both the DXI and FUNI access protocols enable frame relay transmission to be carried over an ATM network. Although both access protocols have a high degree of commonality, there are certain key differences between the two. The primary difference is the use of different hardware and the location of the hardware that provides a frame-to-cell and cell-to-frame translation. Other differences between the two reside in the efficiency and the internetworking capability of each method.

Both DXI and FUNI access protocols require appropriate software operating on a router to generate FUNI or DXI frames. When FUNI is used, FUNI frames flow directly to a serial port, as illustrated in Figure 10-16a. The ATM switch then performs the segmentation and reassembly (SAR) process, converting frames to cells and vice versa. When a DXI access protocol is used for the connection, an ATM CSU/DSU performs the frame-to-cell and cell-to-frame conversion process. This is illustrated in Figure 10-16b.

In examining Figure 10-16, note that the use of the DXI access protocol results in variable-length frames being segmented into ATM fixed-length cells at the subscriber's premises. Because each ATM 53-byte cell includes 5 bytes of overhead, in almost all transmissions the use of the FUNI access protocol will be a more efficient transport mechanism into an ATM network. For example, a 75-byte frame would fill the 48-byte payload of one cell and result in the use of a second cell with only 27 bytes in its 48-byte payload being filled. Because ingress and egress lines into and out of the ATM network normally

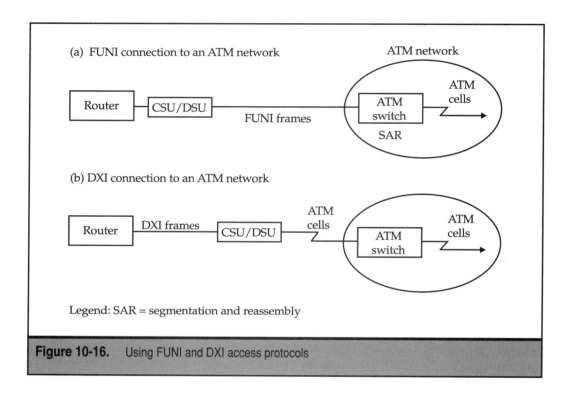

Figure 10-16. Using FUNI and DXI access protocols

operate at a much lower rate than the backbone of the ATM network, the inefficiency of the DXI access protocol may not be suitable for some latency-critical applications, including voice over frame relay.

APPLICATION NOTE: When considering frame relay versus ATM, the use of the FUNI access protocol should be considered over the use of the DXI access protocol when ingress and egress occur at relatively slow data rates. This is because the FUNI access protocol transmits more-bandwidth-efficient frames to the ATM switch, while the DXI access protocol uses less-bandwidth-efficient fixed 53-byte cells.

Frame Structure

The basic structure of the DXI, FUNI, and frame relay frames is the same. It differs only in the composition of the frame relay header field. The top portion of Figure 10-17 illustrates the structure of the basic frame that is common to all three access protocols. The lower portion of Figure 10-17 illustrates the associated header changes for each access protocol.

When FUNI and DXI frames are segmented into cells, their frame addresses are mapped into ATM VPI/VCI (virtual path identifier/virtual connection identifier) using similar procedures. Those mapping procedures result in certain values in the FUNI and DXI frame fields being placed in appropriate fields within the ATM cell header. For example, the Congestion Notification (CN) bit is mapped to the CN bit position in the ATM

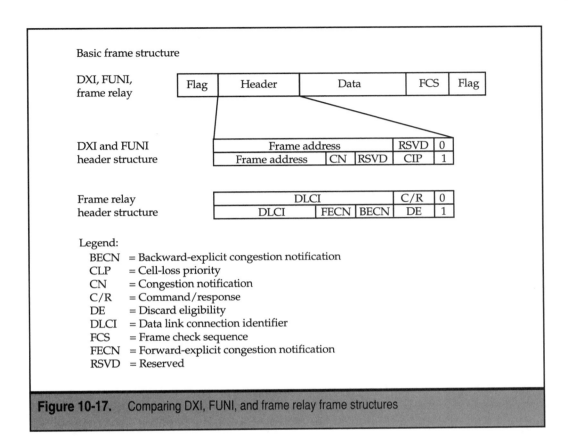

Figure 10-17.   Comparing DXI, FUNI, and frame relay frame structures

cell header. The CN bit performs the same function as the frame relay FECN bit, being set by the network during periods of congestion. However, neither FUNI nor DXI provides for a BECN bit setting. This is because there is no function similar to BECN in ATM. The DXI and FUNI access protocols use the cell loss priority (CLP) bit to perform the same function as the frame relay discard-eligible (DE) bit. Last but not least, the frame relay C/R does not have a corresponding bit in either the FUNI or SXI header.

Mapping Issues

FUNI mandates the support of ATM Adapter Layer 5 (AAL5) in the ATM switch that supports the connection; however, AAL3/4 is optional. It is important to note that although neither AAL3/4 nor AAL5 supports the constant bit rate required for voice circuit emulation, the ATM network operator can specify cell loss and latency for either adaption layer that provides a guaranteed level of delivery. Thus, depending on the latency and loss guarantees specified for mapping frames into variable-bit-rate cells, the use of

ATM in this manner may be suitable for connecting frame relay devices or networks transporting voice. Similarly, DXI frames are segmented into AAL5 cells.

One problem that results from the use of AAL5 is the fact that it performs error detection and correction, which adversely affects the transmission of real-time voice. Thus, another consideration prior to using ATM to transport frame relay, which in turn is transporting digitized voice, is to determine the error rate of the ATM backbone and the ability of equipment to discard retransmitted cells. Obviously, if you can obtain a mapping to AAL1's constant bit rate designed to transport voice, all of the previously mentioned problems are eliminated.

Economics

One of the possible questions in the mind of many readers is the economic trade-offs associated with the use of pure frame relay versus frame relay over ATM. For many economic comparisons, you must examine the cost components of each service. However, this can be a bit difficult with certain comparisons because most frame relay operations do not include a cost component based on the quantity of data transmitted, which most ATM operators do. Another problem associated with comparing the use of the two technologies involves comparing frames and cells when changes are based on the quantity of data transmitted. To provide you with an indication of how you can compare apples and oranges by costing frame relay and ATM service usage, this author will use prices that Quest publicized in early 2000 for frame relay non-discard-eligible, ATM variable-bit-rate, real-time, and ATM available-bit-rate services. Those prices per Mbyte of traffic transmitted were $.04, $.03, $.012, and $.0055, respectively.

Frame Relay Cost An 8-Kbps voice-digitization rate, we noted earlier in this book, is equivalent to 3.6 Mbytes/hour. If we assume voice is packetized in 70-byte segments, the 3.6 Mbytes equates to 51,429 frames (3.6 Mbytes/70 bytes per frame). Because each frame has 6 bytes of overhead, total overhead becomes 51,429 frames × 6 bytes/frame, or 308,574 bytes. Thus, an hour of 8Kbps digitized voice results in 3.6 Mbytes plus 308 Kbytes of overhead, for a total of 3.9 Mbytes. Based on Quest's rate of $.04 and $.03 per Mbyte for non-discard-eligible and discard-eligible transmission, the cost of an hour of digitized voice is either $.156 or $.117!

ATM cost Prior to computing the cost of ATM, we need to determine the number of cells and cell overhead. Because we assumed voice was digitized into 70-byte frames, each frame must be converted into two 53-byte cells. Thus, the 51,429 frames become 102,858 cells. Since each cell is 53 bytes, the total amount of data, including cell overhead, payload, and cell pads, becomes 102,858 cells × 53 bytes/cell, or 5.45 Mbytes. Because we assumed the per-Mbyte cost of ATM to be $.012 for VBR-rt and $.0055 cents for ABR, the cost per hour for each type of ATM class of traffic becomes 5.45 Mbytes × $.012/Mbyte, or $.0654, and 5.45 Mbytes × $.0055/Mbyte, or $.03. Thus, in this example, if we do not consider the cost of equipment and focus our attention on transport charges, the use of an ATM backbone can result in significant savings.

Summary

As indicated in this section frame relay over ATM provides another mechanism for network managers and LAN administrators to consider. Not only can you use low-cost customer premise equipment at the edge of an ATM network but in addition you can obtain the QoS capability of ATM to interconnect geographically separated locations.

CHAPTER 11

Management Issues

In this concluding chapter we will turn our attention to several management issues related to the ability of an organization to transport voice over IP or voice over frame relay. Instead of focusing attention on traditional management issues, such as charge-back, equipment selection, and acquisition, we will turn our attention to a key area you need to explore prior to considering implementation of a voice-over-data network—knowledge about its practicality. In other words, will it work?

Recognizing the Rosetta stone of latency introduced in Chapter 1, we can predetermine the potential success or failure of a voice-over-data network if we can determine the latency or delay through a network. Once we determine the routing delay, we can use that information with ingress and egress delays and codec delays. The total delay would then be used to determine whether the end-to-end delay will be under 250 ms. If it is not, we would more than likely be wasting our time and money in attempting to implement a voice-over-data network facility. However, because in certain situations a lengthy routing delay may be modified to a lower value by the selection of another path or replacement of a bottleneck in the path, we also need to examine the path of the route from the ingress point to the egress point. Fortunately, the TCP/IP protocol stack contains some built-in management tools that can be used to facilitate our examination of both delay and route structure.

USING TCP/IP APPLICATIONS

The TCP/IP protocol suite includes several utility application programs that provide network users with the ability to test and troubleshoot a variety of potential problems. From a voice-over-data-network perspective, there are two key TCP/IP utility application programs you can consider using to obtain information about networking devices, network usage, and the general state of a network. Those utilities are PING and TRACEROUTE.

If you are working in a Windows 2000 environment you have the ability to use a proprietary, souped-up version of TRACEROUTE named PATHPING. In this section we will examine the operation and utilization of each of these three TCP/IP utility application programs.

PING

PING represents a TCP/IP utility application program that is included in every protocol stack this author has worked with. Those protocol stacks include NetManage and FTP software TCP/IP protocol stacks developed for use with Windows 3.1 and several versions of Unix, Windows 95, Windows 98, Windows NT, and the current version of Windows 2000.

The name PING can be traced to two separate historical developments. First, according to some persons, because PING operates by generating an Internet Control Message Protocol (ICMP) Echo Message that results in the destination address issuing an Echo Response message, it is similar to radar and shipboard sonar. That is, a ship generates a sonar signal and uses the ping response from the target to obtain a bearing and distance to the target by timing the delay between transmitting the signal and receiving its pinged reflection. That time represents the round-trip delay. Thus, dividing the time by 2 represents the time to the target. Because the velocity of propagation of a sonar wave is known, the time to the target multiplied by the velocity of propagation equals the distance to the target. According to legend, the Echo Response generated by an Echo Request was viewed similarly to sonar pinging, thus resulting in the term PING being used for this TCP/IP application.

A second commonly told tale concerning the name of the PING application program maintains that the name resulted as an acronym for Packet Internetwork Groper. In this tale, the Echo Request message generated by the client is used to elicit, or "grope," a response from the destination. Because the application was designed to cross network boundaries, it supported internetworking. Thus the name PING was used as an acronym for Packet Internetwork Groper.

Regardless of how the name PING evolved, this application utility program typically represents the first tool commonly used to test and troubleshoot network devices. For example, you can use PING to determine whether a workstation, server, or another TCP/IP network device recently connected to a network is correctly cabled to the network and whether the protocol stack is operational and correctly configured. That is, after a device is connected to a network and its TCP/IP protocol stack is configured, you can test the cabling to the network and the operational state of the protocol stack by PINGing the device from another station on the network or from one located on a different network. If the PING elicits a response, the response informs you that the device is correctly cabled to the network and that the TCP/IP protocol stack is operational. Based on the fact that PING will also return the round-trip delay between source and destination, this application can also provide you with indirect information about network activity. That is, if round-trip delay time is relatively long in comparison to previously obtained round-trip delay times, the cause would normally be attributable to network congestion.

Another little known use of PING that briefly warrants attention is its ability to test IP software without having to worry about the state of communications hardware or software drivers associated with the hardware. To accomplish this task you would use PING with the special Class A loopback address of 127.0.0.1. Later in this section we will note the use of this special IP Class A address with the PING utility application program.

For planning a voice-over-data network implementation, the round-trip delay caused by the use of PING provides the information necessary to determine the delay from a workstation doing the PINGing to the destination by dividing the delay by 2. However, a word of caution is warranted concerning the time reported by PING. If you enter a host name instead of an IP address, the first PING will require a name to address resolution because all routing is via IP addresses. Thus, this action will add to the round-trip delay time. In addition, because the characteristics of a network dynamically change during the day, it's best to use PING periodically throughout the day to examine changes in the round-trip delay.

If you receive a timeout in response to issuing a PING that denotes the absence of an Echo Response message after a predefined period of time, you should not automatically assume the destination is not available or improperly cabled to a network. If the destination is on a different network, it's possible that one or more intermediary devices, such as a router or gateway, is not operational and the PING cannot reach its destination. As we will note later in this section, if you obtain a timeout response to the use of PING when attempting to access an address on a different network, you should turn to the use of the TRACEROUTE utility program prior to concluding that the destination is not operational. In addition, if the round-trip delay appears excessive, you can turn to TRACEROUTE to determine the route and the delays along the route. If you can identify a bottleneck, it may be possible to request your Internet Service Provider (ISP) to upgrade equipment or a transmission facility. If this is not possible, the ISP might be able to reroute your connection or arrange for a different peering agreement where ISP traffic flows between networks to facilitate obtaining a lower latency.

Command Format

Although most modern operating systems are based on the use of a graphic user interface (GUI), most implementations of the PING utility program require the use of a command-line interface. One common command-line interface is as follows:

```
PING [-q] [-v] [-c count] [-s size] {Host_name|IP_Address}
```

where the command-line parameters are as follows:

- ▼ -q implements the quiet mode, resulting only in the display of summary lines at startup and completion.

- ■ -v implements verbose mode, resulting in the listing of ICMP packets received in addition to each Echo Response message.

- ■ -c specifies the number of Echo Requests (count) sent prior to concluding the test.

- ■ -s specifies the number of data bytes (size) transmitted with each Echo Request message.

- ▲ Host_name or IP_Address identifies the target to be PINGed.

To illustrate an example of the use of the PING utility, assume you just installed a computer with the IP address 198.78.46.8 and the host name gil.feds.gov. To test the cabling of the computer to the network and the operation of the protocol stack on the computer, you could go to another station on the network and enter the following command:

```
PING -c 5 gil.feds.gov
```

The preceding command line would result in the transmission of five Echo Request messages to the host address gil.feds.gov. Let's assume the response was five PING timeout messages, each indicating the lack of a response from the destination host. Although you might assume that there is a problem with the destination, it's important to note that routing is based on the use of IP addresses. Because TCP/IP-compliant devices respond to IP addresses, it's quite possible the host is up and operational but, for some reason, an appropriate entry was not made into the local DNS server to enable the host name to be resolved into an appropriate IP address. Recognizing this potential problem, you would then retry PING using the IP address of the destination as follows:

```
PING -c 5 198.78.46.8
```

The response to the PING would appear similar to the following:

```
72 bytes from 198.78.46.8 time = 12.2 ms
72 bytes from 198.78.46.8 time = 12.2 ms
72 bytes from 198.78.46.8 time = 12.2 ms
72 bytes from 198.78.46.8 time = 12.2 ms
72 bytes from 198.78.46.8 time = 12.2 ms
```

If you do not specify the number of data bytes to be used in each Echo Request, a default of 64 is used, with an 8-byte ICMP header resulting in the count of Echo Response packet size being displayed as 72 bytes. Although the round-trip time for each Echo Request to Echo Response message is shown as being the same at 12.2 ms, it's important to note that this is not always true. When you use a host address instead of an IP address, it's important to note that most times the issuance of multiple Echo Requests will result in the first round-trip delay having a longer time value than succeeding times. As previously explained, the reason for this is the fact that the first PING is delayed by the address resolution process. During the address resolution process, the host name is converted into an IP address to enable the Echo Request to be routed toward its destination. If you are using PING to obtain a precise round-trip delay time, such as to determine if a network structure can support voice over IP, which requires a very low level of latency, it is important to recognize the potential of the first round-trip delay time to significantly exceed succeeding delay times. Thus, it's a good and recommended procedure to issue at least two separate PING commands, permitting the first one to obtain the host IP address via the Address Resolution Protocol (ARP) and place that entry in the ARP table maintained by the PINGing device. Then the second PING command to the same destination will reflect round-trip delay times that do not include the address resolution process.

In concluding our examination of the use of the PING application utility program, we will turn our attention to the format and utilization of the program supported by Microsoft Corporation in Windows 95, Windows 98, and Windows NT/2000. Figure 11-1 illustrates the command format for PING displayed in response to entering the command without any command-line parameters.

In examining the command-line entries supported by Microsoft's implementation of PING, note the -t option. The use of that option results in a continuous PINGing of the destination until the CTRL-BREAK key combination is pressed to interrupt the program.

Continuous PINGing requires a continuous response from the destination. Because the destination must suspend what it is doing to respond to the PING, its ability to perform its

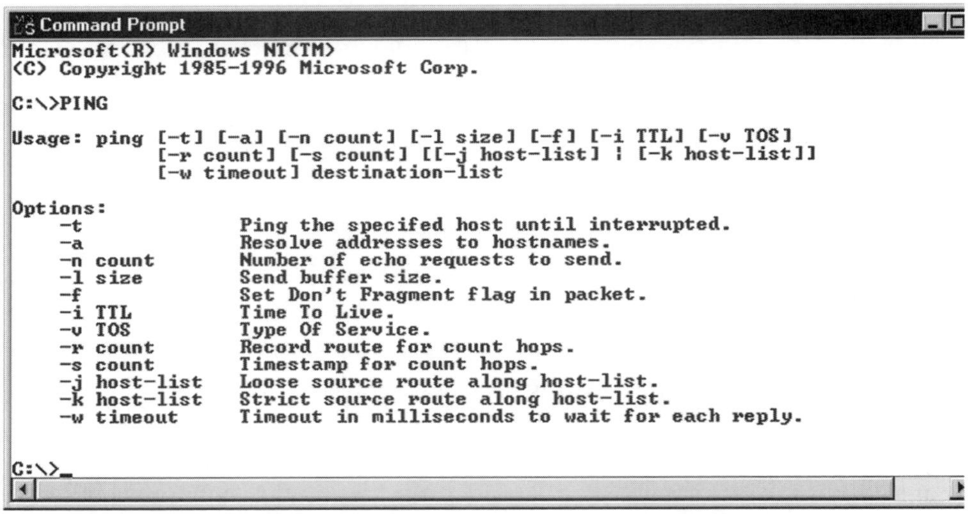

```
Command Prompt                                                    _ □
Microsoft(R) Windows NT(TM)
(C) Copyright 1985-1996 Microsoft Corp.

C:\>PING

Usage: ping [-t] [-a] [-n count] [-l size] [-f] [-i TTL] [-v TOS]
            [-r count] [-s count] [[-j host-list] ¦ [-k host-list]]
            [-w timeout] destination-list

Options:
    -t              Ping the specifed host until interrupted.
    -a              Resolve addresses to hostnames.
    -n count        Number of echo requests to send.
    -l size         Send buffer size.
    -f              Set Don't Fragment flag in packet.
    -i TTL          Time To Live.
    -v TOS          Type Of Service.
    -r count        Record route for count hops.
    -s count        Timestamp for count hops.
    -j host-list    Loose source route along host-list.
    -k host-list    Strict source route along host-list.
    -w timeout      Timeout in milliseconds to wait for each reply.

C:\>_
◄                                                                   ►
```

Figure 11-1. Microsoft Windows 95/98 PING utility program command-line format

intended function, such as a Web server responding to HTTP queries, is degraded. In effect, continuous PINGing is often considered as an unsophisticated hacker denial-of-service attack.

Another area of concern with respect to PING is that it can be used to discover network devices. That is, a sophisticated programmer can use PING within a shell program in an attempt to learn the addresses of network devices and then target those devices. To protect against the adverse effects of PING, some organizations either use a firewall or program a router's access list to prevent PINGs from reaching their network. If a router's access list is programmed, the denial of PINGs is usually expressed on a universal basis. However, it is possible for the router access list administrator to reprogram the list to enable you to issue PINGs for determining the round-trip delay to a predefined destination. Because a firewall has a more sophisticated filtering capability, many network managers and LAN administrators will configure the firewall to permit up to five PINGs per source IP address within a predefined period of time. This action permits a reasonable level of PINGing for a legitimate purpose to occur from users outside the organization's network while precluding denial-of-service attacks based on the use of PINGs. An alternate option some network managers use is to program a firewall to allow employees behind the firewall to issue outbound PINGs while barring inbound PINGs. The problem with this solution is the fact that if every organization implemented it, nobody would be able to legitimately use PING. Based on the preceding, you should consider coordinating the use of PING with your firewall or router administrator. If PING is barred, you can usually use a workstation to issue PINGs. All you need is the IP address of the workstation and about two minutes to enable a change to a firewall policy or perhaps a few minutes longer to change a router access list.

Operation

Figure 11-2 illustrates the use of PING twice to PING the server at American University. If you examine the round-trip delay times shown in Figure 11-2, you will note they vary significantly between each issue of PING that occurred only a half minute from one another. This illustrates a key problem with the use of the public Internet for voice over IP: its variable and unpredictable delay makes it currently unsuitable for many business applications.

Because PING by default simply displays its results, its use to gather statistics would be tedious. As this author is like many readers and prefers to use automation whenever possible instead of a manual effort, he will share with you a trick you can use to automate PING statistics.

Automating PING

As we noted earlier in this book when we discussed the flow of packets through a network, their flow can be affected by many random events. Those events can include the route between source and destination, the traffic offered to each router in the path between source and destination, and the presence or absence of the use of a priority scheme by each router in the path with respect to processing packets. When the latter occurs, the priority, if

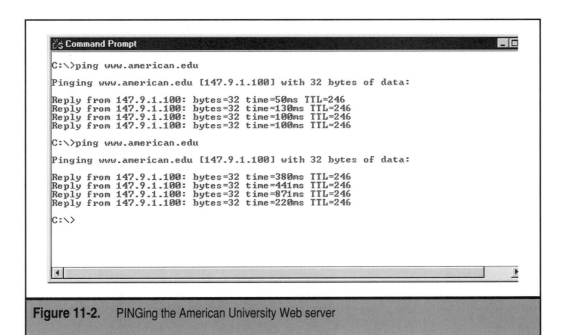

Figure 11-2.   PINGing the American University Web server

any, assigned to packets transporting digitized voice also must be considered. Thus, instead of simply issuing a PING command and observing a few results it would be much better to record the results of the use of this utility application program over a period of time. To do so you can direct the output of the command into a file. To accomplish this task you could issue the PING command with either the -n or -t parameter and direct the resulting display generated by the command onto a file via the use of the greater than (>) symbol.

Figure 11-3 illustrates the use of the PING command and its -t option to generate continuous PINGing of the IP address 198.78.46.9. Note the use of the greater than (>) symbol followed by the filename "pingfile" selected by this author. This command -line entry will run PING until it is terminated and direct the output onto the file named "pingfile." Since forever is a long time, this author used the three-finger salute of CTRL-ALT-DEL to bring up the Close Program dialog box under Windows 95/98 to close the program. Next, the DOS DIR command was used with the filename previously selected to show that the file indeed was saved.

Because this author's old professor was fond of stating "the proof of the pudding is in the eating" and that statement rubbed off on this author, let's examine the file we created and its potential use. Figure 11-4 shows the use of the DOS TYPE command to display the

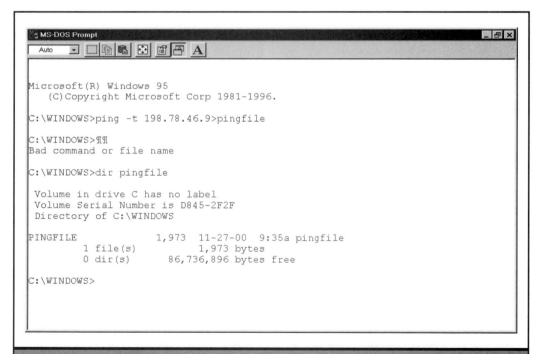

Figure 11-3. By using the PING -t option and directing output onto a file it becomes possible to gather statistics.

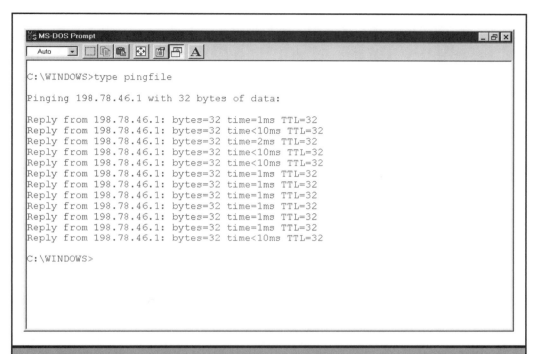

Figure 11-4. Using the DOS TYPE command to display the contents of the previously created file named "pingfile"

contents of the file "pingfile" to the console. Note that the replies from the PINGed IP address are listed row by row, which makes it possible to import the file into the spreadsheet program of your choice. Once this is accomplished you can then use the statistical functions of the spreadsheet to analyze the result of your PINGing effort over a prolonged period of time.

Figure 11-5 illustrates the use of Microsoft's Excel spreadsheet program to open the previously created file named "pingfile." Note that when you select File | Open Excel recognizes that the file is not in the format associated with the program. Thus, Excel automatically displays its Text Import Wizard shown in the near-middle portion of Figure 11-5. There are three steps associated with the Text Import Wizard. The second step provides you with the ability to create, delete, or move break lines. By using this feature it becomes possible to place the round-trip delay times in a specific column. Once this occurs you can use the Replace function from the Edit menu to remove such entries as "time," "=," "<," and "ms" to obtain a column of values. After using the Replace function, you would then obtain the ability to use the spreadsheet's built-in statistical functions to analyze the series of PINGs issued to the target IP address.

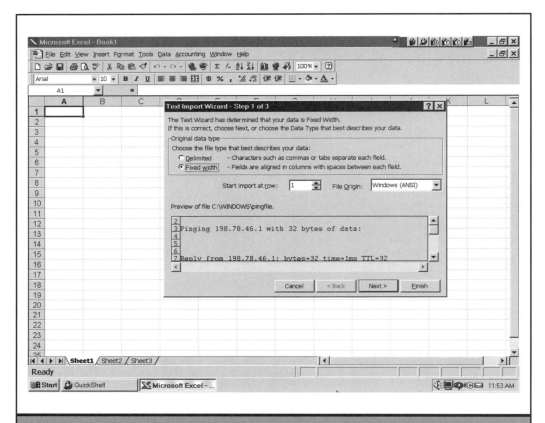

Figure 11-5. Once PINGs are captured onto a file they can be imported into Excel or another spreadsheet program for further analysis.

APPLICATON NOTE: To verify your computer's TCP/IP protocol stack is operating correctly, use PING with the loopback IP address of 127.0.0.1. If the round-trip delay reflects the minimum time PING is capable of computing, the protocol stack is more than likely operating correctly.

Other PING Considerations

Prior to moving forward let's turn our attention to a few additional items concerning the use of PING. First, it may be important to consider using the -1 option that can alter the size of packets transmitted. Although some implementations of PING use a default of 64 bytes of data, a quick look at Figures 11-2 and 11-4 indicates that the Microsoft default is 32 bytes of data. Because routers can be programmed to provide a preference for traffic expediting based upon packet length you may wish to adjust the size of your PINGs to reflect the length of packets transporting digitized voice. In doing so you should consider referencing applicable vendor literature that describes packet length as a function of the voice coder selected. This will enable you to determine if there are delay differences worth noting and select a different coder based upon observed differences in round-trip delay times.

> **APPLICATION NOTE:** You should vary the length of each series of PINGs to reflect the length of packets when different voice coders are used. This can assist you in determining if one or more specific coders are more suitable for use than other coders.

A second item concerning the use of PING that warrants attention concerns the use of the loopback 127.0.0.1 address. While certainly not common, this author was once able to track an abnormal delay to software that apparently was corrupted due to a disk problem. Due to this you should consider using PING with the software loopback address to make sure your TCP/IP protocol stack is functioning correctly. That is, the round-trip delay time when PINGing the loopback address should always represent the minimum delay time the PING implementation is capable of denoting. In a Windows environment the round-trip delay when PINGing the loopback address should always be "<10 ms."

Now that we have an appreciation for the operation and utilization of PING, let's turn our attention to the utility program that should be considered as a follow-up to its use—the TRACEROUTE program.

TRACEROUTE

The TRACEROUTE command invokes a program by that name that provides information about the route that packets take from source to destination. If you PING a device on a different network, the lack of a response via a timeout condition does not necessarily mean that the destination device is not operational. Instead, it's possible that one or more intermediary routers or gateways that provide a connection between networks are not operational. In addition, if PING shows a lengthy round-trip delay, it does not indicate whether most or a portion of the delay results from a single bottleneck whose removal might enable an unworkable solution to become doable. Thus, TRACEROUTE can be used to examine the route from source to destination to determine whether a path to the destination is available and, if so, the delays along the path.

Operation

TRACEROUTE operates by transmitting a series of User Datagram Protocol (UDP) datagrams to an invalid port address at the destination device. First, the program transmits three datagrams in sequence, with each datagram having its IP Time-to-Live (TTL) field value set to unity. This action results in each datagram timing out as soon as it is processed by the first router in the path between the source and destination networks.

After the first router decrements the TTL field value by one and notes its value is zero, it generates an ICMP Time Exceeded Message (TEM) response, which indicates the datagram expired, and sends it to the great bit bucket in the sky. The TEM response enables the device issuing the TRACEROUTE to compute the round-trip delay to the router for each of the three datagrams in the sequence. Next, the issuing device sets the TTL field value to 2 and transmits another sequence of three UDP datagrams. This sequence of three datagrams flows through the first router, which decrements their TTL field values to 1. However, the next router decrements their TTL field values to zero and returns TEM responses to each datagram, enabling the round-trip delay to the second router in the path to be determined.

The previously described process continues until the datagrams either reach their destination or encounter a broken path. If the datagrams reach their destination, the use of an invalid port number in the UDP header results in the destination device generating an ICMP Destination Unreachable message in response to each of the three datagrams. This message tells the TRACEROUTE program that the destination was reached and that it should terminate its operation. In the event an open path is encountered due to a router or gateway failure, TRACEROUTE uses a timer and will terminate its operation when the timer expires. By noting that the last router that responded was not connected to the destination network, you can observe that a wide area network connection or router failure is the reason behind the inability to access a destination device.

Command Format

The following is a commonly used TRACEROUTE command-line format:

```
TRACEROUTE [-m ttl] [-q packets] {IP_Address| Host_name}
```

where:

▼   -m represents the maximum allowable TTL value, which is the number of hops allowed prior to the program terminating. The typical default TTL value used by most implementations of TRACEROUTE is 30.

▲   -q represents the number of UDP packets transmitted with each time-to-live setting. The common default value is 3.

Figure 11-6 illustrates the Microsoft command format for the implementation of TRACEROUTE, which is referred to as TRACERT. Note that you can use the -w option to specify a longer timeout value if the use of the utility results in timeouts. Also note that if the default maximum of 30 hops prevents the use of the program to determine the full path, you can use the -h option to change the default to a higher value.

Figure 11-7 illustrates the use of TRACERT from a Windows NT client to determine the path to the American University Web server. It should be noted that the author used a commercial Internet service provider's network to access the Internet. From the access point, a total of 10 hops were required to reach the destination. In examining the entries in Figure 11-7, you will note that typically routers are configured to return a description in addition to the TEM message. That description typically provides an indication of the location and organization operating the router. For example, hop 4 and hop 5 represent

```
Command Prompt

C:\>TRACERT

Usage: tracert [-d] [-h maximum_hops] [-j host-list] [-w timeout] target_name

Options:
    -d                      Do not resolve addresses to hostnames.
    -h maximum_hops         Maximum number of hops to search for target.
    -j host-list            Loose source route along host-list.
    -w timeout              Wait timeout milliseconds for each reply.

C:\>_
```

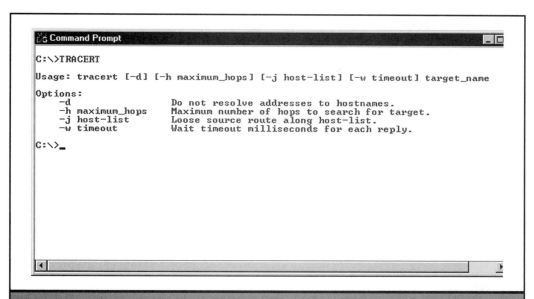

Figure 11-6. The Microsoft TRACERT command format

```
Command Prompt

C:\>tracert www.american.edu

Tracing route to www.american.edu [147.9.1.100]
over a maximum of 30 hops:

  1    <10 ms   <10 ms   <10 ms  205.131.176.1
  2    331 ms    80 ms    70 ms  s11-0-0-28.atlanta1-cr3.bbnplanet.net [4.0.156.1
  3     70 ms    50 ms    40 ms  p2-1.atlanta1-nbr1.bbnplanet.net [4.0.5.114]
  4     50 ms   100 ms   100 ms  p3-3.vienna1-nbr3.bbnplanet.net [4.0.5.141]
  5     70 ms   120 ms   100 ms  p1-0.vienna1-nbr2.bbnplanet.net [4.0.5.45]
  6    130 ms   110 ms   101 ms  p4-1.washdc3-ba1.bbnplanet.net [4.24.4.189]
  7     80 ms    70 ms    90 ms  p0-0-0.washdc3-cr5.bbnplanet.net [4.24.4.110]
  8    110 ms   100 ms   131 ms  4.0.147.6
  9     70 ms   110 ms    80 ms  147.9.5.3
 10    110 ms    90 ms    90 ms  ausolaris1.american.edu [147.9.1.100]

Trace complete.

C:\>
```

Figure 11-7. Using TRACERT to examine the path to the American University Web server

routers operated by BBN Planet in Vienna, Virginia. Because BBN Planet was acquired over two years ago by GTE, it's also apparent that the description of the router was not updated.

Although the Microsoft implementation of TRACEROUTE uses a default of three datagrams to obtain three round-trip delay times per hop, it does not average those times. Other implementations of TRACEROUTE provide a statistical summary of round-trip response time. Similarly, other implementations may have one or more optional parameters that permit you to control timeout delays. Because all versions of TRACEROUTE work in a similar manner, it is important to note that regardless of implementation, its use provides you with the ability to determine whether a path is available to the destination. In addition, it also provides you with the ability to note the delays associated with different points on the route to the destination.

APPLICATION NOTE: You can use the TRACEROUTE utility program to determine the delays associated with a path between source and destination addresses.

PATHPING

Earlier in this section it was mentioned that Windows 2000 includes a new utility program named PATHPING. In concluding this section focused on the use of utility application program we will examine this new program.

Overview

PATHPING can be considered as a souped-up version of Microsoft's TRACERT program. Through the use of PATHPING you can determine which routers or router hops are causing delays or other problems on the path between two IP addresses much better than with the use of TRACERT. The reason for the greater ability of PATHPING is the fact that this utility program uses a greater sampling interval than TRACERT. By default PATHPING will send 100 queries to each host along the path from source to destination. In addition, PATHPING will provide information about packet loss on each hop on the route between source and destination, which facilitates isolating where network problems and abnormal latency is occurring.

Parameters

Figure 11-8 illustrates the use of the PATHPING command without any parameters to generate a display of the parameters it supports. As we review each of the parameters supported by this program we will note how it considerably improves upon TRACERT while retaining many familiar options.

The -n option is used to inform the program not to resolve addresses to host names. The -h option is used to specify the number of hops to search for a target. As noted in Figure 11-8, the default is 30, which is the same for both PING and TRACERT utility programs.

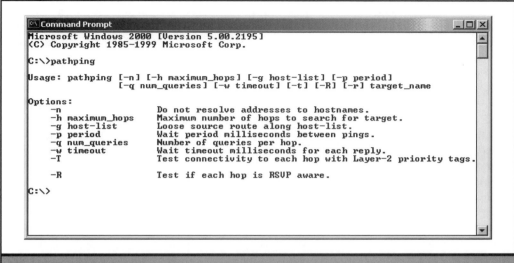

Figure 11-8. PATHPING was added under Windows 2000 and represents an enhanced version of TRACERT.

The -g option when set enables a loose source route along the host-list following the option letter. This option allows intermediate gateways along the host-list separate from consecutive computers.

The -p option is used to specify the time in milliseconds (ms) between PINGs. The default is 250 ms or 0.25 seconds.

The -q option is used to specify the number of queries to send to each host along the route. As previously mentioned the default value is 100.

The -p and -q defaults combined with the number of hops between source and destination can result in a significant waiting period. For example, assume 20 hops between source and destination. With a default of .25 seconds between PINGs and 100 queries sent to each computer, it will require 25 seconds to gather statistics for each hop. Thus, an analysis of 20 hops will require 500 seconds or 8.3 minutes. Although we will soon note the extended capability of PATHPING over TRACERT, the old adage that "patience is a virtue" is true when using this program.

The -w option specifies the number of milliseconds to wait for each reply. The default is 3000 ms or 3 seconds and can further increase the delay we previously computed.

The -T option results in the attachment of an 802.1p priority tag at layer 2 of the frame containing the PING. This tag is sent to each network device along the route as a mechanism to determine which network devices have or do not have layer 2 priority enabled. Because layer 2 priority is only applicable for LANs, the use of this option requires careful consideration. Although the origin and endpoint LANs would obviously prefer to be 802.1p-compliant, when your packets hit a routing center where they pass from one router to another via a LAN it is also desirable to have 802.1p compliance at that location.

The -R option represents another valuable addition under PATHPING. Its use enables you to determine which routers along the path from source to destination support RSVP. This lets you determine if it is possible for bandwidth to be reserved without having to upgrade a router as well as what routers, if any, need to be upgraded to provide an RSVP end-to-end bandwidth reservation capability. You can use this information when negotiating with an ISP for a VPN or conventional Internet access. For example, if five out of six hops along a VPN path support RSVP, you might consider informing the ISP of your organization's need for guaranteed bandwidth for certain applications as a condition of your negotiation effort.

The last PATHPING parameter is the `target_name`. The `target_name` denotes the destination address either as a host name or by the use of an IP address. Similar to our earlier discussion concerning PING and host names, if you use a host name be aware of the fact that if it is a new host name not previously resolved into an IP address, some additional time will be required to perform the host-to-IP address resolution. However, since the default of PATHPING is 100 queries per host, this extra delay is negligible over the extended number of queries in comparison to PING and TRACERT.

APPLICATION NOTE: You can use PATHPING to determine support for RSVP by each router in the path between source and destination as well as if network devices have IEEE 802.1p priority configured.

Operation

Figure 11-9 illustrates the use of PATHPING to trace the route from the author's home to the Yale University Web server. In this example the default of 100 queries was changed to 10 through the use of the -q 10 option. In examining Figure 11-9 note the Round Trip Time (RTT) is specified for each hop. In addition the number of lost packets out of 10 sent to each node and the percentage of lost packets are summarized. Thus, PATHPING provides a considerable amount of additional information beyond TRACERT that can let you know if a path is viable for a voice-over-IP application or, if there are problems, where those problems reside.

TRAFFIC PRIORITIZATION CONSIDERATIONS

As mentioned several times in this book, the key to operating an effective voice-over-data network facility is to obtain predictability and reliability. If your organization is considering implementing voice over IP or voice over frame relay, you need to consider the equipment along the path between source and destination. In concluding this book, we will briefly discuss some additional items you may wish to investigate to ensure that your implementation is headed for success.

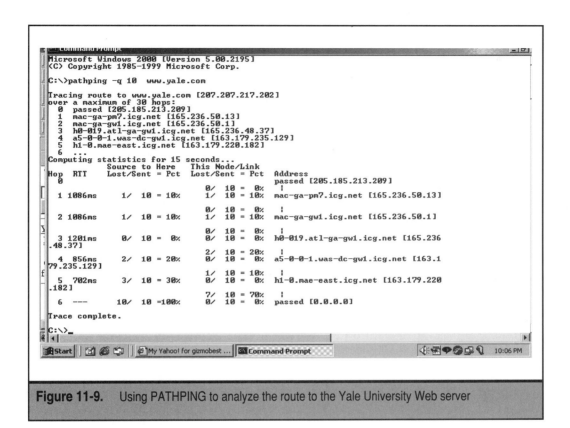

Figure 11-9. Using PATHPING to analyze the route to the Yale University Web server

Shared vs. Switched Operations

If organizational users that will take advantage of voice are LAN-based, you more than likely should consider the use of switches instead of a shared media environment. A shared media Ethernet represents an unpredictable technology, as its basic mode of operation is one of contention for the media. This means that unless LAN utilization is very low, the ability of digitized voice packets to reach a router can be expected to have a considerable variance in delays. If your organization has or can obtain a switch-based infrastructure that supports IEEE 802.1p, you obtain the ability to make packets transporting voice a higher priority than packets transporting data, thus obtaining predictability at the edge of the network.

Ingress Operations

There are three functions necessary to enable packets transporting digitized voice to correctly flow into a packet network. First, due to the effect of long packets upon packets transporting voice, the size or length of all packets must be adjusted. This adjustment is necessary to ensure that when a long packet is processed, its time onto the network does not adversely affect to a significant degree the latency of the packet behind it transporting digitized voice. Second, the ingress device must prioritize packets transporting voice over those carrying data. This will also minimize delays. Last but not least, the line access rate that provides egress to the network must be carefully examined. As we noted earlier in this book, a T1 line results in an insignificant delay for transporting a packet carrying digitized voice, while the use of a 56-Kbps access line results in a significant degree of latency.

Egress Operations

Although it may appear that because data flows into the destination network as a serial bit stream, equipment at the egress location does not require the functionality of ingress equipment. This is not true. Remember that voice conversations are bi-directional. Thus, the three key functions mentioned earlier in this section for ingress equipment are also necessary for egress equipment.

Summary

By carefully examining the features and operational characteristics of potential equipment in conjunction with the use of utility programs, you can obtain a reasonable expectation of the potential success or failure of a voice-over-data application implementation. Although the techniques mentioned in this chapter require a bit of time (no pun intended), it may result in valuable information for implementing voice over data. The best advice this author can provide is that you should use the information in this book to be prepared.

APPENDIX A

Frame Relay Fragmentation Implementation Agreement FRF.12

Frame Relay Forum Technical Committee

Note: The user's attention is called to the possibility that implementation of the Frame Relay Implementation Agreement contained herein may require the use of inventions covered by patent rights held by third parties. By publication of this Frame Relay Implementation Agreement the Frame Relay Forum makes no representation that the implementation of the specification will not infringe on any third party rights. The Frame Relay Forum take no position with respect to any claim that has been or may be asserted by any third party, the validity of any patent rights related to any such claims, or the extent to which a license to use any such rights may not be available.

Editor: Andrew G. Malis

For more information contact:

The Frame Relay Forum
Suite 307
39355 California Street
Fremont, CA 94538 USA

Phone: +1 (510) 608-5920
FAX: +1 (510) 608-5917
E-Mail: frf@frforum.com

TABLE OF CONTENTS

TABLE OF FIGURES

REVISION HISTORY

Version	Change	Date
1.0	Approved	December 1997

FRAME RELAY FRAGMENTATION

IMPLEMENTATION AGREEMENT

1. INTRODUCTION

1.1 Purpose

This document is a Frame Relay Fragmentation Implementation Agreement. It provides transmitting Frame Relay DTEs and DCEs with the ability to fragment long frames into a sequence of shorter frames, which will then be reassembled into the original frame by the receiving peer DTE or DCE. Frame fragmentation is necessary to control delay and delay variation when real-time traffic such as voice is carried across the same interfaces as data.

This agreement supports three applications of fragmentation:

1. Locally across a Frame Relay UNI interface between the DTE-DCE peers.

2. Locally across a Frame Relay NNI interface between the DCE peers.

3. End-to-End between two Frame Relay DTEs interconnected by one or more Frame Relay networks. When used end-to-end, the fragmentation procedure is transparent to Frame Relay network(s) between the transmitting and receiving DTEs.

The agreements herein were reached in the Frame Relay Forum, among vendors and suppliers of Frame Relay network products and services, and are based on the relevant Implementation Agreements and standards referenced in Section 2.

This document may be submitted to bodies involved in ratification of implementation agreements and conformance testing to facilitate multi-vendor interoperability, and to standards bodies for inclusion in international standards.

1.2 Terminology

Must, Shall, or Mandatory—The item is an absolute requirement of this IA.

Should—The item is highly desirable.

May or Optional—The item is not compulsory, and may be followed or ignored according to the needs of the implementer.

Not Applicable—The item is outside the scope of this IA.

1.3 Acronym List

AAL5	ATM Adaptation Layer 5
ATM	Asynchronous Transfer Mode
DCE	Data Communications Equipment
DLCI	Data Link Connection Identifier
DTE	Data Terminal Equipment
FCS	Frame Check Sequence
FR	Frame Relay
IA	Implementation Agreement
IWF	Interworking Function
MTU	Maximum Transmission Unit
NLPID	Network Layer Protocol Identifier
NNI	Network-to-Network Interface
PDU	Protocol Data Unit
PPP	Point-to-Point Protocol
PVC	Permanent Virtual Connection
SVC	Switched Virtual Connection
UI	Unnumbered Information
UNI	User-to-Network Interface
VC	Virtual Connection
VoFR	Voice over Frame Relay

2. RELEVANT STANDARDS

The following is a list of standards and IAs on which this Frame Relay Fragmentation IA is based:

[1]	FRF.1.1	D. Sinicrope (ed.), User-to-Network Implementation Agreement (UNI), January 19, 1996.
[2]	FRF.2.1	L. Greenstein (ed.), Frame Relay Network-to-Network Interface Implementation Agreement, July 10, 1995.
[3]	FRF.3.1	R. Cherukuri (ed.), Multiprotocol Encapsulation Implementation Agreement, June 22, 1995.
[4]	FRF.5	D. O'Leary (ed.), Frame Relay/ATM PVC Network Interworking Implementation Agreement, December 20, 1994.
[5]	FRF.8	D. O'Leary (ed.), Frame Relay/ATM PVC Service Interworking Implementation Agreement, April 14, 1995.
[6]	FRF.9	D. Cantwell (ed.), Data Compression Over Frame Relay Implementation Agreement, January 22, 1996.
[7]	FRF.11	K. Rehbehn, R. Kocen, T. Hatala (eds.), Voice Over Frame Relay Implementation Agreement, March 1997.
[8]	ITU Recommendation Q.922	ISDN Data Link Layer Specification For Frame Mode Bearer Services, 1992.
[9]	RFC 1490	T. Bradley, C. Brown, A. Malis, Multiprotocol Interconnect over Frame Relay, July 26, 1993.
[10]	RFC 1990	K. Sklower, B. Lloyd, G. McGregor, D. Carr, T. Coradetti, The PPP Multilink Protocol (MP), August 1996.

3. OVERVIEW

To properly support voice and other real-time (delay-sensitive) data on lower-speed UNI [1] or NNI [2] links, it is necessary to fragment long frames that share the same UNI or NNI link with shorter frames so that the shorter frames are not excessively delayed. Fragmentation enables interleaving delay-sensitive traffic on one VC with fragments of a long frame on another VC utilizing the same interface.

This Implementation Agreement:

▼ Allows real-time and non-real-time data to share the same Frame Relay UNI or NNI link.

■ Allows the fragmentation of frames of all formats.

■ Defines a fragmentation procedure that can be used by other protocols or IAs, such as FRF.11 [7]. See Section 9.2 for further details.

▲ Defines three fragmentation models, locally across a UNI, locally across a NNI, and End-to-End. All three models share common fragmentation procedures.

The fragmentation procedure previously specified in FRF.3.1 [3] and RFC 1490 [9] is no longer recommended, and is replaced by this IA.

Fragmentation may be used with all Q.922 [8] address formats. For clarity, the figures that include address octets will only illustrate the two-octet address.

4. FRAGMENTATION MODELS

4.1 UNI Fragmentation

UNI (DTE-DCE) fragmentation is used in order to allow real-time and data frames to share the same UNI interface between a DTE and the Frame Relay Network. Fragmentation is strictly local to the interface, and the fragment size can be optimally configured to provide the proper delay and delay variation based upon the *logical* speed of the DTE interface (the logical speed of an interface may be slower than the physical clocking rate if a channelized physical interface is used).

Since fragmentation is local to the interface, the network can take advantage of the higher internal trunk speeds by transporting the complete frames, which is more efficient than transporting a larger number of smaller fragments.

UNI fragmentation is also useful when there is a speed mismatch between the two DTEs at the ends of a VC. It also allows the network to proxy for a DTE that does not implement End-to-End fragmentation.

For UNI fragmentation, the DTE and DCE interfaces act as fragmentation and reassembly peers, as shown in Figure 4-1.

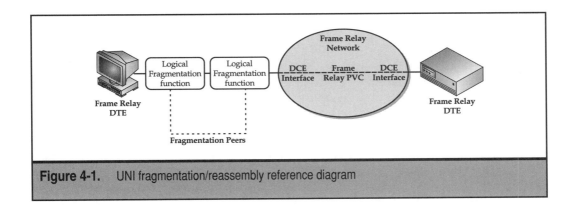

Figure 4-1. UNI fragmentation/reassembly reference diagram

Note that the functionality specified in this Implementation Agreement has been illustrated as a standalone "Logical Fragmentation Function", but it is expected to be implemented in the DTE and DCE interfaces shown in the diagram.

UNI fragmentation is provisioned on an interface-by-interface basis. When UNI fragmentation is used on an interface, then all frames on all DLCIs (including DLCI 0, PVCs, and SVCs) are preceded by the fragmentation header (described in Section 5.1).

Note that UNI and NNI fragmentation formats and procedures are identical.

4.2 NNI Fragmentation

NNI interfaces may also act as fragmentation and reassembly peers, as shown in Figure 4-2.

Fragmentation on slower NNI links allows delay sensitive traffic on one NNI VC to be interleaved with fragments of a long data frame on another VC using the same NNI. As in Figure 4-1, the fragmentation and reassembly functionality has been illustrated as a

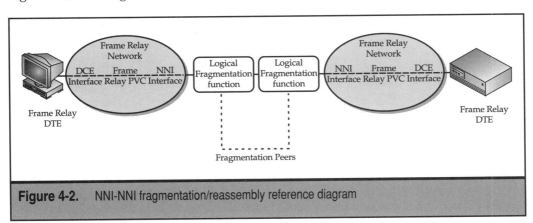

Figure 4-2. NNI-NNI fragmentation/reassembly reference diagram

standalone "Logical Fragmentation Function", but it is expected to be implemented in the NNI interfaces shown in the diagram.

NNI fragmentation is also provisioned on an interface-by-interface basis. When fragmentation is in use on an NNI interface, then all frames on all DLCIs (including DLCI 0, PVCs, and SVCs) are preceded by the fragmentation header (described in Section 5.1).

Note that UNI and NNI fragmentation formats and procedures are identical.

4.3 End-to-End Fragmentation

End-to-End fragmentation is used between peer DTEs, and is restricted to use on PVCs only. It is most useful when peer DTEs wish to exchange both real-time and non-real-time traffic using slower interface(s), but either one or both UNI interfaces does not support UNI fragmentation. Alternatively, they may be on separate FR networks and the path interconnecting the networks may be slow enough to require fragmentation to properly support the real-time traffic, but fragmentation is not supported over the NNI.

When used between DTEs, as shown in Figure 4-3, the fragmentation procedure is transparent to Frame Relay network(s) between the transmitting and receiving DTEs. The transmitting Frame Relay DTEs fragment long frames into a sequence of shorter frames, which will then be reassembled into the original frame by the receiving DTE. Again, the functionality specified in this Implementation Agreement has been illustrated as a standalone "Logical Fragmentation Function", but in this case it is expected to be implemented in the DTEs shown in the diagram.

Unlike UNI and NNI fragmentation, which fragments all frames on an interface, End-to-End fragmentation is limited to fragmenting frames on selected PVCs. When End-to-End fragmentation is provisioned on a particular PVC, frames that exceed the configured maximum fragment size must conform to the fragmentation format. Because DLCI 0 is never carried end-to-end, it is never fragmented using End-to-End fragmentation.

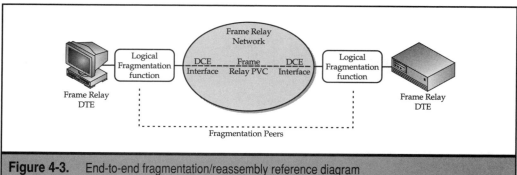

Figure 4-3. End-to-end fragmentation/reassembly reference diagram

5. DATA FRAGMENTATION FORMATS

There are two fragmentation formats, one for Interface (UNI and NNI) fragmentation and the other for End-to-End fragmentation.

5.1 Interface Fragmentation Format

For Interface (UNI and NNI) fragmentation, a two-octet fragmentation header precedes the Frame Relay header.

The format for each fragment of a data frame is shown in Figure 5-1.

The (B)eginning fragment bit is a one-bit field set to 1 on the first data fragment derived from the original frame and set to 0 for all other fragments from the same frame.

The (E)nding fragment bit is a one-bit field set to 1 on the last data fragment and set to 0 for all other data fragments. A data fragment may have both the (B)eginning and (E)nding fragment bits set to 1.

The (C)ontrol bit is set to 0 in all fragments. It is reserved for future control functions.

The sequence number is a 12-bit binary number that is incremented modulo 2^{12} for every data fragment transmitted on a VC. There is a separate sequence number maintained for each DLCI across the interface.

	8	7	6	5	4	3	2	1
Fragmentation	B	E	C	Seq. # high 4 bits				1
header	Sequence # low 8 bits							
Frame Relay	DLCI high six bits						C/R	0
header	DLCI low 4 bits				F	B	DE	1
	Fragment Payload							
	FCS							
	(two octets)							

Figure 5-1.   UNI and NNI data fragment format

Note that the low order bit of the first octet of the fragmentation header is set to 1. This allows the fragmentation header to be distinguishable from the Frame Relay header. This allows a fragmentation entity (UNI or NNI) to detect the misconfiguration of its peer, since the peers must be configured identically to use or not use fragmentation across an interface. If a peer is configured for UNI or NNI fragmentation and receives frames that do not contain the fragmentation header, those frames are discarded. If a peer is not configured for UNI or NNI fragmentation and receives frames with the fragmentation header, those frames will be discarded due to the violation of the Q.922 header format.

The fragment payload and fragmentation procedures are described in Section 6.1.

5.2 End-to-End Fragmentation Format

For End-to-End fragmentation, a two-octet fragmentation header follows the FRF.3.1 [3] multiprotocol encapsulation header. The Network Layer Protocol ID (NLPID) 0xB1 has been assigned to identify this fragmentation header format. See FRF.3.1 for further details concerning NLPIDs and multiprotocol encapsulation over FR.

The format for each fragment of a data frame is shown in Figure 5-2.

The (B)eginning fragment bit is a one-bit field set to 1 on the first data fragment derived from the original frame and set to 0 for all other fragments from the same frame.

The (E)nding fragment bit is a one-bit field set to 1 on the last data fragment and set to 0 for all other data fragments. A data fragment may have both the (B)eginning and (E)nding fragment bits set to 1.

	8	7	6	5	4	3	2	1
Frame Relay	DLCI high six bits						C/R	0
header	DLCI low 4 bits				F	B	DE	1
UI (0x03)	0	0	0	0	0	0	1	1
NLPID (0xB1)	1	0	1	1	0	0	0	1
Fragmentation	B	E	C	Seq. # high 4 bits				0
header	Sequence # low 8 bits							
	Fragment Payload							
	FCS							
	(two octets)							

Figure 5-2. End-to-end data fragment format

The (C)ontrol bit is set to 0 in all fragments. It is reserved for future control functions.

The sequence number is a 12-bit binary number that is incremented modulo 2^{12} for every data fragment transmitted on a PVC. There is a separate sequence number maintained for each fragmented PVC between DTE peers.

The fragment payload and fragmentation procedures are described in Section 6.1.

6. PROCEDURES

6.1 Fragmentation Procedure

This fragmentation procedure is based upon RFC 1990 [10]. Note that the procedure is identical for both header formats.

A series of data fragments is created by taking a frame, removing the leading flag and Q.922 address octets, removing the original FCS and trailing flag following the data, and sending the remaining octets, in their original order, as a series of data fragments. If an FRF.3.1-encapsulated frame is being fragmented, then the Q.922 control, optional pad, and NLPID octets of the original multiprotocol frame are only contained in the first data fragment. The resulting fragments must be transmitted in the same sequence as they occurred in the frame prior to being fragmented. Fragments from multiple VCs may be interleaved with each other on one interface (this is the principal objective of fragmentation).

The first data fragment in the series has the B bit set, and the final data fragment has the E bit set. Every fragment in the series contains the same address octets that were on the original unfragmented frame, including the Frame Relay congestion bits (FECN, BECN, DE).

The first fragment sent on a VC (following a VC becoming active) may have the sequence number set to any value (including zero), and the sequence number must subsequently be incremented by one for each fragment sent. The sequence number is incremented without regard to the original frame boundaries; if the last fragment in one frame used sequence number "N", then the first fragment of the following frame will use sequence number "N+1".[1] This allows lost fragments (and bursts of lost fragments) to be easily detected. Each VC has its own fragmentation sequence number sequence, independent of all other VCs.

If sufficient fragments are sent on an active VC, the sequence number will wrap from all ones to zero, and will eventually also wrap past the original sequence number sent on the VC after it became active. This wrapping may or may not occur on an original frame boundary (it is transparent to frame boundaries).

NNI interfaces are allowed to further fragment End-to-End fragments. The End-to-End fragments are encapsulated within the NNI interface fragments. Thus, the

---

1   For FRF.12 UNI, NNI, and End-to-End fragmentation, the sequence number is incremented modulo 2^{12}; for FRF.11 sub-frame data payload, the sequence number is incremented modulo 2^{13}.

NNI fragments containing the End-to-End fragments can easily be reassembled by the receiving NNI interface, and the End-to-End fragments sent to the destination DTE for final reassembly.

Similarly, UNI interfaces may also further fragment End-to-End fragments, as described in the previous paragraph.

See Section 8 for fragmentation examples.

6.2 Reassembly Procedure

The reassembly procedures are identical for both header formats.

For each VC, the receiver must keep track of the incoming sequence numbers and maintain the most recently received sequence number. The receiver detects the end of a reassembled frame when it receives a fragment bearing the (E)nding bit. Reassembly of the frame is complete if all sequence numbers up to that fragment have been received.

Note that the Frame Relay congestion bits (FECN, BECN, DE) must be logically ORed for all fragments, and the results included in the reassembled frame.

The receiver detects lost fragments when one or more sequence numbers are skipped. When a lost fragment or fragments are detected on a VC, the receiver must discard all currently unassembled and subsequently received fragments for that VC until it receives the first fragment that bears the (B)eginning bit. The fragment bearing the (B)eginning bit is used to begin accumulating a new frame.

In the event of an error (e.g., one or more fragments lost due to transmission error or reassembly buffer overflow), fragments which cannot be reconstructed back into the original frame must be discarded by the receiver.

7. FRAGMENT AND FRAME SIZES

This IA does not recommend any specific fragment size. The fragment size is configured in the transmitter, and two peer transmitters need not use the same fragment size.

For End-to-End fragmentation, the maximum fragment size used on a particular PVC depends on both the access line speeds at the two ends of the connection, the speeds of any NNI interfaces on the path, and the delay and delay variation requirements of the particular applications in use on the PVC. The maximum fragment size to send should be configured on a per-PVC basis. If one or more of the Frame Relay network(s) interconnecting the DTEs are implemented using FRF.5 [4] FR/ATM Network Interworking, then it may be advantageous to make the End-to-End fragment size (not including the FCS) an even multiple of the underlying ATM cell payload size in order to optimize the performance at the ATM layer. See FRF.5 [4] for additional details on how the end-to-end fragments are carried in ATM cells.

For UNI and NNI fragmentation, the maximum fragment size should be configured on a per-interface basis. The optimal fragment size for these interfaces is the result of a tradeoff between the efficiencies of large frames, the MTU size of the constituent networks, and the required delay and delay variation characteristics of the applications.

Receivers must be able to reassemble complete frames up to at least 1600 octets in length.

8. FRAGMENTATION AND OTHER FRF IMPLEMENTATION AGREEMENTS

8.1 FRF.11 VoFR Encapsulation of Data Fragments

Annex C of FRF.11 [7] defines a format allowing encapsulation of fragmented data frames in a Voice over Frame Relay (VoFR) sub-frame data payload. PVCs that utilize the VoFR sub-frame data payload for non-voice frames must use the Data Transfer Syntax Payload Format defined in Annex C of FRF.11, instead of formats indicated in this IA. Such PVCs shall use procedures described in Section 6 of this IA, except as indicated within this section.

Fragmentation using the FRF.11 Annex C format provides End-to-End fragmentation, enabling interleaving of delay-sensitive traffic on one sub-channel with fragments of a long frame on another sub-channel within the PVC.

VoFR data fragments are first created using the procedure in Section 6.1, except as follows:

▼ Each fragment is preceded by an FRF.11 header that identifies it as a primary payload, using sub-channel(s) provisioned to carry data.

■ The data fragments are interleaved with voice sub-frames on difference sub-channels of the same PVC.

■ The order of the data fragments must be maintained within each sub-channel.

■ Because the sequence numbers of FRF.11 Annex C frames use a 13-bit sequence number, the sequence numbers procedures will increment modulo 2^{13}.

▲ UNI or NNI interfaces are allowed to further fragment frames that have been fragmented using the Data Transfer Syntax Payload Format of FRF.11 Annex C. See Section 6.1 of this IA.

8.2 Fragmentation Interaction with FRF.3.1 [3] Multiprotocol Encapsulation and non-Multiprotocol Encapsulated Data

If a VC is configured to carry data other than FRF.3.1, the specification of the relevant protocol applies to the reassembled frame.

If a VC is configured to carry FRF.3.1 multiprotocol encapsulated data, the procedures of FRF.3.1 apply to the reassembled frame. In addition, the procedures of FRF.9 may apply if data compression is enabled for the VC. See Section 8.3.

8.3 Fragmentation Interaction with FRF.9 [6] Compression

Because FRF.9 compressed frames are multiprotocol-encapsulated according to FRF.3.1, compressed frames may also be fragmented. When compression is being used on a PVC, compression is first applied to the frame, then the compressed frame is fragmented using the procedure in Section 6.1. Like any other data frame, Interface and/or End-to-End fragmentation may be used.

On reception, the fragments are first reassembled, and the resulting compressed frame is then uncompressed as per FRF.9.

8.4 End-to-End Fragmentation Interaction with FRF.8 [5] FR/ATM Service Interworking

End-to-End fragmentation must be terminated in the FR side of the FRF.8 Service Interworking Function (IWF). In the FR to ATM direction, fragments will be reassembled by the IWF to form complete frames, which are then passed to the ATM side of the IWF according to the FRF.8 procedures. In the ATM to FR direction, AAL5 PDUs will be passed to the FR side according to the FRF.8 procedures, and then fragmented according to this IA.

Note that the Frame Relay congestion bits (FECN, BECN, DE) must be logically ORed for all fragments and the result passed with the reassembled frame to the ATM side of the interworking unit. In the ATM to Frame Relay direction the ATM congestion bits (EFCN and CLP) are mapped to each Frame Relay frame fragment.

9. FRAGMENTATION EXAMPLES

9.1 Interface Fragmentation Example

An example of the Interface (UNI and NNI) fragmentation procedure, using an FRF.3.1-encapsulated frame as the data to be fragmented, is diagrammed in Figure 9-1. The octets in white indicate the data portion of the original frame that is split into fragments (three fragments in this example). While this example uses an FRF.3.1 frame for illustration purposes, any arbitrary frame contents may be fragmented. For this example, the starting sequence number of 42 was chosen at random. Note that when fragmenting FRF.3.1 data, the control octet, the optional pad (if present), and the NLPID of the original frame are transported in the first frame fragment and are part of the reassembled frame.

9.2 End-to-End Fragmentation Example

An example of the End-to-End fragmentation procedure, using an FRF.3.1-encapsulated frame as the data to be fragmented, is diagrammed in Figure 9-2. The octets in white indicate the data portion of the original frame that is split into fragments (three fragments in

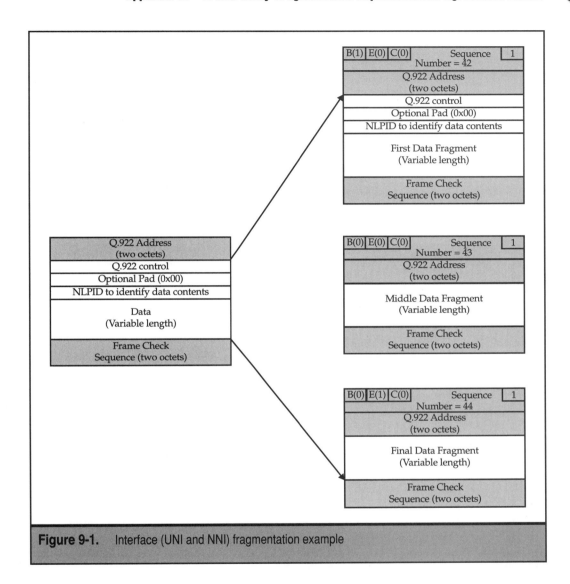

Figure 9-1.   Interface (UNI and NNI) fragmentation example

this example). While this example uses an FRF.3.1 frame for illustration purposes, any arbitrary frame contents may be fragmented. For this example, the starting sequence number of 42 was chosen at random. Note that when fragmenting FRF.3.1 data, the control octet, the optional pad (if present), and the NLPID of the original frame are transported in the first frame fragment and are part of the reassembled frame.

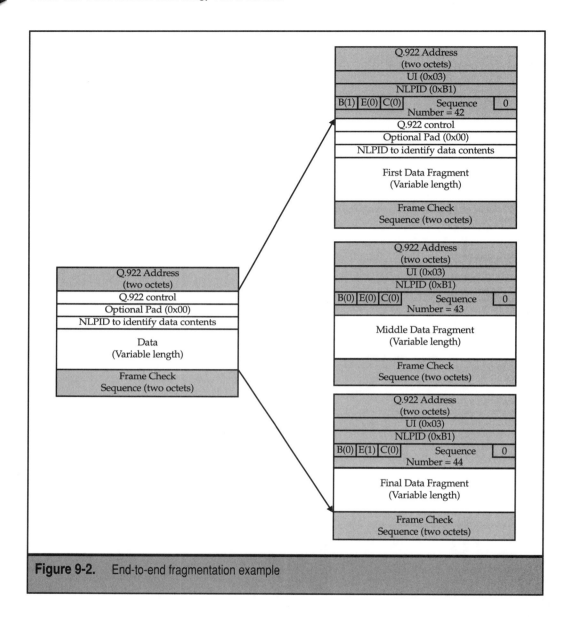

Figure 9-2. End-to-end fragmentation example

9.3 FRF.11 Fragmentation Example

An example of the FRF.11 fragmentation procedure, using an FRF.3.1-encapsulated frame as the data to be fragmented, is diagrammed in Figure 9-3. The octets in white indicate the data portion of the original frame that is split into fragments (three fragments in this example). While this example uses an FRF.3.1 frame for illustration purposes, any arbitrary frame contents may be fragmented. For this example, the starting sequence number of 42 was chosen at random. Note that when fragmenting FRF.3.1 data, the control octet, the optional pad (if present), and the NLPID of the original frame are transported in the first frame fragment and are part of the reassembled frame.

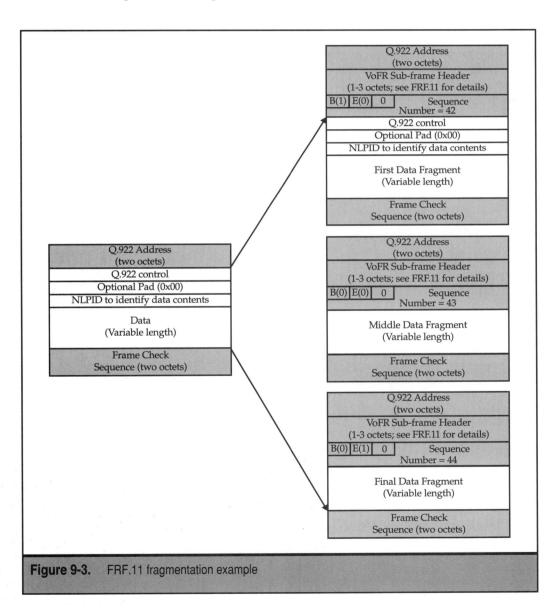

Figure 9-3.   FRF.11 fragmentation example

CONSIDERATIONS FOR CHOOSING THE END-TO-END FRAGMENTATION SIZE

(Informative)

Section 7 mentions that for the case of End-to-End fragmentation, one or more of the network(s) that interconnect the DTEs may use FRF.5 [4] FR/ATM Network Interworking and use ATM cell-based transport. In such cases, the choice of fragment size will have an effect on how efficiently the fragments may be packed into the ATM cells. Presented below is a figure that illustrates the efficiency of different choices for the fragment payload size for the case of networks implementing FRF.5. The efficiency is calculated as follows:

Length of PDU to convert to cells = [FR Header (2 octets)
+ Fragmentation Header (4 octets)
+ Fragmentation Payload Size (**P** octets)
+ AAL5 trailer (8 octets)]

Number of cells needed to carry PDU (**N** cells) = CEILING [Length of PDU /48 octets/cell]

Efficiency (%) = 100 * (**P** octets) / [**N** cells * 53 octets/cell]

Figure A-1 shows Efficiency (%) on the vertical axis versus Fragmentation Payload Size (**P** octets). This does not represent a requirement or a recommendation of FRF.12.

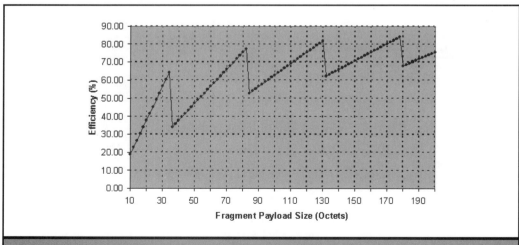

Figure A-1.   Efficiency of different fragment payload sizes when used with FRF.5

INDEX

▼ G

H

I

J - K

L

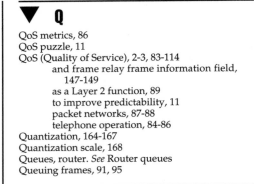

U

 V

INTERNATIONAL CONTACT INFORMATION

AUSTRALIA
McGraw-Hill Book Company Australia Pty. Ltd.
TEL +61-2-9417-9899
FAX +61-2-9417-5687
http://www.mcgraw-hill.com.au
books-it_sydney@mcgraw-hill.com

CANADA
McGraw-Hill Ryerson Ltd.
TEL +905-430-5000
FAX +905-430-5020
http://www.mcgrawhill.ca

GREECE, MIDDLE EAST,
NORTHERN AFRICA
McGraw-Hill Hellas
TEL +30-1-656-0990-3-4
FAX +30-1-654-5525

MEXICO (Also serving Latin America)
McGraw-Hill Interamericana Editores S.A. de C.V.
TEL +525-117-1583
FAX +525-117-1589
http://www.mcgraw-hill.com.mx
fernando_castellanos@mcgraw-hill.com

SINGAPORE (Serving Asia)
McGraw-Hill Book Company
TEL +65-863-1580
FAX +65-862-3354
http://www.mcgraw-hill.com.sg
mghasia@mcgraw-hill.com

SOUTH AFRICA
McGraw-Hill South Africa
TEL +27-11-622-7512
FAX +27-11-622-9045
robyn_swanepoel@mcgraw-hill.com

UNITED KINGDOM & EUROPE
(Excluding Southern Europe)
McGraw-Hill Education Europe
TEL +44-1-628-502500
FAX +44-1-628-770224
http://www.mcgraw-hill.co.uk
computing_neurope@mcgraw-hill.com

ALL OTHER INQUIRIES Contact:
Osborne/McGraw-Hill
TEL +1-510-549-6600
FAX +1-510-883-7600
http://www.osborne.com
omg_international@mcgraw-hill.com